A
COMPLETE GCSE
MATHEMATICS

GENERAL COURSE

Third Edition

11071 ✓
£11-00
29-9-94
W H Smith

11074

A COMPLETE GCSE MATHEMATICS

GENERAL COURSE

A. Greer

Formerly Senior Lecturer,
Gloucestershire College of Arts and Technology

THIRD EDITION

Stanley Thornes (Publishers) Ltd

First published in 1986 by
Stanley Thornes (Publishers) Ltd
Ellenborough House
Wellington Street
CHELTENHAM GL50 1YD

Reprinted 1987 (three times)
Reprinted 1988
Second Edition 1989
Reprinted 1989
Reprinted 1990
Reprinted 1991
Reprinted 1992
Third Edition 1993
Reprinted 1994

British Library Cataloguing in Publication Data

A catalogue record for this book is available from the British Library

ISBN 0-7487-1599-1

Typeset by Tech-Set, Gateshead, Tyne & Wear in 10/12pt Century.
Printed and bound in Great Britain at The Bath Press, Avon.

Contents

Preface to the Third Edition

This book is one of a series of three books intended to cover all the topics prescribed by the National Curriculum at three levels. It is suitable for those students doing the National Curriculum Course at Levels 5, 6, 7 and 8.

Because this is a revision book, the sections on Arithmetic, Algebra, Geometry, Graphical Work and Statistics have been dealt with separately. The emphasis is on a simple approach and a teacher and the class may work, if desired, through the book chapter by chapter.

A large number of examples and exercises has been included which supplement and amplify the text. Many practical examples of mathematics in the real world have been used.

I would like to thank my son David for working out the answers to the exercises.

A. Greer   Gloucester 1993

Coursework

Coursework tasks which are separately assessed at the examination centre may be set by the Examiner or the choice may be left to the teaching staff and the individual candidate. The aim of the three sections of 'Assignments' is to provide a variety of challenges for Intermediate Tier candidates and to supply the teacher with a varied selection of coursework ideas. The assignments are designed to enable the student to develop the skills necessary for completing a comprehensive programme of study.

Coursework should support and extend the GCSE core syllabus, by 'using and applying' the topics covered under Attainment Targets 2–5. These aims have been given special emphasis in this third edition. Coursework 1 is correlated with Chapters 1–12 and does not require knowledge of topics covered in later chapters. Coursework 2 extends the horizons, including topics covered to Chapter 28. Coursework 3 is weighted towards Chapters 29–36, but also requires knowledge of topics covered in earlier chapters.

An effort has been made to offer interesting and stimulating projects. The more extended tasks give considerable guidance so that students can recognise the different branches of research and investigation which are possible before assembling their own folder of coursework. As well as aiming to fulfil the requirements of Ma1, a number of the assignments incorporate several related topic areas from the syllabus, enabling further consolidation of other Attainment Targets.

B.T. Hopkins   Gloucester 1993

Acknowledgements

The author and publishers are grateful to the following who provided material for inclusion:

Cheltenham and Gloucester Omnibus Co. Ltd. for the timetable (1980), page 107.

British Rail for the timetable, page 103.

Operations in Arithmetic

Types of Numbers

Counting numbers or **natural numbers** are the numbers 1, 2, 3, 4, 5, etc.

Whole numbers are the numbers 0, 1, 2, 3, 4, 5, etc.

Positive numbers are numbers which are greater than zero. They either have no sign in front of them or a + sign. Some examples of positive numbers are 5, +9, 8, +21, +293 and 564.

Negative numbers have a value less than zero. They are always written with a − sign in front of them. Some examples of negative numbers are −4, −15, −168 and −783.

Directed numbers or **integers** are whole numbers but they include negative numbers. Some examples of integers are −28, −15, 0, 2, +9, 24, +371 and −892.

Directed numbers are used on thermometers to show Celsius temperatures above

Fig. 1.1

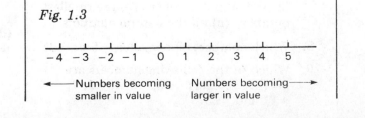

and below freezing point (Fig. 1.1). They are also used on graphs (Fig. 1.2).

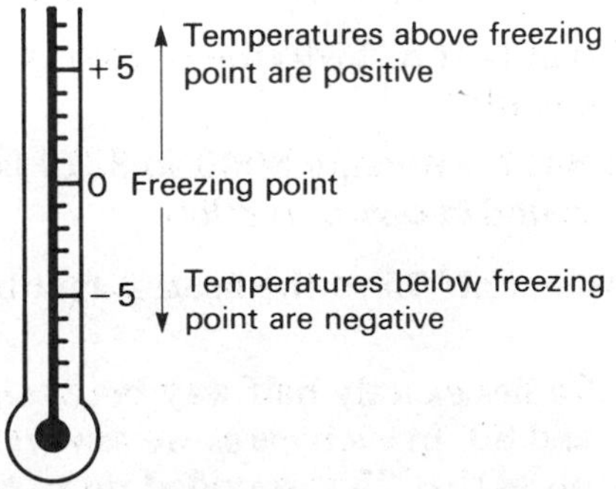

Fig. 1.2

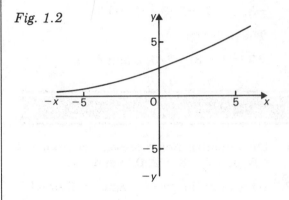

It is also possible to have negative time, for example −12 seconds to blast off.

An overdraft at the bank can be regarded as negative money, for instance −£200 would represent an overdraft of £200 (i.e. the customer owes the bank £200).

The Number Line

Directed numbers may be represented on a number line (Fig. 1.3). Numbers to the right of zero are positive whilst those to the left of zero are negative numbers.

Fig. 1.3

```
 ─┼──┼──┼──┼──┼──┼──┼──┼──┼──┼──
 −4 −3 −2 −1  0  1  2  3  4  5
```
←──── Numbers becoming smaller in value   Numbers becoming larger in value ────→

Negative numbers become smaller in value the further we move to the left of zero. For instance −8 is smaller in value than −3 and −56 is smaller in value than −27.

Positive numbers, however, become greater in value the further we move to the right of zero. For instance 12 has a greater value than 7 and +89 has a greater value than +34.

Example 1

Rearrange in order of size, smallest first, the numbers

$$-3, 5, -7, -2, 8, -4 \text{ and } 3$$

The order is:

$$-7, -4, -3, -2, 3, 5 \text{ and } 8$$

Exercise 1.1

1. On a number line represent the numbers −6, 3, +7, −8, −2, 0, 5 and −4.

2. Which has the greater value, −6 or +3?

3. Write down all the positive numbers from the following: −2, 7, +5, −6, 3, +2 and −3.

4. Write down all the whole numbers from the following set: 7, −4, 3, −6, +2, 0, −9, 5 and −2.

5. Rearrange in order of size, smallest first, the numbers −3, −5, −1, −9 and −4.

6. Rearrange in order of size, largest first, the numbers 3, −8, −15, +25, 18, −6, −2 and 12.

7. From the set of numbers −6, 8, +5, −10, 0, 12, −8 and +7, pick out (a) the largest number, (b) the smallest number, (c) all the natural numbers.

8. Which number is the smaller, −5 or −3?

9. Which of the following numbers are positive integers: +5, −3, 0, −1, 7, +11 and −15?

10. On a Celsius thermometer a temperature of −8 degrees is recorded. Is the temperature below the freezing point of water?

Rounding Off

Calculators save a lot of time and effort when used for addition, subtraction, multiplication and division. We do, however, need to be able to estimate the size of the answer we expect as a check on our use of the calculator.

One way of doing this is to round off the numbers involved. It is generally good enough to round off

> a number between 10 and 100 to the nearest number of tens
>
> a number between 100 and 1000 to the nearest number of hundreds
>
> a number between 1000 and 10 000 to the nearest number of thousands

and so on.

Example 2

(a) Round off the number 67 to the nearest number of tens.

> 67 is roughly 7 tens so 67 is rounded up to 70

(b) Round off 362 to the nearest number of hundreds.

> 362 is roughly 4 hundreds so 362 is rounded up to 400

(c) Round off 8317 to the nearest number of thousands.

> 8317 is roughly 8000 so 8317 is rounded down to 8000

(d) Round off 75 to the nearest number of tens.

> 75 lies exactly half way between 70 and 80. In such cases we always round up so that 75 is rounded up to 80

Exercise 1.2

Round off the following numbers to the nearest number of tens:

1. 42   2. 63   3. 85

4. 92   5. 76

Round off the following numbers to the nearest number of hundreds:

6. 698   7. 703   8. 886

9. 350   10. 885

Round off the following numbers to the nearest number of thousands:

11. 2631   12. 5378   13. 8179

14. 8500   15. 5765

Adding Numbers

+ is the addition sign, meaning plus. Thus $5 + 3 = 8$, which is read as 'five plus three equals eight'. The result obtained by adding numbers is called their **sum**. Thus the sum of 5, 6 and 8 is $5 + 6 + 8 = 19$. *The order in which numbers are added does not affect their sum.*

$$4 + 6 + 8 = 6 + 4 + 8$$
$$= 8 + 6 + 4$$
$$= 18$$

This rule, which is known as the **commutative law,** is very useful when adding numbers mentally:

$$15 + 8 + 5 = (15 + 5) + 8$$
$$= 20 + 8$$
$$= 28$$

It is also useful for checking the result of an addition:

```
  315 ↑        315 |
   17           17
  692 +        692 +
 ─────        ─────
 1024 |       1024 ↓
```

   Add upwards   Check downwards

More often than not a calculator will be used to add numbers greater than 10.

Example 3

Using a calculator find the sum of 278, 89 and 3924.

For a rough estimate we will round off the numbers to

$$300 + 90 + 4000 = 4390$$

Input	Display
278	278.
+	278.
89	89.
+	367.
3924	3924.
=	4291.

The rough estimate shows that the answer, it is 4291, is sensible because the rough estimate and the answer are of the same size.

The whole object of doing arithmetic is to produce correct answers. It is pointless to produce wrong answers. Therefore all calculations must be checked. When using a calculator we make use of the commutative law and input the numbers in reverse order.

Input	Display
3924	3924.
+	3924.
89	89.
+	4013.
278	278.
=	4291.

Since the same answer has been produced twice we can be confident that

$$278 + 89 + 3924 = 4291$$

Questions 1 to 5 inclusive should be done mentally.

1. Find the sum of 3, 7 and 6.

2. Add 6, 8 and 4.

3. Find the total of 12, 7 and 8.

4. Add 3, 6, 7 and 4.

5. What is the sum of 2, 5, 8 and 15?

Work out the following and check the answers:

6. $96 + 348 + 18$

7. $219 + 69 + 3487 + 935$

8. $35\,078 + 21\,009 + 807 + 1288 + 43$

9. $23\,589 + 7\,986\,232 + 245\,899 + 7892 + 335\,681$

10. $17\,438 + 1344 + 2636 + 107\,924 + 11\,358$

Adding Integers

To add integers we make use of the number line. Thus in Fig. 1.4 to find the value of $+3 + 4$, which is usually written $3 + 4$, we measure 3 units to the right of 0 to represent $+3$ and from this point we measure 4 units to the right to represent $+4$. We see from the diagram that

$$3 + 4 = 7$$

Fig. 1.4

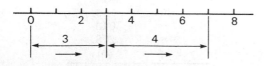

To find the value of $-3 + (-2)$, which is usually written $-3 - 2$, we measure 3 units to the left of 0 (Fig. 1.5) to represent -3.

Then from this points we measure 2 units to the left to represent -2. We see from the diagram that

$$-3 - 2 = -5$$

Fig. 1.5

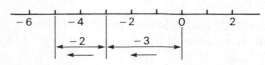

We see that the rules for the addition of integers having the same sign are as follows:

(1) If all the numbers to be added are positive then add them together and place a $+$ sign (or no sign) in front of the sum.

(2) If all the numbers to be added are negative then add the numbers together and place a $-$ sign in front of the sum.

Example 4

(a) $3 + 5 + 6 + 2 = 16$

(b) $-4 - 6 - 9 - 3 = -22$

To find $4 - 6$ we proceed as shown in Fig. 1.6 and we see that

$$4 - 6 = -2$$

Fig. 1.6

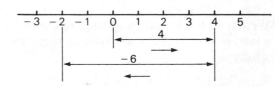

When the signs are different:

(1) For two numbers find the difference between them and place the sign of the larger number in front of the difference.

(2) If there are more than two numbers add the positive numbers together. Next add the negative numbers together. The set of numbers is thereby reduced to two numbers, one positive and the other negative. We then proceed as for two numbers.

Example 5

(a) $8 - 5 = 3$

(b) $-5 + 8 = 3$

(c) $2 - 6 = -4$

(d) $-9 + 3 = -6$

Example 6

$-5 + 8 - 7 + 3 + 6 - 9 + 4 + 1 - 2$

$\quad = (8 + 3 + 6 + 4 + 1)$

$\quad \quad - (5 + 7 + 9 + 2)$

$\quad = 22 - 23$

$\quad = -1$

When using a calculator we can input positive and negative numbers as they occur. There is no need to add the positive and negative numbers separately.

Example 7

Use a calculator to work out $72 - 43 - 59 + 34 + 112 - 89$.

Rough estimate, $70 - 40 - 60 + 30 + 110 - 90 = 20$

Input	Display
72	72.
−	72.
43	43.
−	29.
59	59.
+	−30.
34	34.
+	4.
112	112.
−	116.
89	89.
=	27.

The rough estimate shows that the answer, 27, is sensible.

The answer produced by the machine can be checked by making use of the commutative law and inputting the numbers in reverse order.

Input	Display
−	0.
89	89.
+	−89.
112	112.
+	23.
34	34.
−	57.
59	59.
−	−2.
43	43.
+	−45.
72	72.
=	27.

Since the same answer has been produced by two different methods we can be confident that

$$72 - 43 - 59 + 34 + 112 - 89 = 27$$

Exercise 1.4

Questions 1 to 10 inclusive should be done mentally. Find the value of each of the following:

1. $8 + 5$ 2. $-6 - 3$

3. $-14 - 18$ 4. $-7 - 6 - 3$

5. $8 - 12$ 6. $9 - 17$

7. $8 - 7 - 15$ 8. $-6 + 10$

9. $-3 + 5 + 7 - 4 - 2$

10. $6 + 4 - 3 - 5 - 7 + 2$

Use a calculator to work out the following, in each case making a rough estimate of the answer:

11. $-40 - 23 + 72 + 15 - 18$

12. $324 - 78 - 169 - 89 + 54 + 178 - 26$

13. $-63 - 98 + 37 - 75 + 19 - 28 + 68 - 87$

14. $152 - 78 + 43 - 81 - 128 + 57 - 63$

15. $278 + 452 + 96 + 73 - 156 - 238 - 417 - 82 + 175$

Subtracting Numbers

— is the subtraction sign meaning minus. Thus $8 - 3 = 5$, which is read as eight minus three equals five.

The difference between two numbers is the larger number minus the smaller number.

Example 8

(a) Subtract 7 from 11.

$$11 - 7 = 4$$

(b) Take 2 from 8.

$$8 - 2 = 6$$

(c) Find the difference between 5 and 12.

$$\text{Difference} = 12 - 5 = 7$$

Example 9

Subtract 87 from 395.

Rough estimate, $400 - 90 = 310$

$$\begin{array}{r} 395 \\ 87\ - \\ \hline 308 \\ \hline \end{array}$$

The rough estimate shows that the answer is sensible.

The calculation can be checked by adding the bottom two numbers. If this sum equals the top number then the calculation is correct.

$$\begin{array}{r} 87 \\ 308\ + \\ \hline 395 \\ \hline \end{array}$$

Most subtractions will be done on a calculator as shown in Example 10.

Example 10

Take 2987 from 12 613.

Rough estimate,
$13\,000 - 3000 = 10\,000$

Input	Display
12613	12613.
—	12613.
2987	2987.
=	9626.

Hence $12\,613 - 2987 = 9626$.

The rough estimate shows that the answer is sensible.

To check the answer we add 2987 and 9626 to see if this sum equals 12 613.

Input	Display
2987	2987.
+	2987.
9626	9626.
=	12613.

Subtracting Integers

Suppose we want to find the value of $-(+5) - 4$. To represent $+5$ we measure 5 units to the right of 0. Therefore to represent $-(+5)$ we must reverse direction and measure 5 units to the left of 0 (Fig. 1.7).

Hence $-(+5)$ is the same as -5.

Therefore

$$-(+5) - 4 = -5 - 4$$
$$= -9$$

Fig. 1.7

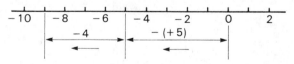

To subtract a directed number change its sign and add the resulting number.

Example 11

(a) $5 - (-3) = 5 + 3 = 8$

(b) $-7 - (+4) = -7 - 4 = -11$

(c) $8 - (+5) = 8 - 5 = 3$

Exercise 1.5

Questions 1 to 5 inclusive should be done mentally.

1. Subtract 5 from 9.

2. Take 3 from 8.

3. Find the difference between 7 and 4.

4. Work out $8 - 2$.

5. How many less is 4 than 9?

Use a calculator to work out each of the following, in each case making a rough estimate of the answer:

6. $247 - 89$ 7. $1253 - 357$

8. $2788 - 1359$ 9. $11\,913 - 9876$

10. $234\,582 - 98\,687$

Questions 11 to 15 inclusive should be done mentally. Find the value of each of the following:

11. $7 - (+6)$ 12. $-5 - (-6)$

13. $5 - (-3)$ 14. $-3 - (-4)$

15. $-10 - (-4)$

Multiplication

$\times$ is the multiplication sign, meaning multiplied by, or times. Thus $6 \times 8 = 48$ which is read as six times eight equals forty-eight.

Multiplication is a quick way of adding equal numbers.

$$7 + 7 + 7 + 7 + 7 = 35$$

$5 \times 7 = 35$ (i.e. five sevens equal thirty-five)

35 is called the **product** of 5 and 7.

The product of 4, 5 and 8 is $4 \times 5 \times 8 = 160$.

The order in which the numbers are multiplied does not affect the product. Thus

$$3 \times 4 \times 6 = 6 \times 4 \times 3$$
$$= 4 \times 3 \times 6$$
$$= 72$$

This rule is called the **commutative law for multiplication.** It may be used:

(1) To simplify multiplication when done mentally:

$$4 \times 9 \times 25 = (4 \times 25) \times 9$$
$$= 100 \times 9$$
$$= 900$$

(2) To check a product by interchanging numbers:

$$28 \times 73 = 2044$$
$$73 \times 28 = 2044$$

More often than not a calculator will be used for multiplication because of the time taken when using pen and paper.

Example 12

Use a calculator to find the product of 356 and 98.

Rough estimate, $350 \times 100 = 35\,000$

Input	Display
356	356.
$\times$	356.
98	98.
=	34888.

Hence $356 \times 98 = 34\,888$.

The rough estimate shows that the answer is sensible.

To check the product produced by the machine multiply in the reverse order, i.e. 98×356.

Example 13

Work out $27 \times 15 \times 54 \times 29$.

Rough estimate, $30 \times 20 \times 50 \times 30 = 30 \times 30 \times 1000 = 900\,000$

Input	Display
27	27.
×	27.
15	15.
×	405.
54	54.
×	21870.
29	29.
=	634230.

Although there is a big difference between the answer and the rough estimate the rough estimate shows the answer is 634 230 and not 63 423 or 6 342 300.

Hence $27 \times 15 \times 54 \times 29 = 634\,230$.

This product may be checked by multiplying in the reverse order, i.e. $29 \times 54 \times 15 \times 27$.

Multiplication of Directed Numbers

$$4 + 4 + 4 = 12$$
$$3 \times 4 = 12$$

Hence two positive numbers multiplied together give a positive product.

$$-4 - 4 - 4 = -12$$
$$3 \times (-4) = -12$$

Hence a positive number multiplied by a negative number gives a negative product.

Suppose we wish to find the value of $(-3) \times (-4)$:

$$(-3) \times (4 - 4) = (-3) \times 0$$
$$= 0$$
$$(-3) \times 4 + (-3) \times (-4) = 0$$
$$-12 + (-3) \times (-4) = 0$$
$$\therefore \qquad (-3) \times (-4) = 12$$

Hence a negative number multiplied by a negative number gives a positive product.

Therefore the rule for multiplication is:

(1) Numbers with like signs give a positive product.

(2) Numbers with unlike signs give a negative product.

Example 14

(a) $2 \times 3 = 6$

(b) $(-2) \times 3 = -6$

(c) $2 \times (-3) = -6$

(d) $(-2) \times (-3) = 6$

(e) $(-3) \times (-5) \times (-8) = 15 \times (-8)$
$$= -120$$

(f) $(-2) \times 3 \times (-4) = (-6) \times (-4) = 24$

Exercise 1.6

Questions 1 to 5 inclusive should be done mentally.

1. Find values for
   (a) $2 \times 3 \times 5$
   (b) $4 \times 7 \times 5$

2. Multiply 5 by 8.

3. Find the products of
   (a) 3, 4 and 5
   (b) 5, 7 and 8

4. Work out $2 \times 3 \times 5 \times 7$.

5. What is the product of 2, 25, 5 and 4?

Using a calculator work out the following:

6. 42×75

7. 235×39

8. $18 \times 27 \times 35$

9. $8 \times 12 \times 15 \times 22$

10. $9 \times 11 \times 14 \times 16 \times 5$

Questions 11 to 18 inclusive should be done mentally.

Find values for each of the following:

11. $8 \times (-3)$ **12.** $(-5) \times 2$

13. $6 \times (-3)$ **14.** $(-2) \times (-4)$

15. $(-2) \times (-5) \times (-6)$

16. $4 \times (-3) \times (-2)$

17. $2 \times (-5) \times 4$

18. $(-3) \times (-4) \times 5$

Division

$\div$ is the division sign, meaning divided by.

There are several ways of writing division problems. 6 divided by 3 may be written:

(1) $6 \div 3$, which reads six divided by three.

(2) $\frac{6}{3}$, which reads six over three.

(3) $3\overline{)6}$, which reads three into six.

Division will generally be done by using a calculator.

Example 15

Divide 1554 by 42.

Rough estimate, $1600 \div 40 = 40$

Input	Display
1554	1554.
$\div$	1554.
42	42.
=	37.

Hence $1554 \div 42 = 37$.

The rough estimate shows that the answer is sensible.

The result of a division can always be checked by multiplying back. Thus if

$$1554 \div 42 = 37$$

then

$$42 \times 37 = 1554$$

We can check that this is so by using a calculator in the usual way.

Remainders

Consider $17 \div 3$.

3 will not divide into 17 exactly and the answer is 5 remainder 2.

Note that $3 \times 5 + 2 = 15 + 2 = 17$.

For more difficult numbers the remainder may be found by using a calculator as shown in Example 16.

Example 16

Find the remainder when 10 798 is divided by 29.

Input	Display	
10798	10798.	
$\div$	10798.	
29	29.	
$-$	372.34482	(this is the answer to the division)
372	372	
$\times$	0.34482	(this is the remainder in decimal form)
29	29.	
=	9.99978	(this is the required remainder, see note below)

Hence $10\,798 \div 29 = 372$ remainder 10.

The remainder must be a whole number but because the calculator is limited to 8 digits calculations of this kind are never exactly right, and we expect small errors to occur. 9.999 78 is so near to 10 that this must be the remainder. Note that

$$372 \times 29 + 10 = 10\,798$$

Division of Directed Numbers

The rules for division must be similar to the rules for multiplication. Since

$$3 \times (-4) = -12$$

$$\frac{-12}{3} = -4$$

and

$$\frac{-12}{-4} = 3$$

Therefore the rule is:

(1) Numbers with like signs, when divided, give a positive answer.

(2) Numbers with unlike signs, when divided, give a negative answer.

Example 17

(a) $\dfrac{8}{4} = 2$ (b) $\dfrac{8}{-4} = -2$

(c) $\dfrac{-8}{4} = -2$ (d) $\dfrac{-8}{-4} = 2$

Combined Multiplication and Division

Example 18

Work out $\dfrac{722 \times 420}{19 \times 168}$

Rough estimate, $\dfrac{700 \times 400}{20 \times 200} = 70$

Input	Display
722	722.
×	722.
420	420.
÷	303240.
19	19.
÷	15960.
168	168.
=	95.

Hence $\dfrac{722 \times 420}{19 \times 168} = 95.$

The rough estimate shows that the answer is sensible.

Again we can check the calculation by multiplying back.

If the correct answer is 95, then

$$722 \times 420 = 19 \times 168 \times 95$$

which can be verified by using a calculator.

Example 19

Work out $\dfrac{4 \times (-8) \times 3}{(-2) \times 6 \times (-4)}$

$$\dfrac{4 \times (-8) \times 3}{(-2) \times 6 \times (-4)} = \dfrac{-96}{48}$$

$$= -2$$

Exercise 1.7

Using a calculator work out the following. In each case make a rough estimate of the answer:

1. $1064 \div 8$ 2. $255 \div 17$

3. $567 \div 21$ 4. $32\overline{)512}$

5. $71\overline{)3053}$ 6. $61\overline{)3599}$

7. $\dfrac{378}{21}$ 8. $\dfrac{16\,308}{453}$

9. $\dfrac{49\,182}{2342}$

10. $2\,233\,700 \div 6382$

Questions 11 to 18 inclusive should be done mentally.

11. $8 \div (-2)$ 12. $(-9) \div 3$

13. $(-8) \div (-2)$ 14. $(-14) \div (-7)$

15. $10 \div (-5)$ 16. $(-12) \div (-4)$

17. $18 \div (-6)$ 18. $(-24) \div (-8)$

Work out each of the following:

19. $\dfrac{-16}{(-2) \times (-4)}$

20. $\dfrac{2 \times (-8) \times 6}{(-3) \times (-4) \times (-2)}$

21. $\dfrac{15 \times (-3) \times 2}{(-5) \times (-6)}$

22. $\dfrac{(-9) \times (-5) \times (-4)}{3 \times 15 \times (-2)}$

23. $\dfrac{648 \times 252}{108 \times 28}$ 24. $\dfrac{651 \times 252}{27 \times 217}$

25. $\dfrac{79\,380 \times 30\,429}{49 \times 27}$

26. $\dfrac{90\,343 \times 16\,695}{43 \times 45}$

Work out each of the following and state the remainder:

27. $15 \div 2$ 28. $23 \div 3$ 29. $19 \div 4$

30. $11 \div 6$ 31. $33 \div 5$ 32. $47 \div 8$

33. $302 \div 11$ 34. $601 \div 19$

35. $5631 \div 21$

Sequence of Arithmetical Operations

In working out problems such as

$$5 + (3 + 6)$$
or $$15 - (7 - 2)$$
or $$2 \times 4 + 2$$
or $$7 + 6 \div 2$$

a certain sequence of operations must be observed.

(1) Work out the contents of any brackets first:

$$4 \times (3 + 5) = 4 \times 8$$
$$= 32$$
$$15 - (7 - 2) = 15 - 5$$
$$= 10$$

We get the same answer and sometimes make the work easier if we expand the bracket as shown below:

$$8 \times (14 + 29) = 8 \times 14 + 8 \times 29$$
$$= 112 + 232$$
$$= 344$$

Sometimes the multiplication sign is omitted.

$$5(8 - 4) \text{ means } 5 \times (8 - 4)$$
$$= 5 \times 4$$
$$= 20$$

(2) Multiply and divide before adding and subtracting:

$$3 \times 6 + 5 = 18 + 5$$
$$= 23$$
$$5 - 6 \div 2 = 5 - 3$$
$$= 2$$

Example 20

Find the value of $11 - 12 \div 4 + 3 \times (6 - 2)$.

$$11 - 12 \div 4 + 3 \times (6 - 2)$$
$$= 11 - 12 \div 4 + 3 \times 4$$

(by working out the contents of the bracket)

$$= 11 - 3 + 12$$

(by multiplying and dividing)

$$= 20$$

(by adding and subtracting)

Exercise 1.8

Work out the value of each of the following:

1. $7 + 4 \times 3$
2. $2 \times 6 - 3$
3. $5 \times 4 - 3 \times 6 + 5$
4. $8 \times 5 - 15 \div 5 + 7$
5. $3 + 6 \times (3 + 2)$
6. $9 - 4 \times (3 - 2)$
7. $10 - 12 \div 6 + 3(8 - 3)$
8. $15 \div (4 + 1) - 9 \times 3 + 7(4 + 3)$
9. $35 \div (25 - 20)$

10. $24 \div 4 - 24 \div 6$

11. $3(2 - 5) + 4$

12. $25 \div (-5) - 7$

13. $3 - (-12) \times 4 - 15 \div (-3)$

14. $36 - (-4) + 36 \div (-6)$

15. $4(2 - 8) + (-12) \div (-3)$

Some Important Facts

(1) When zero is added to any number the sum is the number.

$$873 + 0 = 873 \quad \text{and} \quad 0 + 74 = 74$$

(2) When any number is multiplied by 1 the product is the number.

$$9483 \times 1 = 9483 \quad \text{and} \quad 1 \times 534 = 534$$

(3) When a number is multiplied by zero the product is zero.

$$178 \times 0 = 0 \quad \text{and} \quad 0 \times 2398 = 0$$

(4) It is impossible to divide a number by zero. Thus, for instance, $8 \div 0$ is meaningless.

(5) When zero is divided by any number the result is zero.

$$0 \div 79 = 0 \quad \text{and} \quad 0 \div 7 = 0$$

Exercise 1.9

This exercise should be done mentally.

Work out the value of each of the following:

1. $768 + 0$ 2. 256×1

3. 0×50 4. $0 \times 4 \times 5$

5. $3 \times 6 \times 1$ 6. $7 \times 0 \times 6$

7. $16 - 8 \times 0 - 3$

8. $12 - 24 \div 6 + 0 \times (7 - 3)$

9. $0 \div 32$

10. $(5 - 5) \times 3$

11. $7 \times (3 - 9) + 0 \div 7$

12. $24 \times 0 + 3(7 - 7) - 5$

Sequences

A set of numbers connected by some definite law is called a **sequence of numbers**. Each of the numbers in the sequence is called a **term** in the sequence.

Example 21

(a) Write down the next two terms of the sequence $4, 8, 12, 16, \ldots$

Each term of the sequence is formed by adding 4 to the previous term. Therefore

$$5\text{th term} = 16 + 4$$
$$= 20$$
$$6\text{th term} = 20 + 4$$
$$= 24$$

(b) Find the next two terms of the sequence $5, 2, -1, -4, \ldots$

Each term of the sequence is 3 less than the previous term. Therefore

$$5\text{th term} = -4 - 3$$
$$= -7$$
$$6\text{th term} = -7 - 3$$
$$= -10$$

(c) In the sequence $1, 3, 9, ?, 81, ?$ write down the terms denoted by a question mark.

Each term is formed by multiplying the previous term by 3. Therefore

$$4\text{th term} = 9 \times 3$$
$$= 27$$
$$6\text{th term} = 81 \times 3$$
$$= 243$$

(d) Find the next two terms of the sequence 1, 2, 3, 5, 8, ...

Each term is formed by adding the two previous terms. Therefore

$$6\text{th term} = 5 + 8$$
$$= 13$$
$$7\text{th term} = 8 + 13$$
$$= 21$$

(e) Write down the next two terms of the sequence 2, 7, 17, 37, ...

Each term is formed by multiplying the previous term by 2 and adding 3 to the product. Therefore

$$5\text{th term} = 37 \times 2 + 3$$
$$= 74 + 3$$
$$= 77$$
$$6\text{th term} = 77 \times 2 + 3$$
$$= 154 + 3$$
$$= 157$$

Exercise 1.10

Write down the next two terms for each of the following sequences:

1. 2, 10, 50, ...   2. 1, 5, 9, 13, ...

3. 3, 9, 15, 21, ...   4. 176, 88, 44, ...

5. 1, 3, 6, 10, ...

6. 1, 1, 2, 3, 5, 8, 13, ...

7. 3, 8, 18, 38, ...   8. 5, 9, 21, 57, ...

9. 2, 0, −2, −4, ...

10. −5, −3, −1, 1, ...

Write down the terms denoted by the question marks in the following sequences:

11. 0, 2, 4, ?, ?, 10, 12, 14

12. 3, 5, 8, 13, ?, 34, 55, ?

13. 2, 4, 10, 28, ?, ?, 730

14. 5, 11, 29, 83, ?, 731, ?

15. 243, 81, 27, ?, 3, ?

Miscellaneous Exercise 1

1. What is the difference between the two numbers 45 and 72?

2. Work out the value of $8 + 5 \times (4 - 2)$.

3. Calculate the value of $23 \times 0 + 5(3 - 3)$.

4. What is the remainder when 33 is divided by 5?

5. Write down the terms denoted by ? in the following sequence of numbers: 0, 2, 4, ?, ?, 10, 12, 14, ?

6. The instructions for cooking a turkey are 'cook for 20 minutes per pound plus an additional 25 minutes'. How long will it take to cook a turkey weighing 8 pounds?

7. Write down the next three terms of the sequence 21, 17, 13 ...

8. If the temperature rises from −5 °C to 3 °C, what is the increase in temperature?

9. Write down the next three terms of the following sequences:
   (a) 21, 17, 13, ...
   (b) 40, 32, 24, ...

10. Subtract one hundred thousand from one million.

11. $53 + 29 = 37 + ?$ What is the missing number?

12. A car travels 9 kilometres on 1 litre of petrol. How many litres will be needed for a journey of 324 kilometres?

13. Write down the next two numbers of the sequence 1, 4, 9, 16, 25, ...

14. Find the value of
    (a) $(-5) \times (-8)$ (b) $-5-8$
    (c) $8 \div (-2)$ (d) $30 \times (-10)$

15. Write down in order of size, smallest first, $-3, 0, -5$ and 3.

16. A wall is 342 bricks long. It has 23 courses of bricks. How many bricks have been used in building the wall?

17. Which is the larger and by how much: $(5+2) \times (7-4)$ or $5 + 2(7-4)$?

18. Fill in the blanks in the following sequence: ?, ?, 12, 24, 48, ?

19. Calculate the answers to the following:
    (a) $(7-3) \times (7+4)$
    (b) $7 - 3(7+4)$

20. In each of the following insert either a $+$ sign or a $-$ sign to make the statement correct:
    (a) $27 + 5 - 12 = 27 + (5 \ldots 12)$
    (b) $13 - 4 + 9 = 13 \ldots (4-9)$

Mental Test 1

Try to answer the following questions without writing anything down except the answer.

1. Add $3, 6, 7$ and 4.

2. Find the sum of $2, 4, 6$ and 8.

3. What is the difference between 17 and 28?

4. Take 14 away from 23.

5. Multiply 14 by 5.

6. Find the product of 15 and 6.

7. Divide 72 by 8.

8. Find the value of $2 \times 3 + 5$.

9. Work out $17 - 2(6-4)$.

10. Work out $3 \times 5 - 12 \div (3+1)$.

11. Find the value of $-6 - 3$.

12. Work out $-8 - 7 - 15$.

13. Subtract $+6$ from 4.

14. Find the value of $-12 - (-15)$.

15. Find the product of $-3, -4$ and -5.

16. Work out $(-16) \div (-4)$.

17. Find the value of $18 \div (2-5)$.

18. Find the next term of the series $2, 6, 18, \ldots$

19. Find the next term of the following pattern of numbers:
    $$32, -16, 8, -4, \ldots$$

20. Find the next term of the sequence $6, 4, 2, 0, \ldots$

2 Factors and Multiples

Odd and Even Numbers

Counting in twos, starting with 2, the sequence is 2, 4, 6, 8, 10, 12, 14, ... Continuing the sequence, we discover that each number in it ends in either 0, 2, 4, 6 or 8. Such numbers are called **even numbers**. Thus 90, 282, 654, 8736 and 29 348 are all even numbers. If we now start at 1 and count in twos the sequence is 1, 3, 5, 7, 9, 11, 13, 15, 17, 19, ... Carrying on the sequence, we discover that each number in it ends in either 1, 3, 5, 7 or 9. Such numbers are called **odd numbers**. Thus 91, 103, 825, 9867 and 23 659 are all odd numbers. Note that any number which when divided by 2 has a remainder of 1 is an odd number.

Powers of Numbers

The quantity $5 \times 5 \times 5$ is usually written as 5^3, and 5^3 is called the third **power** of 5. The number 3 which indicates the number of fives to be multiplied together is called the **index** (plural: indices).

$$2^5 = 2 \times 2 \times 2 \times 2 \times 2$$
$$= 32$$
$$9^2 = 9 \times 9$$
$$= 81$$

When a number is multiplied by itself it is called the **square** of the number. Thus the square of a number is the number raised to the power of 2.

$$\text{Square of } 3 = 3^2$$
$$= 3 \times 3$$
$$= 9$$
$$\text{Square of } 15 = 15^2$$
$$= 15 \times 15$$
$$= 225$$

The cube of a number is the number raised to the power of three.

$$\text{Cube of 7 is } 7^3 = 7 \times 7 \times 7$$
$$= 343$$
$$\text{Cube of 85 is } 85^3 = 85 \times 85 \times 85$$
$$= 614\,125$$

Powers of numbers can easily be calculated using a calculator.

Example 1

Using a calculator find the value of 4^5.

Input	Display
4	4.
××	4.
=	16.
=	64.
=	256.
=	1024.

Hence $4^5 = 1024$.

Square Roots of Numbers

The square root of a number is the number whose square equals the given number. Since $5^2 = 25$, the square root of 25 is 5. The sign $\sqrt{}$ is used to denote a positive square root and we write $\sqrt{25} = 5$.

Similarly, since $7^2 = 49$, $\sqrt{49} = 7$.

Most calculators possess a square-root key which is used in the way shown in Example 2.

Example 2

Using a calculator find $\sqrt{529}$.

Input	Display
529	529.
$\sqrt{}$	23.

Hence $\sqrt{529} = 23$.

To check, note that $23^2 = 529$.

Square Root of a Product

The square root of a product is the product of the square roots of the individual numbers. Thus

$$\sqrt{4 \times 9} = \sqrt{4} \times \sqrt{9}$$
$$= 2 \times 3$$
$$= 6$$

Example 3

Find the square root of $16 \times 25 \times 49$.

$$\sqrt{16 \times 25 \times 49} = \sqrt{16} \times \sqrt{25} \times \sqrt{49}$$
$$= 4 \times 5 \times 7$$
$$= 140$$

Cube Roots of Numbers

The cube root of a number is that number whose cube equals the given number. Since $2^3 = 2 \times 2 \times 2 = 8$ the cube root of 8 is 2.

Since $15^3 = 15 \times 15 \times 15 = 3375$, $\sqrt[3]{3375} = 15$.

The cube root of 2 is usually written $\sqrt[3]{2}$ and the cube root of 47 is written $\sqrt[3]{47}$ etc.

The cube root of a number may be found by trial and improvement as shown in Example 4 (see also page 148).

Example 4

Using trial and improvement find $\sqrt[3]{29\,791}$.

Try 30: $30^3 = 27\,000$ which is too small.

Try 32: $32^3 = 32\,768$ which is too large.

So the value of $\sqrt[3]{29\,791}$ lies somewhere between 30 and 32.

Try 31: $31^3 = 29\,791$ which is correct and so $\sqrt[3]{29\,791} = 31$.

Exercise 2.1

Questions 1 to 20 inclusive should be done mentally. State whether the numbers given below are even or odd:

1. 95
2. 936
3. 852
4. 5729
5. 8887
6. 63 752
7. 456 310
8. 9367
9. From the set of numbers 17, 36, 59, 98, 121, 136 and 259 write down (a) all the odd numbers, (b) all the even numbers.
10. Consider the numbers 16, 19, 35, 68, 89, 137 and 342. Write down all the odd numbers.

Find values for each of the following:

11. The square of 6
12. The cube of 5

13. (a) 7^2 (b) 4^3

14. The square root of 64

15. 9 squared

16. (a) $\sqrt{16}$ (b) $\sqrt{144}$

17. The square root of 9×36

18. $\sqrt{64 \times 64}$

19. $\sqrt{9 \times 25 \times 36}$

20. $\sqrt{64 \times 36 \times 49}$

21. Find the value of each of the following:
    (a) 3^4 (b) 2^5 (c) 4^6

22. Find the sum of 2^6 and 6^2.

23. Find the product of 2^4 and 3^3.

24. Find the difference between 2^5 and 5^2.

25. Find the square root of the following numbers:
    (a) 169   (b) 625
    (c) 1024   (d) 7569
    (e) 15 625

26. By using trial and improvement find the cube roots of the following numbers:
    (a) 343 using initial guesses of 5 and 9.
    (b) 1331 using initial guesses of 9 and 13.
    (c) 15 625 using initial guesses of 20 and 30.

The Digit Sum of a Number

The digits of a number are the single figures which make up the number. For instance the digits of the number 312 are 3, 1 and 2.

The digit sum of a number is found by adding the digits of the number together.

Thus the digit sum of 312 is $3 + 1 + 2 = 6$.

Sometimes the digit sum contains more than one digit. In such cases the digits of the sum are added together until a single figure is obtained.

Example 5

Find the digit sum of the number 9587.

Adding the digits of 9587:

$$9 + 5 + 8 + 7 = 29$$

Adding the digits of 29:

$$2 + 9 = 11$$

Adding the digits of 11:

$$1 + 1 = 2$$

Tests for Divisibility

The following tests for divisibility are useful for work on factors, which follow later in this chapter.

If a number is even, then 2 will divide into it without leaving a remainder, i.e. the number is divisible by 2. Thus 38, 40, 56, 84 and 930 are all divisible by 2.

Numbers which are divisible by 2 are called multiples of 2. Thus 20, 42, 54, 136 and 978 are all multiples of 2.

A number is divisible by 3 if it divides exactly into the digit sum of the number without leaving a remainder.

Example 6

Is the number 4317 divisible by 3?

Adding the digits of 4317:

$$4 + 3 + 1 + 7 = 15$$

Adding the digits of 15:

$$1 + 5 = 6$$

Therefore the digit sum of 4317 is 6. Since 3 divides into 6 without leaving a remainder, the number 4317 is divisible by 3.

Numbers which are divisible by 3 are called multiples of 3. Thus 6, 9, 15, 24 and 36 are all multiples of 3.

A number is divisible by 5 if the number ends in 0 or 5. Hence 20, 35, 95, 8000 and 16 340 are all divisible by 5. The numbers 132, 984, 7906 and 8173 are not divisible by 5 because they do not end in 0 or 5.

Numbers which are divisible by 5 are called multiples of 5. Thus 15, 20, 30, 50 and 85 are all multiples of 5.

Numbers which end in 0 are divisible by 10. Thus 60, 9730, 27 000 and 16 876 320 are all divisible by 10.

Exercise 2.2

Find the digit sum of the following numbers:

1. 305   2. 171   3. 324
4. 976   5. 6283   6. 9457
7. 9207   8. 9875

Which of the following numbers are divisible by 2?

9. 12   10. 39   11. 82
12. 1986 13. 6032 14. 26 735
15. 29 008

Which of the following numbers are multiples of 2?

16. 92   17. 870   18. 1368
19. 7357 20. 9651

Which of the following numbers are divisible by 3?

21. 417   22. 484   23. 2740
24. 9354 25. 18 567

State which of the following numbers are multiples of 3:

26. 144   27. 286   28. 264
29. 5874 30. 29 739

Which of the following numbers are divisible by 5?

31. 105   32. 1237 33. 9180
34. 3415 35. 28 735

State which of the following numbers are multiples of 5:

36. 81   37. 180   38. 965
39. 4329 40. 2695

Which of the following numbers are divisible by 10?

41. 992   42. 1500 43. 4983
44. 9830 45. 26 325 410

Factors

A number is a factor of another number if it divides into that number without leaving a remainder.

Since 63 is divisible by 7, 7 is a factor of 63.

63 has other factors, namely 1, 3, 9, 21 and 63 since each of these numbers divides into 63 without leaving a remainder.

Exercise 2.3

Questions 1 to 5 should be done mentally.

1. Is 6 a factor of
   (a) 12   (b) 31   (c) 42
   (d) 66   (e) 86

2. Is 5 a factor of
   (a) 8   (b) 12   (c) 25
   (d) 32   (e) 50

3. Is 9 a factor of
   (a) 30   (b) 45   (c) 72
   (d) 128 (e) 135

4. Which of the following numbers are factors of 42?

   (a) ②   (b) 3   (c) 4
   (d) 5   (e) ⑥   (f) ⑦
   (g) 8   (h) 9

5. Which if the following numbers are factors of 72?

   (a) ②   (b) ③   (c) 4
   (d) 5   (e) ⑥   (f) 7
   (g) ⑧   (h) ⑨   (i) 11
   (j) ⑫   (k) 13   (l) 14

6. Consider the numbers 15, 31, 35, 49, 62, 84 and 109. Three of these numbers have a common factor. What is it?

Prime Numbers

Every number has itself and 1 as factors. If a number has no other factors it is said to be a **prime number**. It is useful to learn all of the prime numbers up to 100. They are

   2, 3, 5, 7, 11, 13, 17, 19, 23, 29, 31,
   37, 41, 43, 47, 53, 59, 61, 67, 71, 73,
   79, 83, 89, 97

Notice that, with the exception of 2, all the prime numbers are odd numbers. Also note that although 1 is a factor of all other numbers it is not regarded as being a prime number.

Prime Factors

A factor which is a prime number is called a **prime factor**. In the statement $63 = 7 \times 9$, 7 is a prime factor of 63 but 9 is not a prime factor of 63.

Example 7

Find the prime factors of 63.

$$63 = 3 \times 3 \times 7$$

Since $3 \times 3 = 3^2$ we can write

$$63 = 3^2 \times 7$$

and 3^2 and 7 are the prime factors of 63.

The prime factors of any number can be found by successive division as shown in Example 8. Note that we start by dividing by the smallest of the prime factors and end by dividing by the largest of them.

Example 8

Find the prime factors of 420.

2	420
2	210
3	105
5	35
7	7
	1

Hence

$$420 = 2 \times 2 \times 3 \times 5 \times 7$$
$$= 2^2 \times 3 \times 5 \times 7$$

Example 9

Find all the factors of 60.

The factors are:

1, 2, 3, 4, 5, 6, 10, 12, 15, 20, 30, 60,

If we pair off the factors from each end as shown above and multiply them together we get

$1 \times 60 = 60$	$2 \times 30 = 60$
$3 \times 20 = 60$	$4 \times 15 = 60$
$5 \times 12 = 60$	$6 \times 10 = 60$

We see that each pair gives a product of 60. We can use this method of pairing to check that all the factors have been found.

Lowest Common Multiple (LCM)

The multiples of 3 are 3, 6, 9, 12, 15, 18, 21, 24, 27, 30, 33, 36, 39, 42, etc.

The multiples of 4 are 4, 8, 12, 16, 20, 24, 28, 32, 36, 40, 44, etc.

We see that 12, 24 and 36 are multiples which are common to both 3 and 4. We say that 12 is the LCM of 3 and 4.

Note that 12 is the smallest number into which 3 and 4 will divide exactly.

Example 10

Find the LCM of 4, 10 and 12.

$$4 = 2 \times 2$$
$$= 2^2$$
$$10 = 2 \times 5$$
$$12 = 2 \times 2 \times 3$$
$$= 2^2 \times 3$$

The LCM is the product of the highest power of each prime factor. Hence

$$LCM = 2^2 \times 3 \times 5$$
$$= 60$$

Note that 60 is the smallest number into which 4, 10 and 12 will divide exactly.

Highest Common Factor (HCF)

The factors of 30 are 1, 2, 3, 5, 6, 10, 15 and 30.

The factors of 40 are 1, 2, 4, 5, 8, 10, 20 and 40.

We see that 1, 2, 5 and 10 are factors which are common to both 30 and 40. The highest of these common factors is 10 and we say that the HCF of 30 and 40 is 10. Note that 10 is the highest number which will divide exactly into 30 and 40.

Example 11

Find the HCF of 42, 98 and 112.

$$42 = 2 \times 3 \times 7$$
$$98 = 2 \times 7 \times 7$$
$$= 2 \times 7^2$$
$$112 = 2 \times 2 \times 2 \times 2 \times 7$$
$$= 2^4 \times 7$$

We see that only 2 and 7 are factors of all three numbers. Hence

$$HCF \text{ of } 42, 98 \text{ and } 112 = 2 \times 7$$
$$= 14$$

Note that 14 is the highest number which will divide exactly into 42, 98 and 112.

Exercise 2.4

1. Write down all the factors of
    (a) 15   (b) 64   (c) 48

2. Write down the next two prime numbers larger than 19.

3. From the set of numbers 24, 33, 45, 61, 49 and 27 write down (a) an even number, (b) an odd number, (c) a prime number, (d) a multiple of 7.

4. Consider the numbers 11, 21, 31, 77 and 112.

    (a) Two of these numbers are prime. What is the number which lies exactly half-way between them?

    (b) Three of these numbers have a common factor. What is it?

5. Express each of the following numbers as a product of its prime factors:
    (a) 24   (b) 72   (c) 45

6. Find the LCM of the following sets of numbers:

   (a) 6 and 15   (b) 12 and 6

   (c) 3, 5 and 6   (d) 4, 10 and 12

   (e) 3, 5 and 9

7. Write down all the factors of 20 and 24. Hence find the common factors and write down the HCF of 20 and 24.

8. Write down the multiples of 6 and 10 less than 61. Find the common multiples and hence find the LCM of 6 and 10.

9. Find the HCF of the following sets of numbers:

   (a) 10 and 30   (b) 6 and 15

   (c) 4, 10 and 12

10. Consider the numbers 2, 3, 4, 5, 6, 12, 18 and 24.

    (a) Which of them are factors of 12?

    (b) Which of them are multiples of 6?

11. Write down all the factors of

    (a) 12   (b) 36   (c) 60   (d) 100

12. Write down all the factors of 84 and check by pairing.

Miscellaneous Exercise 2

1. Write down two prime numbers whose sum is 16 and whose difference is 6.

2. List the following numbers:

   (a) Multiples of 8 less than 48.

   (b) The next three prime numbers greater than 23.

3. Find the smallest number that is divisible by 3, 4, 6 and 9.

4. Find the value of $13^2 - 12^2$.

5. Work out the values of $\sqrt{16 \times 49}$.

6. Find the sum of all the prime numbers between 10 and 20.

7. Consider the number sequence 1, 3, 6, 10, 15, 21, ... What is the next number in the sequence that has a square root which is a whole number?

8. What is the lowest common multiple of 3, 4 and 6?

9. Given the numbers 49, 55, 60, 63, 81 and 122 write down (a) the multiples of 5, (b) the multiples of 3, (c) the even numbers.

10. Express 360 as a product of prime factors.

11. Find the LCM of the following sets of numbers:
    (a) 10 and 15 (b) 12 and 15

12. Write down (a) all the positive whole numbers which are factors of 42, (b) all the multiples of 5 between 11 and 29.

13. Three prime numbers are 11, 13 and 37.
    (a) Find their sum. Is the sum a prime number?
    (b) Calculate their product. Is the product a prime number?

14. (a) Find the sum of 2^5 and 5^2.
    (b) Calculate the product of 3^2 and 2^3.

15. Consider the sequence 2, 3, 5, 8, 13, 21, 34, 55.
    (a) List all the even numbers in the sequence.
    (b) List all the prime numbers in the sequence.
    (c) Find the sum of the last four numbers.
    (d) Write down the next number in the sequence.

16. List the following numbers:
    (a) The prime factors of 100.
    (b) Multiples of 6 less than 55.
    (c) The next three prime numbers after 25.

A Complete GCSE Mathematics General Course

Mental Test 2

Try to answer the following questions without writing anything down except the answer.

1. Is 128 an even number?

2. Is 347 divisible by 3?

3. What is the square of 8?

4. Find the value of 2^4.

5. Is 529 an even number?

6. What is the square root of 36?

7. Find the value of $\sqrt{25 \times 64}$.

8. Find the difference between 2^3 and 3^2.

9. Is the number 425 divisible by 5?

10. Write down the factors of 24.

11. Write down all the multiples of 7 up to 41.

12. State the prime factors of 18.

13. What is the LCM of 2 and 5?

14. What is the HCF of 15 and 25?

15. Find the sum of the prime numbers in the following set: 3, 6, 8, 9, 11, 15, 19 and 21.

Fractions

Introduction

The circle shown in Fig. 3.1 has been divided into eight equal parts. Each of these parts is called one-eighth of the circle and is written $\frac{1}{8}$. The number below the line shows how many equal parts there are and it is called the **denominator**. The number above the line shows how many of these equal parts are taken and it is called the **numerator**. If five of the eight equal parts are taken then we have taken $\frac{5}{8}$ of the circle.

Fig. 3.1

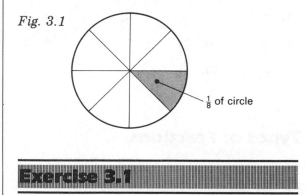

$\frac{1}{8}$ of circle

Exercise 3.1

Fig. 3.2 shows some drawings of circles. Write down the fractions represented by the shaded portions.

Fig. 3.2

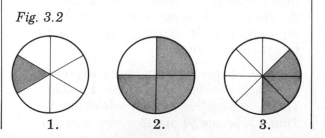

1.   2.   3.

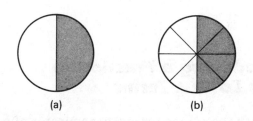

4.   5.

In a similar way to that shown in Fig. 3.2, sketch five circles and shade them to represent:

6. $\frac{7}{8}$ 7. $\frac{2}{5}$ 8. $\frac{5}{12}$ 9. $\frac{3}{10}$

10. $\frac{4}{7}$

Equivalent Fractions

In Fig. 3.3 we have shaded $\frac{1}{2}$ of a circle in diagram (a) and $\frac{4}{8}$ of a circle in diagram (b). From the diagrams we see that the same area has been shaded in both diagrams.

Hence

$$\frac{1}{2} = \frac{4}{8}$$

Fig. 3.3

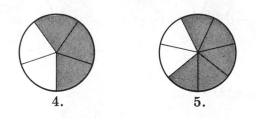

(a)   (b)

Two fractions are **equivalent** if they have the same value. We see that $\frac{1}{2}$ and $\frac{4}{8}$ are equivalent fractions and that to change $\frac{1}{2}$ into $\frac{4}{8}$ we have multiplied top and bottom by 4.

By drawing more sketches of fractions (Fig. 3.4) we find that if we multiply or divide the top and bottom of a fraction by the same number (provided it is not zero) we do not alter its value.

Fig. 3.4

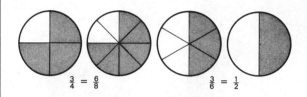

$$\frac{3}{4} = \frac{6}{8} \qquad\qquad \frac{3}{6} = \frac{1}{2}$$

Example 1

(a) $\frac{3}{7}$ is equivalent to $\dfrac{3 \times 5}{7 \times 5} = \dfrac{15}{35}$

(b) $\frac{12}{20}$ is equivalent to $\dfrac{12 \div 4}{20 \div 4} = \dfrac{3}{5}$

Exercise 3.2

Draw suitable sketches to show that

1. $\frac{2}{3} = \frac{6}{9}$ 2. $\frac{4}{5} = \frac{8}{10}$ 3. $\frac{3}{4} = \frac{6}{8}$

4. $\frac{8}{14} = \frac{4}{7}$ 5. $\frac{6}{10} = \frac{3}{5}$

Copy the following fractions and fill in the missing numbers:

6. $\frac{3}{4} = \frac{}{20}$ 7. $\frac{2}{3} = \frac{}{12}$ 8. $\frac{3}{7} = \frac{}{28}$

9. $\frac{3}{8} = \frac{}{32}$ 10. $\frac{1}{6} = \frac{}{24}$ 11. $\frac{9}{12} = \frac{}{36}$

12. $\frac{4}{8} = \frac{}{4}$ 13. $\frac{9}{15} = \frac{}{5}$ 14. $\frac{35}{42} = \frac{}{6}$

15. $\frac{21}{49} = \frac{}{7}$

Reducing a Fraction to its Lowest Terms

When the numerator and denominator of a fraction possess no common factors, the fraction is said to be in its **lowest terms**.

The fraction $\frac{7}{16}$ is in its lowest terms because there is no whole number that will divide exactly into 7 and 16.

The fraction $\frac{6}{8}$ is not in its lowest terms because 2 will divide exactly into 6 and 8 to give the equivalent fraction $\frac{3}{4}$.

Example 2

Reduce $\frac{8}{12}$ to its lowest terms.

$$\frac{8}{12} \text{ is equivalent to } \dfrac{8 \div 4}{12 \div 4} = \dfrac{2}{3}$$

Exercise 3.3

Reduce each of the following fractions to its lowest terms:

1. $\frac{8}{16}$ 2. $\frac{15}{25}$ 3. $\frac{9}{15}$ 4. $\frac{8}{64}$

5. $\frac{12}{18}$ 6. $\frac{42}{48}$ 7. $\frac{22}{33}$ 8. $\frac{18}{27}$

9. $\frac{21}{24}$ 10. $\frac{35}{56}$

Types of Fractions

If the top number of a fraction is less than the bottom number then the fraction is called a **proper fraction**. Thus $\frac{2}{3}, \frac{5}{8}$ and $\frac{9}{20}$ are all proper fractions.

If the top number of a fraction is greater than the bottom number then the fraction is called an **improper fraction** or a **top-heavy fraction**. Thus $\frac{8}{5}, \frac{5}{2}$ and $\frac{19}{7}$ are all top-heavy fractions.

Every top-heavy fraction can be expressed as a whole number and a proper fraction. These are sometimes called **mixed numbers**. Thus $2\frac{3}{5}, 5\frac{7}{8}$ and $9\frac{1}{2}$ are all mixed numbers.

Example 3

Change $\frac{7}{3}$ into a mixed number.

$\frac{7}{3}$ may be written as $\dfrac{6+1}{3} = \frac{6}{3} + \frac{1}{3}$

$$= 2 + \frac{1}{3}$$

$$= 2\frac{1}{3}$$

From this example we see that the rule for changing top-heavy fractions into mixed numbers is:

Divide the numerator (top number) by the denominator (bottom number). The answer becomes the whole-number part of the mixed number. The remainder becomes the numerator (top number) of the fractional part whose denominator is the denominator of the top-heavy fraction.

Example 4

Change $\frac{35}{4}$ into a mixed number.

$$35 \div 4 = 8 \text{ remainder } 3$$

Therefore $\frac{35}{4} = 8\frac{3}{4}$.

All mixed numbers can be changed into top-heavy fractions. The rule is:

To change a mixed number into a top-heavy fraction multiply the whole number by the denominator (bottom part) of the fractional part. Then add the numerator (top part) of the fractional part to this product. This gives the numerator of the top-heavy fraction whose denominator is the denominator of the original fractional part of the mixed number.

Example 5

Change $5\frac{2}{7}$ into a top-heavy fraction.

$$5\frac{2}{7} = \dfrac{(5 \times 7) + 2}{7}$$

$$= \dfrac{35 + 2}{7}$$

$$= \frac{37}{7}$$

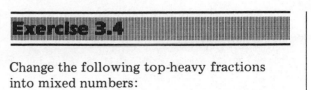
Exercise 3.4

Change the following top-heavy fractions into mixed numbers:

1. $\frac{7}{2}$ 2. $\frac{19}{8}$ 3. $\frac{22}{9}$

4. $\frac{12}{11}$ 5. $\frac{21}{8}$

Change the following mixed numbers into top-heavy fractions:

6. $2\frac{3}{8}$ 7. $5\frac{1}{10}$ 8. $8\frac{2}{3}$

9. $6\frac{5}{7}$ 10. $4\frac{2}{9}$

Comparing the Values of Fractions

When putting fractions in order of size, first express each of the fractions with the same denominator. This common denominator should be the LCM of the denominators of the fractions to be compared.

Example 6

Arrange the fractions $\frac{3}{4}, \frac{5}{8}, \frac{7}{10}$ and $\frac{11}{20}$ in order of size, beginning with the smallest.

The LCM of the denominators 4, 8, 10 and 20 is 40.

We now change each of the given fractions so that it has a denominator of 40.

$\frac{3}{4}$ is equivalent to $\dfrac{3 \times 10}{4 \times 10} = \dfrac{30}{40}$

$\frac{5}{8}$ is equivalent to $\dfrac{5 \times 5}{8 \times 5} = \dfrac{25}{40}$

$\frac{7}{10}$ is equivalent to $\dfrac{7 \times 4}{10 \times 4} = \dfrac{28}{40}$

$\frac{11}{20}$ is equivalent to $\dfrac{11 \times 2}{20 \times 2} = \dfrac{22}{40}$

After the fractions have been expressed with the same denominator, all we have to do is compare their numerators.

Therefore the order is

$$\tfrac{22}{40}, \tfrac{25}{40}, \tfrac{28}{40} \text{ and } \tfrac{30}{40}, \text{ or } \tfrac{11}{20}, \tfrac{5}{8}, \tfrac{7}{10} \text{ and } \tfrac{3}{4}$$

Exercise 3.5

Arrange the following sets of fractions in order of size, beginning with the smallest:

1. $\tfrac{1}{2}, \tfrac{5}{6}, \tfrac{2}{3}$ and $\tfrac{7}{12}$
2. $\tfrac{9}{10}, \tfrac{3}{4}, \tfrac{7}{8}$ and $\tfrac{3}{5}$
3. $\tfrac{3}{4}, \tfrac{5}{8}, \tfrac{9}{16}$ and $\tfrac{17}{32}$
4. $\tfrac{1}{3}, \tfrac{1}{4}, \tfrac{1}{2}$ and $\tfrac{1}{5}$
5. $\tfrac{3}{8}, \tfrac{5}{9}, \tfrac{2}{6}$ and $\tfrac{5}{18}$

Addition of Fractions

The steps when adding fractions are as follows:

(1) Find the LCM of the denominators of the fractions to be added.
(2) Express each of the fractions with this common denominator.
(3) Add the numerators of the new fractions to give the numerator of the answer. The denominator of the answer is the LCM found in (1).

Example 7

Add $\tfrac{2}{7}$ and $\tfrac{3}{4}$.

The LCM of the denominators 7 and 4 is 28.

Expressing each fraction with a denominator of 28 gives

$$\tfrac{2}{7} \text{ is equivalent to } \frac{2 \times 4}{7 \times 4} = \frac{8}{28}$$

$$\tfrac{3}{4} \text{ is equivalent to } \frac{3 \times 7}{4 \times 7} = \frac{21}{28}$$

$$\tfrac{2}{7} + \tfrac{3}{4} = \tfrac{8}{28} + \tfrac{21}{28}$$
$$= \frac{8 + 21}{28}$$
$$= \frac{29}{28}$$
$$= 1\tfrac{1}{28}$$

The work may be set out as follows:

$$\tfrac{2}{7} + \tfrac{3}{4} = \frac{(2 \times 4) + (3 \times 7)}{28}$$
$$= \frac{8 + 21}{28}$$
$$= \tfrac{29}{28}$$
$$= 1\tfrac{1}{28}$$

If mixed numbers are present then add the whole numbers first.

Example 8

Add $5\tfrac{1}{2}$, $2\tfrac{2}{3}$ and $3\tfrac{2}{5}$.

The LCM of the denominators 2, 3 and 5 is 30.

$$5\tfrac{1}{2} + 2\tfrac{2}{3} + 3\tfrac{2}{5}$$
$$= 5 + 2 + 3 + \tfrac{1}{2} + \tfrac{2}{3} + \tfrac{2}{5}$$
$$= 10 + \frac{(1 \times 15) + (2 \times 10) + (2 \times 6)}{30}$$
$$= 10 + \frac{15 + 20 + 12}{30}$$
$$= 10 + \tfrac{47}{30}$$
$$= 10 + 1\tfrac{17}{30}$$
$$= 10 + 1 + \tfrac{17}{30}$$
$$= 11 + \tfrac{17}{30}$$
$$= 11\tfrac{17}{30}$$

Exercise 3.6

Work out:

1. $\frac{1}{2} + \frac{1}{3}$ 2. $\frac{2}{5} + \frac{9}{10}$

3. $\frac{3}{4} + \frac{3}{8}$ 4. $\frac{3}{10} + \frac{1}{4}$

5. $\frac{1}{2} + \frac{3}{4} + \frac{7}{8}$ 6. $\frac{1}{8} + \frac{2}{3} + \frac{5}{12}$

7. $1\frac{3}{8} + 3\frac{9}{16}$ 8. $7\frac{2}{3} + 6\frac{3}{5}$

9. $3\frac{3}{8} + 5\frac{2}{7} + 4\frac{3}{4}$ 10. $4\frac{1}{2} + 3\frac{5}{6} + 2\frac{1}{3}$

11. $7\frac{3}{8} + 2\frac{3}{4} + \frac{7}{8} + \frac{5}{16}$ 12. $7\frac{2}{3} + \frac{2}{5} + \frac{3}{10} + 2\frac{1}{2}$

Subtraction of Fractions

The method is similar to that used in addition. We first find the LCM of the denominators of the fractions and then express each fraction with this common denominator. We then subtract the numerators of the new fractions.

Example 9

Subtract $\frac{2}{5}$ from $\frac{5}{8}$.

The LCM of the denominators 5 and 8 is 40.

$$\frac{5}{8} - \frac{2}{5} = \frac{(5 \times 5) - (2 \times 8)}{40}$$

$$= \frac{25 - 16}{40}$$

$$= \frac{9}{40}$$

Example 10

Take $2\frac{1}{4}$ from $3\frac{7}{10}$.

The LCM of the denominators 4 and 10 is 20.

$$3\frac{7}{10} - 2\frac{1}{4} = 3 - 2 + \frac{7}{10} - \frac{1}{4}$$

$$= 1 + \frac{(7 \times 2) - (1 \times 5)}{20}$$

$$= 1 + \frac{14 - 5}{20}$$

$$= 1 + \frac{9}{20}$$

$$= 1\frac{9}{20}$$

Sometimes we have to 'borrow' from the whole number as shown in Example 11.

Example 11

Subtract $2\frac{3}{4}$ from $4\frac{2}{3}$.

The LCM of 4 and 3 is 12.

$$4\frac{2}{3} - 2\frac{3}{4} = 4 - 2 + \frac{2}{3} - \frac{3}{4}$$

$$= 2 + \frac{2}{3} - \frac{3}{4}$$

Because $\frac{3}{4}$ is larger than $\frac{2}{3}$ we cannot subtract and so we write

$$4\frac{2}{3} - 2\frac{3}{4} = 1 + 1 + \frac{2}{3} - \frac{3}{4}$$

$$= 1 + \frac{(1 \times 12) + (2 \times 4) - (3 \times 3)}{12}$$

$$= 1 + \frac{12 + 8 - 9}{12}$$

$$= 1 + \frac{11}{12}$$

$$= 1\frac{11}{12}$$

'Borrowing' can be avoided by changing mixed numbers into top-heavy fractions before attempting to subtract.

$$4\frac{2}{3} - 2\frac{3}{4} = \frac{14}{3} - \frac{11}{4}$$

$$= \frac{(14 \times 4) - (11 \times 3)}{12}$$

$$= \frac{56 - 33}{12}$$

$$= \frac{23}{12}$$

$$= 1\frac{11}{12}$$

Exercise 3.7

Work out:

1. $\frac{1}{2} - \frac{1}{3}$ 2. $\frac{7}{8} - \frac{5}{6}$ 3. $\frac{1}{3} - \frac{1}{5}$

4. $\frac{2}{3} - \frac{1}{2}$ 5. $3\frac{3}{8} - 1\frac{1}{4}$ 6. $3 - \frac{5}{7}$

7. $5\frac{7}{16} - 3\frac{1}{5}$ 8. $5\frac{3}{8} - 2\frac{9}{10}$ 9. $4\frac{3}{8} - 3\frac{2}{5}$

10. $2\frac{2}{3} - 1\frac{3}{4}$

Combined Addition and Subtraction

When a problem contains both addition and subtraction it is often best to make each of the mixed numbers into a top-heavy fraction.

Example 12

Work out $5\frac{3}{8} - 1\frac{1}{4} + 2\frac{1}{2} - \frac{7}{16}$.

$$5\frac{3}{8} - 1\frac{1}{4} + 2\frac{1}{2} - \frac{7}{16}$$

$$= \frac{43}{8} - \frac{5}{4} + \frac{5}{2} - \frac{7}{16}$$

$$= \frac{(43 \times 2) - (5 \times 4) + (5 \times 8) - 7}{16}$$

$$= \frac{86 - 20 + 40 - 7}{16}$$

$$= \frac{(86 + 40) - (20 + 7)}{16}$$

$$= \frac{126 - 27}{16}$$

$$= \frac{99}{16}$$

$$= 6\frac{3}{16}$$

Exercise 3.8

Work out:

1. $2\frac{1}{2} + 3\frac{1}{4} - 4\frac{3}{8}$

2. $5\frac{1}{10} - 3\frac{1}{2} - 1\frac{1}{4}$

3. $4\frac{3}{8} - 2\frac{1}{2} + 5$

4. $6\frac{1}{2} - 3\frac{1}{6} + 2\frac{1}{12} - 4\frac{3}{4}$

5. $1\frac{3}{16} - 2\frac{2}{5} + 3\frac{3}{4} + 5\frac{5}{8}$

6. $12\frac{3}{4} - 6\frac{7}{8} - 2\frac{13}{16} + 5\frac{1}{2}$

7. $3\frac{9}{20} + 1\frac{3}{8} - 2\frac{7}{10} - \frac{3}{4}$

8. $1\frac{1}{12} - 4\frac{3}{4} - 3\frac{2}{3} + 8\frac{3}{8}$

Multiplication of Fractions

To multiply two or more fractions together, first multiply all the top numbers together and then multiply all the bottom numbers together. Mixed numbers must always be made into top-heavy fractions before attempting to multiply.

Example 13

Multiply $\frac{5}{8}$ by $\frac{3}{7}$.

$$\frac{5}{8} \times \frac{3}{7} = \frac{5 \times 3}{8 \times 7}$$

$$= \frac{15}{56}$$

Example 14

Multiply $1\frac{3}{8}$ by $1\frac{1}{4}$.

$$1\frac{3}{8} \times 1\frac{1}{4} = \frac{11}{8} \times \frac{5}{4}$$

$$= \frac{11 \times 5}{8 \times 4}$$

$$= \frac{55}{32}$$

$$= 1\frac{23}{32}$$

Exercise 3.9

Work out:

1. $\frac{2}{3} \times \frac{4}{5}$ 2. $\frac{3}{4} \times \frac{5}{7}$ 3. $\frac{2}{9} \times 1\frac{2}{3}$

4. $\frac{5}{9} \times 2\frac{3}{4}$ 5. $1\frac{2}{5} \times 3\frac{1}{2}$ 6. $2\frac{1}{2} \times 2\frac{1}{3}$

7. $1\frac{2}{9} \times 1\frac{2}{5}$ 8. $1\frac{7}{8} \times 1\frac{4}{7}$

Cancelling

Example 15

Work out $\frac{2}{3} \times 1\frac{7}{8}$.

$$\frac{2}{3} \times 1\frac{7}{8} = \frac{2}{3} \times \frac{15}{8}$$
$$= \frac{2 \times 15}{3 \times 8}$$
$$= \frac{30}{24}$$
$$= \frac{5}{4}$$
$$= 1\frac{1}{4}$$

The step of reducing $\frac{30}{24}$ to its lowest terms has been achieved by dividing 6 into both the numerator and the denominator. The work can often be made easier by cancelling the common factors before multiplying:

$$\frac{\overset{1}{\cancel{2}}}{\underset{1}{\cancel{3}}} \times \frac{\overset{5}{\cancel{15}}}{\underset{4}{\cancel{8}}} = \frac{1 \times 5}{1 \times 4}$$

$$= \frac{5}{4}$$
$$= 1\frac{1}{4}$$

We have divided 2 into 2 (a top number) and into 8 (a bottom number). Also we have divided 3 into 15 (a top number) and into 3 (a bottom number). You will see that we have divided the top and bottom numbers by the same amount. Note carefully that we can only cancel between a top number and a bottom number.

Example 16

Work out $3\frac{1}{8} \times \frac{7}{10} \times \frac{2}{21}$.

$$3\frac{1}{8} \times \frac{7}{10} \times \frac{2}{21} = \frac{\overset{5}{\cancel{25}}}{\underset{4}{\cancel{8}}} \times \frac{\overset{1}{\cancel{7}}}{\underset{2}{\cancel{10}}} \times \frac{\overset{1}{\cancel{2}}}{\underset{3}{\cancel{21}}}$$

$$= \frac{5 \times 1 \times 1}{4 \times 2 \times 3}$$

$$= \frac{5}{24}$$

Sometimes in calculations involving fractions the word 'of' appears. It should always be taken as meaning multiply.

Example 17

Find $\frac{4}{5}$ of 20.

$$\frac{4}{5} \text{ of } 20 = \frac{4}{\underset{1}{\cancel{5}}} \times \frac{\overset{4}{\cancel{20}}}{1}$$

$$= \frac{4 \times 4}{1 \times 1}$$

$$= \frac{16}{1}$$

$$= 16$$

Exercise 3.10

Work out:

1. $\frac{3}{4} \times 1\frac{7}{9}$ 2. $5\frac{1}{5} \times \frac{10}{13}$

3. $1\frac{5}{8} \times \frac{7}{26}$ 4. $1\frac{1}{2} \times \frac{2}{5} \times 2\frac{1}{2}$

5. $\frac{5}{8} \times \frac{7}{10} \times \frac{2}{21}$ 6. $2 \times 1\frac{1}{2} \times 1\frac{1}{3}$

7. $3\frac{3}{4} \times 1\frac{3}{5} \times 1\frac{1}{8}$ 8. $\frac{15}{32} \times \frac{8}{11} \times 24\frac{1}{5}$

9. $\frac{3}{4}$ of 16   10. $\frac{5}{7}$ of 140

11. $\frac{2}{3}$ of $4\frac{1}{2}$ 12. $\frac{4}{5}$ of $2\frac{1}{2}$

Division of Fractions

To divide by a fraction, invert it and multiply.

Example 18

(a) Divide $\frac{3}{5}$ by $\frac{2}{7}$.

$$\frac{3}{5} \div \frac{2}{7} = \frac{3}{5} \times \frac{7}{2}$$

$$= \frac{3 \times 7}{5 \times 2}$$

$$= \frac{21}{10}$$

$$= 2\frac{1}{10}$$

(b) Divide $1\frac{4}{5}$ by $2\frac{1}{3}$.

$$1\frac{4}{5} \div 2\frac{1}{3} = \frac{9}{5} \div \frac{7}{3}$$

$$= \frac{9}{5} \times \frac{3}{7}$$

$$= \frac{9 \times 3}{5 \times 7}$$

$$= \frac{27}{35}$$

Exercise 3.11

Work out:

1. $\frac{4}{5} \div 1\frac{1}{3}$ 2. $2 \div \frac{1}{4}$ 3. $\frac{5}{8} \div \frac{15}{32}$

4. $3\frac{3}{4} \div 2\frac{1}{2}$ 5. $2\frac{1}{2} \div 3\frac{3}{4}$ 6. $5 \div 5\frac{1}{5}$

7. $3\frac{1}{15} \div 2\frac{5}{9}$ 8. $2\frac{3}{10} \div \frac{3}{5}$

Operations with Fractions

The sequence of operations when dealing with fractions is the same as that used with whole numbers:

(1) Work out the contents of brackets.

(2) Multiply and divide.

(3) Add and subtract.

Example 19

Simplify $(2\frac{1}{2} - 1\frac{1}{3}) \div 1\frac{5}{9}$.

$$(2\frac{1}{2} - 1\frac{1}{3}) \div 1\frac{5}{9} = (\frac{5}{2} - \frac{4}{3}) \div \frac{14}{9}$$

$$= \frac{5 \times 3 - 4 \times 2}{6} \div \frac{14}{9}$$

$$= \frac{7}{6} \div \frac{14}{9}$$

$$= \frac{7^1}{6_2} \times \frac{9^3}{14_2}$$

$$= \frac{1 \times 3}{2 \times 2}$$

$$= \frac{3}{4}$$

Exercise 3.12

Work out:

1. $\frac{1}{4} \div (1\frac{1}{8} \times \frac{2}{5})$ 2. $1\frac{2}{3} \div (\frac{3}{5} \div \frac{9}{10})$

3. $(1\frac{7}{8} \times 2\frac{2}{5}) - 3\frac{2}{3}$ 4. $(2\frac{2}{3} + 1\frac{1}{5}) \div 5\frac{4}{5}$

5. $3\frac{2}{3} \div (\frac{2}{3} + \frac{4}{5})$ 6. $\frac{2}{5} \times (\frac{2}{3} - \frac{1}{4}) + \frac{1}{2}$

7. $2\frac{8}{9} \div (1\frac{2}{3} + \frac{1}{2})$ 8. $(2\frac{1}{2} - 1\frac{3}{8}) \times 1\frac{1}{3}$

Practical Applications of Fractions

The examples which follow are all of a practical nature. They depend upon fractions for their solution.

Example 20

(a) A girl spends $\frac{3}{4}$ of her pocket money and has 90p left. How much did she have to start with?

The whole amount of her pocket money is represented by 1, so the amount left is represented by

$$1 - \frac{3}{4} = \frac{1}{4}$$

$\frac{1}{4}$ represents 90p

1 represents $4 \times 90p = 360p$

The girl had £3.60 to start with.

(b) A group of school children went to a hamburger bar. $\frac{2}{5}$ of them bought hamburgers, $\frac{1}{4}$ bought chips and the remainder bought drinks.

(i) What fraction bought food?

(ii) What fraction bought drinks?

(i)   Fraction who bought food $= \dfrac{2}{5} + \dfrac{1}{4}$

$$= \frac{8 + 5}{20}$$

$$= \frac{13}{20}$$

(ii) The whole group of school children is a whole unit, i.e. 1. Therefore

Fraction who bought drinks

$$= 1 - \frac{13}{20}$$

$$= \frac{20 - 13}{20}$$

$$= \frac{7}{20}$$

Exercise 3.13

1. Calculate $\frac{3}{16}$ of £800.

2. Jane takes $5\frac{3}{4}$ minutes to iron a blouse. How many blouses can she iron in 23 minutes?

3. At a Youth Club $\frac{2}{5}$ of those present were playing darts and $\frac{1}{4}$ were playing other games.
   (a) What fraction were playing games?
   (b) What fraction were not playing games?

4. A watering can holds $12\frac{1}{2}$ litres. It is filled 11 times from a tank containing 400 litres. How much water is left in the tank?

5. A school has 600 pupils. $\frac{1}{5}$ are in the upper school, $\frac{1}{4}$ in the middle school and the remainder in the lower school. How many pupils are in the lower school?

6. A boy spends $\frac{5}{8}$ of his pocket money and has 60p left. How much money did he have to start with?

7. During 'bob a job week' a boy scout decided to earn money by cleaning shoes. It takes him $2\frac{1}{2}$ minutes to clean one pair. At one house he was given 12 pairs to clean. How long did it take him to complete the task?

8. The profits of a business are £29 000. It is shared between two partners A and B. If A receives $\frac{9}{20}$ of the profits, how much money does B receive?

Miscellaneous Exercise 3

1. Express $\frac{3}{8} \times \frac{5}{6} \times \frac{4}{15}$ as a single fraction.

2. Find the value of $\frac{2}{3} - \frac{4}{9}$.

3. Find the next three terms of the sequence $1\frac{3}{4}, 2, 2\frac{1}{4}, \ldots$

4. Work out $6 - 1\frac{7}{8}$.

5. Make $3\frac{1}{5} + 1\frac{5}{8}$ into a single fraction.

6. What is the square root of $\frac{16}{49}$?

7. Which is the smallest of the fractions $\frac{2}{3}, \frac{11}{15}, \frac{7}{10}$ and $\frac{5}{6}$?

8. What is $\frac{2}{3}$ of 60?

9. Write down the next two terms of the sequence $2, 1\frac{2}{3}, 1\frac{1}{3}, \ldots$

10. Write down the next term of the sequence $\frac{1}{2}, \frac{3}{4}, \frac{9}{8}, \ldots$

11. Consider the fractions $\frac{3}{5}, \frac{3}{4}, \frac{7}{10}$ and $\frac{13}{20}$.
    (a) Which fraction is the largest?
    (b) What is the difference between the largest and smallest fractions?
    (c) Calculate the product of $\frac{3}{5}$ and $\frac{3}{4}$.
    (d) Divide $\frac{3}{5}$ by $\frac{13}{20}$.

12. Give the answers to the following in their lowest terms:
    (a) $\frac{3}{5} \times \frac{10}{21}$ (b) $\frac{5}{8} \div 2\frac{3}{4}$
    (c) $\frac{7}{8} - \frac{2}{3}$ (d) $1\frac{1}{3} + 2\frac{3}{4}$

13. Find $\frac{3}{4}$ of $\frac{2}{3}$.

14. Find the next three terms in the sequence $2\frac{1}{4}, 2\frac{5}{8}, 3, \ldots$

15. Find $\frac{5}{8}$ of 32.

16. Write down the next fraction in the sequence $\frac{3}{4}, \frac{9}{16}, \frac{3}{8}, \ldots$

17. Express $\frac{1}{2} - \frac{3}{4} + \frac{5}{8} - \frac{7}{16} + \frac{19}{32}$ as a single fraction.

18. Express as a single fraction $(2\frac{1}{2} + 1\frac{1}{4}) \div 2\frac{1}{2}$.

19. A man left $\frac{3}{8}$ of his money to his wife and half the remainder to his son. If he left £8000, how much did his son receive?

20. Find $\frac{3}{4}$ of $7\frac{1}{3}$.

Mental Test 3

Try to write down the answers to the following without writing anything else.

1. Work out $\frac{3}{8} + \frac{1}{2}$.

2. Find the sum of $\frac{3}{10}$ and $\frac{7}{20}$.

3. Take $\frac{3}{5}$ from $\frac{7}{10}$.

4. Add $\frac{1}{2}$, $\frac{1}{4}$ and $\frac{1}{8}$.

5. Find the product of $\frac{1}{2}$ and $\frac{1}{5}$.

6. Work out $\frac{2}{3} \times \frac{3}{7}$.

7. Work out $\frac{2}{5} \times \frac{5}{8}$.

8. What is $\frac{1}{4}$ of 20?

9. Work out $\frac{1}{5} \div 2$.

10. Work out $\frac{1}{4} \div \frac{1}{2}$.

11. Write down $\frac{12}{15}$ in its lowest terms.

12. Write $5\frac{2}{7}$ as a top-heavy fraction.

4 The Decimal System

Place Value

When we write the number 666 we mean $600 + 60 + 6$. Reading from left to right, each figure 6 is ten times the value of the next 6.

We now have to decide how to deal with fractional quantities, i.e. quantities whose values are less than 1. If we regard 666.666 as meaning $600 + 60 + 6 + \frac{6}{10} + \frac{6}{100} + \frac{6}{1000}$ then the dot, called the **decimal point**, separates the whole numbers from the fractional parts.

Notice that in the fractional or decimal part .666, each figure 6 has ten times the value of the following 6. So $\frac{6}{10}$ has ten times the value of $\frac{6}{100}$ and this has ten times the value of $\frac{6}{1000}$.

Decimals, then, are fractions which have denominators of 10, 100, 1000, etc., according to the position of the figure after the decimal point.

If we want to write six hundred and five we write 605, the zero keeping the place for the missing tens. In the same way, to write $\frac{3}{10} + \frac{5}{1000}$ we write .305, the zero keeping the place for the missing hundredths. Also, $\frac{4}{100} + \frac{7}{1000}$ would be written .047, the zero keeping the place for the missing tenths.

Wh_n there are no whole numbers it is usual t_ _ _t a zero in front of the decimal point _ _t, for instance, .35 would be written

A table similar to the following may be used to illustrate place value.

Hundreds	Tens	Units		Tenths	Hundredths	Thousandths	Number in decimal form
		9	.	6	5		9.65
	8	3	.	2	7	4	83.274
		0	.	3			0.3
		0	.	4	7		0.47
9	7	6	.	0	5	3	976.053

From the table we see that:

the first column after the decimal point represents tenths;

the second column after the decimal point represents hundredths;

the third column after the decimal point represents thousandths and so on.

Example 1

(a) Write down the place value of the figure 7 in the number 54.374.

Since the 7 occurs in the second column after the decimal point, its value is 7 hundredths, i.e. $\frac{7}{100}$.

(b) Find the difference between the actual values of the two sevens in the number 679.73.

The first seven has a value of 70.

The second seven has a value of $\frac{7}{10}$.

The difference between the actual values of the two sevens is
$70 - \frac{7}{10} = 69\frac{3}{10} = 69.3$

Exercise 4.1

Write down the place value of

1. The figure 9 in
    (a) 34.79   (b) 5.369
    (c) 475.92   (d) 89.753

2. The figure 5 in
    (a) 579.2   (b) 0.059
    (c) 28.57   (d) 7.852

3. The figure 8 in
    (a) 8734.9   (b) 0.0083
    (c) 19.84   (d) 77.784

4. What is the numerical difference between the actual values of the two fours in the number 0.404?

5. What is the numerical difference between the actual values of the two nines in the number 594.93?

6. Read off as decimal numbers:
    (a) $\frac{7}{10}$ (b) $\frac{3}{10} + \frac{7}{100}$
    (c) $\frac{5}{10} + \frac{8}{100} + \frac{9}{1000}$ (d) $\frac{9}{1000}$
    (e) $\frac{3}{100}$ (f) $\frac{1}{100} + \frac{7}{1000}$

7. Read off the following with denominators (bottom numbers) 10, 100, 1000:
    (a) 0.2   (b) 4.6
    (c) 3.58   (d) 0.256
    (e) 0.004   (f) 0.036
    (g) 80.29   (h) 0.032

Number Line for Decimals

Fig. 4.1 shows a number line for decimal numbers. The numbers −1.3, 2.4 and 5.8 have been shown.

Fig. 4.1

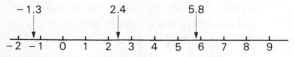

Exercise 4.2

1. Make a copy of the scale shown in Fig. 4.1. Draw arrows and label the following points:
    (a) 7.2   (b) 3.7   (c) 4.2
    (d) 0.7   (e) −0.8   (f) −1.8

2. By looking at the number line (Fig. 4.1) decide which number is the larger:
    (a) 0.4 or 0.8   (b) 3.2 or 4.9
    (c) −2.0 or 2.0   (d) −0.8 or 0.3
    (e) −1.9 or 1.2

3. Rearrange the following numbers in order of size, largest first: −9, −29 and −3.

4. Write down in order of size, smallest first, the numbers −4, 0, −6 and 3.

5. Write the following numbers in order of size, smallest first: 5.205, −52.05, 5.502, 5.025 and −5.052.

The Decimal Point and Following Zeros

It is sometimes useful to add zeros at the end of numbers, particularly when adding and subtracting decimal numbers.

The number 3 means 3 units.

The number 3.0 means 3 units and no tenths.

The number 3.00 means 3 units, no tenths and no hundredths.

The number 8.6 means 8 units and six tenths.

The number 8.60 means eight units, six units and no hundredths.

Sometimes noughts are placed in front of integers. For instance, a bearing could be 036°, and James Bond is known as 00

Adding and Subtracting Decimals

Adding and subtracting decimals is done in exactly the same way as for whole numbers. Care must be taken, however, to write the decimal points underneath one another. This makes sure that all the figures having the same place value fall in the same column. Putting zeros at the end of numbers so that all the numbers to be added have the same number of figures after the decimal point also helps.

Example 2

Add 11.362, 2.63 and 27.1.

$$
\begin{array}{r}
11.362 \\
2.630 \\
27.100\, + \\
\hline
41.092 \\
\end{array}
$$

Therefore
$11.362 + 2.63 + 27.1 = 41.092.$

Most problems involving the addition of numbers will be done using a calculator. The method is as follows:

Input	Display
11.362	11.362
+	11.362
2.63	2.63
+	13.992
27.1	27.1
=	41.092

The addition may be checked by inputting the numbers in reverse order, i.e. $27.1 + 2.63 + 11.362$.

Example 3

Subtract 8.34 from 17.2.

$$
\begin{array}{r}
17.20 \\
8.34\, - \\
\hline
8.86 \\
\end{array}
$$

Therefore $17.2 - 8.34 = 8.86.$

Using a calculator the programme is:

Input	Display
17.2	17.2
−	17.2
8.34	8.34
=	8.86

Subtraction may always be checked by adding the bottom two numbers on the calculation. If this sum equals the top number the subtraction is correct. Thus in Example 3:

$$8.34 + 8.86 = 17.20$$

Hence the answer is correct.

Many problems involve combined addition and subtraction. More often than not a calculator will be used, as shown in Example 4.

Example 4

Work out $9.24 - 35.7 - 58.32 + 6.53 + 75.6$

Input	Display
9.24	9.24
−	9.24
35.7	35.7
−	−26.46
58.32	58.32
+	−84.78
6.53	6.53
+	−78.25
75.6	75.6
=	−2.65

Hence $9.24 - 35.7 - 58.32 + 6.53 + 75.6 = -2.65.$

The work may be checked by inputting the numbers in reverse order.

Exercise 4.3

Find the value of:

1. $2.375 + 0.63$

2. $3.196 + 2.48 + 18.5$

3. $38.7 + 0.05 + 23$

4. $27.418 + 0.97 + 25 + 1.967$

5. $12.48 - 3.5$ 6. $0.867 - 0.039$

7. $5.48 - 0.0691$

8. $87.3 - 15.64 + 3.28 - 56.25$

9. $7.21 - 4.48 - 12.57 - 54.2 + 37.91 + 86.56$

10. $23.65 - 34.09 - 27.5 - 36.23 - 11.33 + 5.7$

Multiplying by Powers of 10

To multiply by 10 move all the figures one place to the left.

$$8.532 \times 10 = 85.32$$

$$0.083 \times 10 = 0.83$$

To multiply by 100 move all the figures two places to the left.

$$97.48 \times 100 = 9748$$

$$7.638 \times 100 = 763.8$$

To multiply by 1000 move all the figures three places to the left.

$$975.364 \times 1000 = 975\,364$$

$$0.007\,34 \times 1000 = 7.34$$

$$2.9 \times 1000 = 2900.0$$

$$= 2900$$

Dividing by Powers of 10

To divide by 10 move all the figures one place to the right.

$$57.48 \div 10 = 5.748$$

$$0.279 \div 10 = 0.0279$$

To divide by 100 move all the figures two places to the right.

$$598.65 \div 100 = 5.9865$$

$$4.738 \div 100 = 0.047\,38$$

To divide by 1000 move all the figures three places to the right.

$$9438 \div 1000 = 9.438$$

$$25.472 \div 1000 = 0.025\,472$$

Exercise 4.4

Multiply the following numbers by
(a) 10, (b) 100 and (c) 1000:

1. 0.45   2. 7.893   3. 0.058

4. 89.2346   5. 8.2643   6. 0.063

7. 0.000 28   8. 0.008

Divide each of the following numbers by
(a) 10, (b) 100 and (c) 1000:

9. 289   10. 28.17   11. 827.34

12. 0.04   13. 0.625   14. 0.0038

15. 6.52

Multiplying and Dividing Decimals

Multiplying and dividing longhand is very time consuming and it is best to use a calculator for these operations.

Example 5

Multiply 18.23 by 9.76.

Input	Display
18.23	18.23
×	18.23
9.76	9.76
=	177.9248

Therefore $18.23 \times 9.76 = 177.9248$.
To check the work, input the numbers in reverse order, i.e. 9.76×18.23.

Example 6

Divide 60.888 by 3.54.

Input	Display
60.888	60.888
÷	60.888
3.54	3.54
=	17.2

Therefore $60.888 \div 3.54 = 17.2$
Many problems occur that involve both multiplication and division together and these are best done using a calculator.

Example 7

Work out $\dfrac{12.88 \times 2.59}{4.6 \times 0.74}$

Input	Display
12.88	12.88
×	12.88
2.59	2.59
÷	33.3592
4.6	4.6
÷	7.252
0.74	0.74
=	9.8

Therefore $\dfrac{12.88 \times 2.59}{4.6 \times 0.74} = 9.8$

Exercise 4.5

Work out each of the following using a calculator:

1. 25.42×29.23
2. 0.361×2.63
3. 0.76×0.38
4. 3.025×2.45
5. 0.043×0.32
6. $44.8 \div 12.8$
7. $3.484 \div 0.26$
8. $3.914 \div 2.06$
9. $11.56 \div 2.72$
10. $0.0144 \div 0.32$
11. $\dfrac{53.04 \times 0.806}{3.4 \times 0.26}$
12. $\dfrac{1.6856 \times 14.9}{0.028}$
13. $\dfrac{3.122 \times 0.703}{11.15 \times 0.037}$
14. $\dfrac{65.178 \times 0.1674}{19.17 \times 0.62}$
15. $\dfrac{0.001\,35 \times 0.008\,37}{0.003 \times 0.027}$

Decimal Places

Example 8

Work out $15.187 \div 3.57$.

Using a calculator

$$15.187 \div 3.57 = 4.254\,061\,6$$

This is the limit of accuracy of the calculator

but it may not be the correct answer. It is the answer correct to 7 decimal places because this is the number of places following the decimal point.

For many purposes numbers can be approximated by stating them to so many decimal places (d.p.).

If the first figure to be discarded is 5 or more the previous figure is increased by 1.

Example 9

$$93.7257 = 93.726 \quad \text{correct to 3 d.p.}$$
$$= 93.73 \quad \text{correct to 2 d.p.}$$
$$= 93.7 \quad \text{correct to 1 d.p.}$$

Notice carefully how zeros must be kept to show the size of the number or to indicate that it is one of the decimal places.

Example 10

(a) $0.007\,362 = 0.007$ correct to 3 d.p.
$$= 0.01 \quad \text{correct to 2 d.p.}$$

(b) $7.601 = 7.60$ correct to 2 d.p.
$$= 7.6 \quad \text{correct to 1 d.p.}$$

Exercise 4.6

Write down the following numbers correct to the number of decimal places stated:

1. 19.372   (a) 2 d.p.   (b) 1 d.p.
2. 0.007 519   (a) 5 d.p.   (b) 3 d.p.
   (c) 2 d.p.
3. 4.9703   (a) 3 d.p.   (b) 2 d.p.

Use a calculator to find the value of

4. $18.89 \div 14.2$ correct to 2 d.p.
5. $0.0396 \div 2.51$ correct to 3 d.p.
6. $7.21 \div 0.038$ correct to 2 d.p.
7. $\dfrac{184.3 \times 0.0063}{11.43 \times 0.7362}$ correct to 4 d.p.
8. $\dfrac{765.8 \times 0.000\,116}{178.2 \times 26.43}$ correct to 7 d.p.

Significant Figures

A second way of approximating a number is to use **significant figures**. In the number 2179 the 2 represents the most significant figure because it has the greatest value. Similarly 2 and 1 are the two most significant figures whilst 2, 1 and 7 are the three most significant figures. The rules regarding significant figures are as follows:

(1) If the first figure to be discarded is 5 or more, increase the previous figure by 1.

$$7.192\,53 = 7.1925 \quad \text{correct to 5 s.f.}$$
$$= 7.193 \quad \text{correct to 4 s.f.}$$
$$= 7.19 \quad \text{correct to 3 s.f.}$$
$$= 7.2 \quad \text{correct to 2 s.f.}$$

(2) Zeros must be kept to show the position of the decimal point or to indicate that zero is a significant figure.

$$35\,291 = 35\,290 \quad \text{correct to 4 s.f.}$$
$$= 35\,300 \quad \text{correct to 3 s.f.}$$
$$= 35\,000 \quad \text{correct to 2 s.f.}$$
$$0.0739 = 0.074 \quad \text{correct to 2 s.f.}$$
$$= 0.07 \quad \text{correct to 1 s.f.}$$
$$18.403 = 18.40 \quad \text{correct to 4 s.f.}$$
$$= 18.4 \quad \text{correct to 3 s.f.}$$

Unnecessary Zeros

A possible source of confusion is in deciding which noughts are necessary and which are not.

(1) Noughts are not needed after the last figure of a number unless the nought is a significant figure.

$$6.300 = 6.3; \quad 4.70 = 4.7; \quad 3.00 = 3$$

The number half way between 5.4 and 5.8 is 5.6.

The number half way between 5.9 and 6.1 is 6.0. In this case the nought is a significant figure and should be left.

(2) Noughts are needed to keep the places for any missing hundreds, tens, units, tenths, hundredths, etc. The noughts in these numbers are needed:

$$70; \quad 7.205; \quad 800; \quad 9.005$$

(3) One zero is usually put before the decimal point if there are no whole numbers, but this is not essential.

.837 is usually written 0.837

.058 is usually written 0.058

Number of Significant Figures in a Calculation

The answer to a calculation should not contain more significant figures than the least number of significant figures used amongst the given numbers.

Example 11

Find the value of $1.3 \times 7.231 \times 1.24$

The least number of significant figures amongst the given numbers is two (for the number 1.3). Hence the product should only be stated to two significant figures.

$$1.3 \times 7.231 \times 1.24 = 12$$
$$\text{correct to 2 s.f.}$$

Exercise 4.7

Write down the following numbers correct to the number of significant figures stated:

1. 24.935 82
    (a) to 6 s.f.   (b) to 4 s.f.
    (c) to 2 s.f.

2. 0.008 357 1
    (a) to 4 s.f.   (b) to 3 s.f.
    (c) to 2 s.f.   (d) to 1 s.f.

3. 17 359 285
   (a) to 5 s.f.   (b) to 2 s.f.
   (c) to 1 s.f.

4. 0.078 03
   (a) to 3 s.f.   (b) to 2 s.f.
   (c) to 1 s.f.

Write the following numbers without unnecessary zeros:

5. 48.90   6. 4.000   7. 0.5000

8. 600.00   9. 108 070

Each number in the following questions is correct to the number of significant figures shown. Use a calculator to work out the answers and state them to the correct number of significant figures:

10. 15.64×19.75 11. 14.6×8.7

12. $13.96 \div 0.42$

13. $43.5 \times 0.87 \times 1.23$

14. $\dfrac{15.76 \times 8.3}{9.725}$

Rough Estimates for Calculations

Before doing a calculation you should always do a rough estimate to make sure that the answer is sensible. You can often see if there is a mistake by looking at the size of the answer. In doing a rough estimate try to select numbers which are easy to add, subtract and multiply. If division is needed try to select numbers which will cancel.

Example 12

(a) Multiply 32.4 by 0.259.

For a rough estimate we will take

$$32 \times 0.25 = 32 \times \tfrac{1}{4}$$
$$= 32 \div 4$$
$$= 8$$

Accurate calculation:

$$32.4 \times 0.259 = 8.39$$
$$\text{(correct to 3 s.f.)}$$

(The rough estimate shows that the answer is not 83.9 or 0.839.)

(b) Add 5.32, 0.925 and 17.81.

For a rough estimate we will take

$$5 + 1 + 18 = 24$$

Accurate calculation:

$$5.32 + 0.925 + 17.81 = 24.055$$

(c) Work out $\dfrac{47.5 \times 36.52}{11.3 \times 2.75}$

For a rough estimate we will take

$$\frac{\cancel{50}^{\,5} \times \cancel{36}^{\,12}}{\cancel{10}_{\,1} \times \cancel{3}_{\,1}} = \frac{5 \times 12}{1 \times 1}$$
$$= 60$$

Accurate calculation:

$$\frac{47.5 \times 36.52}{11.3 \times 2.75} = 55.8$$
$$\text{(correct to 3 s.f.)}$$

Exercise 4.8

Do a rough estimate for each of the following and then, using a calculator, work out the accurate answer:

1. $18.25 + 39.3 + 429.8$

2. $76.815 - 37.23 - 9.63 - 28.27$

3. 22×0.57 (correct to 2 s.f.)

4. 41.35×0.26 (correct to 2 s.f.)

5. $0.732 \times 0.098 \times 2.17$ (correct to 2 s.f.)

6. $92.17 \div 31.45$ (correct to 4 s.f.)

7. $0.092 \div 0.035$ (correct to 2 s.f.)

8. $27.18 \times 29.19 \times 0.030$ (correct to 2 s.f.)

9. $\dfrac{1.456 \times 0.0125}{0.0532}$ (correct to 3 s.f.)

10. $\dfrac{29.92 \times 31.32}{10.89 \times 2.95}$ (correct to 3 s.f.)

Fraction-to-Decimal Conversion

In Chapter 1 we found that the line separating numerator and denominator of a fraction takes the place of a division sign. Thus

$\frac{17}{80}$ is the same as $17 \div 80$

Example 13

(a) Convert $\frac{27}{32}$ to a decimal number.

$\frac{27}{32} = 27 \div 32 = 0.843\,75$

(by using a calculator)

(b) Convert $2\frac{9}{16}$ to a decimal number.

With mixed numbers we need only convert the fractional part to a decimal number. Thus

$2\frac{9}{16} = 2 + \frac{9}{16} = 2 + (9 \div 16)$
$= 2 + 0.5625$
$= 2.5625$

Recurring Decimals

Consider the calculation

$\frac{3}{8} = 3 \div 8 = 0.375 \qquad 8)\overline{3.000}$
$\qquad\qquad\qquad\qquad\qquad 0.375$

By adding three noughts after the decimal point we are able to complete the division and give an exact answer.

Now consider

$\frac{2}{3} = 2 \div 3 = 0.666\,666\ldots \quad 3)\overline{2.000\,000\ldots}$
$\qquad\qquad\qquad\qquad\qquad\qquad 0.666\,666\ldots$

We see that we will continue to obtain sixes for evermore and we say that the 6 **recurs**.

Now consider

$\frac{7}{11} = 7 \div 11 = 0.636\,363\ldots$

$\qquad\qquad\qquad 11)\overline{7.000\,000\ldots}$
$\qquad\qquad\qquad\quad 0.636\,363\ldots$

Here the 63 recurs. Sometimes it is one figure which recurs and sometimes it is a group of figures. If one figure or a group of figures continues to recur we are said to have a **recurring decimal**.

To save writing so many figures the dot notation is used. For example

$\frac{1}{3} = 1 \div 3 = 0.\dot{3}$ (meaning $0.333\,333\ldots$)

$\frac{1}{6} = 1 \div 6 = 0.1\dot{6}$ (meaning $0.166\,666\ldots$)

$\frac{5}{11} = 5 \div 11 = 0.\dot{4}\dot{5}$ (meaning $0.454\,545\ldots$)

Note that when two figures have dots over them they both recur.

$0.4\dot{0}2\dot{7} = 0.402\,702\,702\,7\ldots$

For all practical purposes we never need recurring decimals; what we need is a decimal given to so many significant figures or decimal places. Thus

$\frac{2}{3} = 0.67$ (correct to 2 s.f.)

$\frac{7}{11} = 0.636$ (correct to 3 d.p.)

Converting Decimals into Fractions

It will be remembered that decimals are fractions with denominators (bottom numbers) of 10, 100, 1000, etc. Thus

$0.53 = \frac{53}{100}$ and $0.625 = \frac{625}{1000} = \frac{5}{8}$

Example 14

Find the difference between $2\frac{5}{16}$ and 2.3214.

$$2\frac{5}{16} = 2 + (5 \div 16)$$
$$= 2 + 0.3125$$
$$= 2.3125$$
$$\text{Difference} = 2.3214 - 2\frac{5}{16}$$
$$= 2.3214 - 2.3125$$
$$= 0.0089$$

Exercise 4.9

Using a calculator convert the following fractions to decimals:

1. $\frac{3}{4}$ 2. $\frac{1}{5}$ 3. $\frac{7}{8}$ 4. $\frac{13}{16}$

5. $1\frac{5}{8}$ 6. $2\frac{19}{32}$ 7. $3\frac{15}{64}$

Write down the following recurring decimals correct to 5 d.p.:

8. $0.\dot{5}$ 9. $0.\dot{8}$ 10. $0.1\dot{7}$

11. $0.4\dot{5}$ 12. $0.\dot{3}\dot{5}$ 13. $0.\dot{2}\dot{1}$

14. $0.\dot{4}2\dot{8}$ 15. $0.\dot{5}67\dot{1}$

Convert each of the following into a recurring decimal:

16. $\frac{2}{9}$ 17. $\frac{2}{11}$ 18. $\frac{7}{9}$

19. $\frac{5}{11}$ 20. $\frac{4}{9}$

Write down the following decimals as fractions in their lowest terms:

21. 0.3 22. 0.65 23. 0.375

24. 0.4375 25. 2.62 26. 1.75

27. 9.185 28. 7.36

29. Find the difference between $\frac{3}{16}$ and 0.17.

30. What is the difference between $5\frac{3}{8}$ and 3.627?

31. Work out $\frac{13}{16} - 0.723$.

32. Find the difference between 0.281 35 and $\frac{9}{32}$.

33. Subtract 0.295 from $\frac{19}{64}$.

Estimation

Suppose that you want to put an extension on to your house. You would explain to the builder what you require and ask him to prepare an **estimate**. This estimate is made to give you some idea how much the extension is likely to cost. It may be a little more or a little less than the actual cost. (If the builder tells you the exact cost of the work he is giving you a quotation.)

To make his estimate the builder would have to work out roughly what he would need to charge for building work. His estimate might be as follows:

Building materials	£2500
Electrical work	£300
Plumbing work	£500
Labour	£3600
Final estimate	£6900

His costs would only be approximate (i.e. not correct but nearly correct). The costs given above might be correct to the nearest £100, and we say that the figures have been **rounded** to the nearest £100.

Rounding is another convenient way of approximating numbers. For instance

297 rounded to the nearest 100 is 300

648 rounded to the nearest 50 is 650

82 rounded to the nearest 10 is 80

Exercise 4.10

Round the following numbers to the nearest 100:

1. 698 2. 703 3. 870

4. 3620 5. 4870

Round the following numbers to the nearest 50:

6. 57   7. 237   8. 490

9. 880   10. 17 385

Round the following numbers to the nearest 10:

11. 94   12. 376   13. 693

14. 589   15. 604

Types of Numbers

We have already seen (in Chapter 1) that:

Counting or **natural** numbers are the numbers $1, 2, 3, 4, \ldots$

Whole numbers are the numbers $0, 1, 2, 3, 4, \ldots$

Integers are whole numbers but they include negative numbers. Thus the numbers $\ldots, -2, -1, 0, 1, 2, 3, \ldots$ are all integers.

A number with a sign in front of it is called a **directed** number. Thus -75 is a negative directed number or a negative integer.

$\frac{2}{3}$ is a **fractional** number and $-\frac{5}{6}$ is a negative fraction. 0.325 is a positive **decimal** number whilst -0.874 is a negative decimal number.

Fractional and decimal numbers, both positive and negative, as well as integers can be shown on a number line (Fig. 4.2).

Fig. 4.2

Rational numbers are numbers which can be expressed as a fraction. Thus 0.625 is a rational number because it can be expressed as the fraction $\frac{5}{8}$. Recurring decimals are also rational numbers because they can be expressed as fractions.

Irrational numbers cannot be expressed as fractions. For instance $\sqrt{2} = 1.414\,213\ldots$ and it is impossible to find an exact value for $\sqrt{2}$, and hence $\sqrt{2}$ is an irrational number.

Not all square roots are irrational. For instance

$$\sqrt{9} = 3 \quad \text{and} \quad \sqrt{2.25} = 1.5 = \tfrac{3}{2}$$

and these two numbers are rational.

Exercise 4.11

1. Which of the following numbers are positive integers:

    $5, -8, \frac{2}{3}, 1\frac{1}{4}, 2.75$ and 198?

2. Which of the following are negative integers:

    $8\frac{1}{2}, -9, 7, -\frac{1}{3}, 11$ and $-4\frac{3}{5}$?

3. Which of the following are rational numbers and which are irrational:

    $1.57, \frac{1}{7}, -5.625, \sqrt{16}, \sqrt{15},$ 6.76 and $-3\frac{1}{2}$?

4. Which of the following are whole numbers:

    $\frac{1}{4}, -2, 0, 6, 3.1$ and -0.4?

5. Arrange in order of size, smallest first, the numbers:

    $3, -4.7, \frac{5}{8}, -3\frac{3}{4}, 0, 7, -3.4$ and 2.5.

Finding the Square Root of a Decimal Number

Most calculators have a square-root key. Its use is shown in Example 15.

Example 15

Find the square root of 27.53 correct to 2 decimal places.

Input	Display
27.53	27.53
$\sqrt{}$	5.2469 ...

$\sqrt{27.53} = 5.25$ to 2 decimal places

Exercise 4.12

Find the square root of

1. 1.72
2. 8.61
3. 3.84

4. 12.61
5. 76.22
6. 90.41

7. 128
8. 342

Miscellaneous Exercise 4

1. Work out $0.017 \div 0.027$ correct to 2 decimal places.

2. What is 0.081 778 rounded to the nearest thousandth?

3. By rounding off the numbers to the nearest number of tens, obtain the best estimate for $\dfrac{27.5 \times 60.52}{11.3 \times 20.51}$

4. Write $\frac{9}{11}$ as a recurring decimal.

5. Find the value of $\dfrac{0.8 \times 7 \times 1.3}{2.1 \times 4}$ correct to 1 decimal place.

6. In the number 52.058, find the numerical difference between the actual values of the two digits 5.

7. Write the number 738.0584 correct to
   (a) 2 decimal places
   (b) 1 decimal place

8. Change the following fractions to decimal numbers:
   (a) $\frac{15}{16}$ (b) $9\frac{4}{25}$

9. Find the difference between $3\frac{7}{20}$ and 3.37.

10. Find the sum of 3.98, 0.745 and 52.073.

11. Find the exact value of $\dfrac{4.2 \times 0.0462}{0.077}$

12. Round off 0.049 49
    (a) correct to 3 significant figures
    (b) correct to 3 decimal places

13. Calculate
    (a) 7.3×2.1 (b) $52.16 \div 3.2$

14. Find the exact value of $\dfrac{45}{0.9} + \dfrac{6.6}{0.55}$

15. Express 0.166 66 correct to 3 decimal places.

16. Express 0.026 66 correct to 2 significant figures.

17. Express $27 \div 81$ as
    (a) a fraction in its lowest terms
    (b) a decimal number correct to 3 decimal places

18. Work out
    (a) 87.4×1000 (b) $3.26 \div 100$

19. Work out $4 + 0.7 - 5.38$.

20. Rearrange the numbers 0.3, -0.7, $\frac{2}{3}$, -4.2 and 1.6 in order of size, starting with the greatest value.

21. Find the sum of 3.8, 273.09 and 0.265.

22. Express as decimal numbers
    (a) $\frac{7}{16}$ (b) $\frac{1}{20}$

23. Express as fractions in their lowest terms
    (a) 0.95   (b) 0.32

24. Find the square root of 86.32 correct to 4 significant figures.

Mental Test 4

Try to give the answers to the following
without writing anything else.

1. Add 1.2 and 1.35.

2. Find the sum of 0.23 and 0.032.

3. Subtract 0.73 from 1.85.

4. Multiply 3.2 by 5.

5. Find the product of 0.3 and 0.2.

6. Write down the place value of the digit
    9 in the number 0.394.

7. Read off the number 0.4 as a fraction in
    its lowest terms.

8. Read off $\frac{53}{100}$ as a decimal number.

9. Which number is the larger, -7.6 or
    4.3?

10. Multiply 0.78 by 10.

11. Find the product of 0.357 and 1000.

12. Divide 53.21 by 100.

13. Write the number 2.467 correct to 2
    decimal places.

14. Write the number 53 648 correct to 2
    significant figures.

15. Write the number 17.504 correct to 3
    significant figures.

16. Convert $\frac{4}{5}$ to a decimal number.

17. Write the recurring decimal $0.1\dot{7}$
    correct to 3 significant figures.

18. Convert $4\frac{3}{4}$ to a decimal number.

19. Round off the number 3623 to the
    nearest hundred.

20. Is the number $-\frac{2}{3}$ rational?

5 Measurement

Measurement of Length

In the metric system the standard unit of length is the metre (abbreviation: m). For some purposes the metre is too large a unit and it is therefore split up into smaller units as follows:

$$1 \text{ metre (m)} = 10 \text{ decimetres (dm)}$$
$$= 100 \text{ centimetres (cm)}$$
$$= 1000 \text{ millimetres (mm)}$$

In dealing with large distances the metre is too small a unit and large distances are measured in kilometres such that

$$1 \text{ kilometre (km)} = 1000 \text{ metres (m)}$$

Because the metric system is essentially a decimal system it is easy to convert from one unit to another by multiplying or dividing by 10, 100 or 1000.

Example 1

(a) Convert 28.35 m into millimetres.

$$28.35 \text{ m} = 28.35 \times 1000 \text{ mm}$$
$$= 28\,350 \text{ mm}$$

(b) Convert 879 cm into metres.

$$879 \text{ cm} = 879 \div 100 \text{ m} = 8.79 \text{ m}$$

(c) Convert 734 mm into centimetres.

$$734 \text{ mm} = 734 \div 10 \text{ cm} = 73.4 \text{ cm}$$

(d) Convert 87 652 m into kilometres.

$$87\,652 \text{ m} = 87\,652 \div 1000 \text{ km}$$
$$= 87.652 \text{ km}$$

Exercise 5.1

This exercise should be done mentally.

Convert the following to millimetres:

1. 5 cm
2. 4.8 cm
3. 0.9 cm
4. 2 m
5. 8.3 m
6. 28.1 m
7. 7 dm
8. 19.4 dm

Convert the following to centimetres:

9. 8 m
10. 48 m
11. 0.8 m
12. 28 dm
13. 3.25 dm
14. 800 mm
15. 8964 mm
16. 5.4 mm

Convert the following to metres:

17. 8000 mm
18. 25 634 mm
19. 54.3 mm
20. 300 cm
21. 37 864 cm
22. 9.35 cm
23. 18 km
24. 6.2 km

Convert the following to kilometres:

25. 9000 m
26. 460 m
27. 58 m
28. 1800 cm
29. 750 cm
30. 470 000 mm

In the imperial system lengths are measured in inches, feet, yards and miles.

12 inches (in) =

3 feet (ft

1760 yards (y

Example 2

(a) Change 468 inches into feet.

$$468 \text{ in} = 468 \div 12 \text{ ft} = 39 \text{ ft}$$

(b) Change 1548 inches into yards.

$$1548 \text{ in} = 1548 \div 36 \text{ yd} = 43 \text{ yd}$$

(c) Change 22 352 yards into miles.

$$22\,352 \text{ yd} = 22\,352 \div 1760 \text{ miles}$$
$$= 12.7 \text{ miles}$$

(d) Change 32 yards into inches.

$$32 \text{ yd} = 32 \times 36 \text{ in}$$
$$= 1152 \text{ in}$$

Exercise 5.2

A calculator should be used to work out the answers.

Change to feet:

1. 48 in   2. 30 in   3. 39 in

4. 74.4 in   5. 8 yd   6. 9.2 yd

7. 813 yd   8. 39.4 yd

Change to yards:

9. 216 in   10. 158.4 in

11. 586.8 in   12. 2021.4 in

13. 36 ft   14. 8.7 ft

15. 67.5 ft   16. 1209 ft

Change to inches:

17. 8 ft   18. 23 ft   19. 15.8 ft

20. 31.9 ft   21. 7 yd   22. 5.1 yd

23. 6.8 yd   24. 54.2 yd

Change to miles:

        ?80 yd   26. 11 264 yd

        ?ft   28. 49 104 ft

Measurement of Mass

In the metric system light objects are weighed using grams or milligrams whilst heavier objects are weighed using kilograms or tonnes.

$$1 \text{ gram (g)} = 1000 \text{ milligrams (mg)}$$
$$1 \text{ kilogram (kg)} = 1000 \text{ grams (g)}$$
$$1 \text{ tonne (t)} = 1000 \text{ kilograms (kg)}$$

Example 3

(a) Change 900 g into kilograms.

$$900 \text{ g} = 900 \div 1000 \text{ kg} = 0.9 \text{ kg}$$

(b) Change 8 kg to grams.

$$8 \text{ kg} = 8 \times 1000 \text{ g}$$
$$= 8000 \text{ g}$$

(c) Change 47 000 mg to grams.

$$47\,000 \text{ mg} = 47\,000 \div 1000 \text{ g}$$
$$= 47 \text{ g}$$

(d) Change 43 500 kg to tonnes.

$$43\,500 \text{ kg} = 43\,500 \div 1000 \text{ t}$$
$$= 43.5 \text{ t}$$

Exercise 5.3

This exercise should be done mentally.

Change to kilograms:

1. 5000 g   2. 6980 g

3. 8 t   4. 11.2 t

Change to grams:

5. 9 kg   6. 39 kg

7. 0.75 kg   8. 45 000 mg

9. 1830 mg   10. 85 mg

Change to milligrams:

11. 7 g   12. 4.9 g

13. 0.3 g   14. 0.009 g

Change to tonnes:

15. 50 000 kg   16. 8570 kg

17. 453 kg   18. 45 kg

In the imperial system light objects are weighed in ounces. Heavier objects are measured in pounds whilst very heavy objects are measured in hundredweights or tons.

16 ounces (oz) = 1 pound (lb)

112 pounds (lb) = 1 hundredweight (cwt)

20 hundred-weight (cwt) = 1 ton

1 ton = 2240 lb

Example 4

(a) How many ounces are there in 2 pounds?

$$2 \text{ lb} = 2 \times 16 \text{ oz}$$
$$= 32 \text{ oz}$$

(b) Change 419.2 ounces to pounds.

$$419.2 \text{ oz} = 419.2 \div 16 \text{ lb}$$
$$= 26.2 \text{ lb}$$

(c) Change 817.6 pounds into hundred-weight.

$$817.6 \text{ lb} = 817.6 \div 112 \text{ cwt}$$
$$= 7.3 \text{ cwt}$$

(d) How many pounds are there in 8.6 tons?

$$8.6 \text{ tons} = 8.6 \times 2240 \text{ lb}$$
$$= 19 264 \text{ lb}$$

Exercise 5.4

A calculator should be used to work out the answers.

Change to ounces:

1. $\frac{1}{4}$ lb   2. 5 lb

3. $4\frac{1}{2}$ lb   4. 8.3 lb

Change to pounds:

5. 112 oz   6. 4 oz

7. 12 oz   8. 52 oz

9. 4 cwt   10. 5.2 cwt

11. 6 tons   12. 7.3 tons

Change to hundredweights:

13. 448 lb   14. 436.8 lb

15. 8 tons   16. 12.4 tons

Change to tons:

17. 13 440 lb   18. 1344 lb

19. 80 cwt   20. 184 cwt

Measurement of Capacity

Fluids of various kinds are usually stored in tins, bottles and tanks. The amount of fluid that a container will hold is known as its **capacity**.

In the metric system capacities are measured in litres, centilitres and millilitres such that

$$1 \text{ litre } (\ell) = 100 \text{ centilitres } (c\ell)$$
$$= 1000 \text{ millilitres } (m\ell)$$

Example 5

(a) How many centilitres are there in 0.7 litre?

$$0.7 \ell = 0.7 \times 100 \text{ c}\ell$$
$$= 70 \text{ c}\ell$$

(b) Change 45 000 millilitres into litres.

$$45 000 \text{ m}\ell = 45 000 \div 1000 \ell$$
$$= 45 \ell$$

Exercise 5.5

Change to centilitres:

1. 6 ℓ 2. 4.3 ℓ 3. 0.25 ℓ

4. 500 mℓ 5. 8000 mℓ

Change to millilitres:

6. 2 ℓ 7. 0.3 ℓ

8. 0.005 ℓ 9. 50 cℓ

10. 3 cℓ

Change to litres:

11. 550 cℓ 12. 40 cℓ

13. 8 cℓ 14. 3250 mℓ

15. 90 mℓ

In the imperial system capacities are measured in fluid ounces, pints and gallons such that

$$20 \text{ fluid ounces (fl oz)} = 1 \text{ pint (pt)}$$

$$8 \text{ pints (pt)} = 1 \text{ gallon (gal)}$$

Example 6

(a) Change 7 gallons into pints.

$$7 \text{ gal} = 7 \times 8 \text{ pt}$$

$$= 56 \text{ pt}$$

(b) Change 54 fluid ounces into pints.

$$54 \text{ fl oz} = 54 \div 20 \text{ pt}$$

$$= 2.7 \text{ pt}$$

Exercise 5.6

Change into pints:

1. 60 fl oz   2. 25 fl oz

3. 45 fl oz   4. 9 gal

5. 6.25 gal

Change into fluid ounces:

6. 4 pt   7. 1.3 pt   8. 0.45 pt

9. 2 gal   10. 0.3 gal

Change into gallons:

11. 24 pt   12. 5.2 pt   13. 760 pt

14. 320 fl oz   15. 4000 fl oz

Conversion of Imperial and Metric Units

Conversions from metric units to imperial units and vice versa are often needed. Sometimes an accurate conversion is required but often an approximate conversion is good enough.

Metric Equivalents
1 inch = 2.5 centimetres = $2\frac{1}{2}$ cm approx. (i.e. 2 in $\approx$ 5 cm)
1 foot = 30.48 centimetres = 30 cm approx.
1 yard = 0.91 metres = 1 m approx.
1 kilometre = 0.621 miles = $\frac{5}{8}$ mile approx. (i.e. 5 miles $\approx$ 8 km)
1 kilogram = 2.205 pounds = $2\frac{1}{4}$ lb approx. (i.e. 4 kg $\approx$ 9 lb)
1 fluid ounce = 28.41 millilitres = 30 mℓ approx.
1 litre = 1.760 pints = $1\frac{3}{4}$ pt approx. (i.e. 4$\ell \approx$ 7 pt)
1 gallon = 4.546 litres = $4\frac{1}{2}\ell$ approx. (i.e. 2 gal $\approx$ 9 ℓ)

Example 7

(a) Given that 1 kg = 2.205 lb, find in pounds the weight of a 50 kg sack of potatoes.

$$50 \text{ kg} = 50 \times 2.205 \text{ lb} = 110.25 \text{ lb}$$

(b) 1 kilometre = $\frac{5}{8}$ mile approx. Find the distance in kilometres of 135 miles.

$$135 \text{ miles} = 135 \div \frac{5}{8} \text{ kilometres}$$

$$= \frac{\overset{27}{\cancel{135}}}{1} \times \frac{8}{\underset{1}{\cancel{5}}} \text{ kilometres}$$

$$= 27 \times 8 \text{ kilometres}$$

$$= 216 \text{ kilometres}$$

Exercise 5.7

1. If 1 kg = $2\frac{1}{4}$ lb, find how many pounds there are in 64 kg.

2. Taking 1 fluid ounce to be 30 millilitres, how many fluid ounces are equivalent to 180 millilitres?

3. If 1 kilometre = 0.621 miles, find to the nearest mile the distance equivalent to 32 km.

4. If 8 kilometres = 5 miles approximately, find the distance in kilometres of 25 miles.

5. A woman buys 2 kg of tomatoes. How many pounds of tomatoes does she buy? (Take 4 kg = 9 lb.)

6. A shopper buys 5 lb of apples. How many kilograms of apples does she buy if 1 kg = 2.2 lb?

7. A consignment of steel weighing 5 tonnes is delivered to a factory. Taking 1 kg = 2.205 lb, find the weight of the steel in tons.

8. Taking 1 litre = $1\frac{3}{4}$ pints, convert 49 pints into litres.

Addition and Subtraction of Metric Quantities

Metric quantities are added and subtracted in the same way as decimal numbers. However, it is important that all the quantities should be in the same units.

Example 8

Add 15.2 m, 39.2 cm and 150.2 mm. State the answer in metres.

$$15.2 \text{ m} + 39.2 \text{ cm} + 150.2 \text{ mm}$$

$$= 15.2 \text{ m} + 0.392 \text{ m} + 0.1502 \text{ m}$$

$$= 15.7422 \text{ m}$$

Example 9

Subtract 158 mm from 7.3 m. State the answer in metres.

$$7.3 \text{ m} - 158 \text{ mm} = 7.3 \text{ m} - 0.158 \text{ m}$$

$$= 7.142 \text{ m}$$

Exercise 5.8

Add the following lengths, giving the answer in metres:

1. 47 cm, 5.83 m and 15 mm

2. 93 km, 462 m and 5 cm

3. 0.185 m, 7.36 cm and 8.2 mm

4. Add together 792 g, 15 000 mg and 1.265 kg, giving the answer in kilograms.

5. A greengrocer starts the day with 127 kg of apples. He sells $3\frac{1}{2}$ kg, 450 g and 25 kg. What weight of apples has he left?

6. A length of ribbon 2.5 m long has the following lengths cut from it: 25 cm, 863 mm and 70 cm. What length of ribbon, in centimetres, remains?

7. Subtract 15.2 cm from 0.78 m, giving the answer in millimetres.

8. Subtract 25 mm from 7.5 cm, giving the answer in centimetres.

Multiplication and Division of Metric Quantities

Metric quantities are multiplied and divided in exactly the same way as decimal numbers.

Example 10

57 lengths of wood each 95 cm long are required by a builder. What total length of wood, in metres, is needed?

$$57 \times 95 \text{ cm} = 57 \times 0.95 \text{ m}$$

$$= 54.15 \text{ m}$$

Hence total length of wood needed is 54.15 m.

Example 11

Frozen peas are packed in bags containing 450 grams. How many full bags can be obtained from 2000 kg of peas?

$$450 \text{ grams} = 0.45 \text{ kilograms}$$

$$\text{Number of bags filled} = 2000 \div 0.45$$

$$= 4444$$

Exercise 5.9

1. 95 lengths of steel bar, each 127 cm long, are required by a toy manufacturer. Find the total length of bar, in metres, needed.

2. 209 lengths of cloth each 135 cm long have to be cut from a bale containing 300 m. What length of cloth, in metres, remains?

3. How many lengths of string, each 53 cm long, can be cut from a ball containing 25 m? What length of string remains?

4. How many pieces of ribbon each 36 cm long can be cut from a reel containing 30 m?

5. Butter is packed in 250-gram packs. How many packs can be obtained from 8 tonnes of butter?

6. A certain spice is packed in jars containing 32 grams. How many jars can be filled from 15 kg of spice?

7. How many doses of 5 millilitres can be obtained from a full medicine bottle with a capacity of 24 centilitres?

8. A certain type of tablet has a weight of 8.2 milligrams. How much, in kilograms, do 5000 of these tablets weigh?

Money

The British system of currency uses the pound as the basic unit. The only sub-unit used is pence, such that

$$100 \text{ pence} = 1 \text{ pound}$$

The abbreviation p is used for pence and £ is used for pounds. A decimal point is used to separate the pounds from the pence, for instance £3.58 means 3 pounds and 58 pence. pence.

There are two ways of expressing amounts less than £1. For example 74 pence may be written as 74p or £0.74.

5 pence may be written as 5p or £0.05.

Adding and Subtracting Sums of Money

The addition and subtraction of sums of money is done in exactly the same way as the addition and subtraction of decimal numbers.

Example 12

(a) Add £8.94, £3.28 and 95p.

When amounts are given in pence it is best to convert these to pounds.

$$£8.94 + £3.28 + 95\text{p} = £8.94 + £3.28$$
$$+ £0.95$$

$$= £13.17$$

(b) Subtract 47p from £5.00.

$$£5.00 - 47p = £5.00 - £0.47$$
$$= £4.53$$

Exercise 5.10

1. Add £2.15, £7.28 and £6.54.

2. Add 27p, 82p, 97p and 31p.

3. Add £15.36, £10.42, 75p, 86p and £73.25.

4. Subtract £5.49 from £12.62.

5. Subtract 79p from £3.27.

6. A lady spends amounts of £43.64, £59.76 and £87.49 in a departmental store. How much did she spend altogether?

7. A man deposited £540 in his bank account but withdrew amounts of £138.26, £57.49 and £78.56 before depositing a further £436. How much money has he in his account?

8. Emma had £10. She spent amounts of £3.58, £4.29 and £1.35. How much money had she left?

Multiplication and Division with Money

The methods used when multiplying and dividing sums of money are very similar to those used with decimal numbers.

Example 13

(a) Find the cost of 35 articles if each costs 82p.

Since 82p = £0.82

cost of 35 articles = £0.82 × 35

= £28.70

(b) If 37 similar articles cost £34.78, how much does each article cost?

Cost of each article = £34.78 ÷ 37

= £0.94

Exercise 5.11

1. Find the cost of 12 articles if each costs 12p.

2. How much do 87 articles at 6p each cost?

3. Calculate the cost of 23 articles if each costs £9.46.

4. Bread rolls cost 15p each. 140 are needed for a party. How much will they cost?

5. 12 similar articles cost £6.24. How much does each cost?

6. 41 similar articles cost £102.09. How much does each cost?

7. I bought 3 loaves at 72p each, $\frac{1}{2}$ kg of butter at £1.80 per kg and $\frac{1}{4}$ kg of cheese at £9.60 per kg. How much was the bill?

8. An electricity bill was £74.36 for a quarter (which is 13 weeks). How much was the cost electricity per week?

Miscellaneous Exercise 5

1. What is the total cost of renting a television set for six months at £18.32 per month?

2. Calculate the total cost of three pairs of jeans at £18.83 per pair and four pairs of socks at 84p per pair.

3. How much change would I get from £5 when I buy 4 packets of coffee at 92p per packet.

4. (a) Change 12.34 kg into grams.
   (b) Change 123.4 cm into metres.
   (c) Change 1234 mℓ to litres.

5. How many pieces of string, each 30 cm long, can be cut from a piece of string 4.2 m long.

6. If 4 kg = 9 lb approximately, work out in pounds the weight of a bag of flour weighing 3 kg.

7. A large piece of cheese weighs $2\frac{1}{4}$ kg. Smaller pieces weighing 525 g, 485 g and 370 g are cut from it. What weight of cheese, in grams, is left?

8. A bottle contains 20 cℓ of medicine. How many 5 mℓ doses can be obtained from it?

9. 1 km = $\frac{5}{8}$ of a mile approximately. How far, in miles, is a distance of 160 km?

10. Express each of the following in metres:
    (a) 2 km   (b) 2 km 56 m
    (c) 672 cm   (d) 28 400 mm

11. How many 25-millilitre doses can be obtained from a bottle containing $1\frac{1}{2}$ litres.

12. How many (a) milligrams in 3 grams, (b) centimetres in 9 metres, (c) litres in 3500 millilitres?

13. How many lengths of tape, each 75 cm long, can be cut from a reel 10 m long? What is the length of the piece left over?

14. Express 45 cm as a fraction of 1 m. Express the fraction in its lowest terms.

15. Which is the heaviest: 80 g, 0.8 kg, 800 mg or 0.88 g?

16. A bank delivers £50 worth of 2 pence pieces to a supermarket. Calculate the weight of these coins in kilograms, if each coin weighs 6.92 grams.

17. Calculate the exact cost of 375 toys at £3.21 each.

18. Find the cost of taking 34 children on an outing at £2.65 each.

19. A cup costs 48p and a saucer 31p. How much change should be obtained from a £10 note after purchasing 8 cups and 8 saucers?

Mental Test 5

Try to answer the following questions without writing anything down except the answer.

1. Convert 4 m into centimetres.

2. Change 5 m into millimetres.

3. Change 20 mm into centimetres.

4. Change 7 cm into millimetres.

5. Change 6000 m into kilometres.

6. Convert 9.2 km into metres.

7. Change 36 inches into feet.

8. Change 72 in into yards.

9. How many feet are there in 1 mile?

10. Change 4000 g into kilograms.

11. How many grams are there in 8 kg?

12. Convert 7500 mg into grams.

13. Convert 9.2 g into milligrams.

14. How many ounces are there in 20 lb?

15. How many pounds are there in 3 cwt?

16. Change 80 fluid ounces into pints.

17. How many fluid ounces are there in 5 pints?

18. How many centilitres are there in 4 litres?

19. Change 6 gallons into pints.

20. If 1 in = $2\frac{1}{2}$ cm, how many centimetres are there in 4 inches?

21. If 1 fl oz = 30 mℓ, calculate the number of fluid ounces in 90 millilitres.

22. Add 500 cm and 8 m, giving the answer in metres.

23. Add 7 cm and 50 mm, giving the answer in centimetres.

24. Subtract 50 mm from 2 metres, giving the answer in millimetres.

25. How many doses of 5 millilitres can be obtained from a medicine bottle holding 1 litre?

26. Add £3.35 and 40p.

27. Subtract 27p from £1.50.

28. Find the cost of ten articles if each one costs 57p.

29. If 20 articles cost £10, how much does each one cost?

Ratio and Proportion

Ratio

Concrete is made by mixing sand and cement in the ratio $3:1$ (three to one) and then adding water. You could put three sacks of sand to one sack of cement or three bucketfuls of sand to one bucketful of cement or three cups of sand to one cup of cement. These ratios are all $3:1$.

As long as sand and cement are kept in the same proportion when mixed with water they will make concrete. The actual amount of sand and cement used only affects the amount of concrete made.

Example 1

Blackcurrant jam is made by mixing 4 kg of blackcurrants to 2 ℓ of water and 6 kg of sugar. This will make 8 kg of jam.

(a) If 4 kg of jam are required, work out the amount of each ingredient needed.

(b) If 12 kg of blackcurrants are used, how much sugar is required?

(c) If 4 ℓ of water is used how much sugar and how many blackcurrants are needed?

> **(a)** Since half the amount of jam is required, we need only half of the ingredients. Therefore 2 kg of blackcurrants, 1 ℓ of water and 3 kg of sugar are needed.
>
> **(b)** Since three times as many blackcurrants are used we need three times as much sugar: 18 kg.

> **(c)** Since twice as much water is used we need twice as much of the other ingredients. Hence we need 8 kg of blackcurrants and 12 kg of sugar.

Model trains are made to a scale of $1:72$. Every measurement on the model is $\frac{1}{72}$nd of the real measurement. The model measurements and the real measurements are in the ratio of $1:72$.

Exercise 6.1

1. A recipe for a cake to serve four people is as follows:

   400 g butter   320 g castor sugar

   1 egg separated   800 g self-raising flour

   200 g currants   60 g mixed peel

   1 pinch of salt

   (a) Write down the ingredients needed to make a cake to serve eight people.

   (b) If 1200 g of butter is used, how much sugar is needed?

   (c) If 3200 g of flour is used, how much mixed peel is needed?

   (d) If 800 g of currants are used, how much flour is required?

2. A mortar mixture is made of cement, sand and water in the ratio $1:2:4$. If 80 kg of cement is used, how much sand and water is required?

3. A metal alloy is made by mixing copper and zinc in the ratio of $5:1$. If 8 kg of zinc is used, how much copper is required?

4. Solder is made by mixing tin and lead in the ratio of $3:1$. If 5 kg of lead is used, work out

   (a) the weight of tin required

   (b) the total weight of solder made

5. A suitable batter for pancakes uses 120 grams of flour, 1 egg and 250 millilitres of milk. This is sufficient for 12 pancakes.

   (a) Work out the amount of each ingredient if 24 pancakes are to be made.

   (b) If 360 grams of flour is used, how many eggs should be used?

   (c) If 1 litre of milk is used, how many grams of flour should be used?

Simplifying Ratios

Problems are often made easier by putting the ratios into their simplest terms.

The ratios $1:3, 3:7$ and $9:4$ are in their simplest terms because there is no number which will divide exactly into both sides.

The ratio $8:6$ is not in its simplest terms because 2 will divide into both sides to give $4:3$, which is the same as $8:6$.

Example 2

Put the ratio $72:84$ in its simplest terms.

   12 divides into 72 and 84, hence

   $72:84$ is the same as
   $72 \div 12 : 84 \div 12$. That is
   $72:84$ is the same as $6:7$,
   which is the ratio $72:84$
   in its simplest terms.

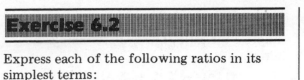

Express each of the following ratios in its simplest terms:

1.	$5:20$	2.	$4:12$
3.	$42:49$	4.	$35:42$
5.	$64:56$	6.	$60:48$
7.	$7500:5000$	8.	$50:60:70$
9.	$25:30:45$	10.	$20:25:30:35$

Exercise 6.2 should bring out the similarities between ratios and fractions. Compare, for instance

$$42:49 = 6:7 \quad \text{with} \quad \tfrac{42}{49} = \tfrac{6}{7}$$

and $\quad 5:20 = 1:4 \quad \text{with} \quad \tfrac{5}{20} = \tfrac{1}{4}$

Exercise 6.3

Express the following ratios as fractions in their lowest terms:

1.	$9:7$	2.	$5:10$	3.	$14:7$
4.	$12:15$	5.	$16:24$	6.	$80:100$
7.	$3:9$	8.	$15:18$	9.	$21:24$
10.	$45:81$				

To simplify a ratio such as $1\tfrac{1}{4}:\tfrac{1}{3}$ change the mixed number into a top-heavy fraction and then express each fraction with the same denominator (bottom number). The common denominator should be the LCM of the original denominators.

$\quad 1\tfrac{1}{4}:\tfrac{1}{3}$ is the same as $\tfrac{5}{4}:\tfrac{1}{3}$

The LCM of 4 and 3 is 12. Expressing each fraction with a denominator of 12 gives

$\quad \tfrac{5}{4}:\tfrac{1}{3}$ is the same as $\tfrac{15}{12}:\tfrac{4}{12}$

We now multiply each side of the ratio by 12:

$\quad \tfrac{15}{12}:\tfrac{4}{12}$ is the same as $15:4$

This is the ratio $1\tfrac{1}{4}:\tfrac{1}{3}$ expressed in its simplest terms.

Example 3

Express the ratio $1\frac{1}{3}:2\frac{1}{2}$ in its simplest terms.

$$1\frac{1}{3}:2\frac{1}{2} \;=\; \frac{4}{3}:\frac{5}{2} \;=\; \frac{8}{6}:\frac{15}{6} \;=\; 8:15$$

Exercise 6.4

Simplify the following ratios:

1. $\frac{3}{4}:\frac{1}{3}$ 2. $\frac{2}{5}:\frac{3}{10}$ 3. $\frac{1}{2}:\frac{3}{8}$

4. $\frac{1}{6}:\frac{5}{12}$ 5. $2\frac{1}{4}:\frac{1}{4}$ 6. $3\frac{1}{2}:1\frac{1}{2}$

7. $\frac{5}{6}:1\frac{2}{3}$ 8. $2\frac{3}{4}:1\frac{1}{2}$ 9. $1\frac{1}{3}:3\frac{1}{4}$

10. $2\frac{2}{5}:5\frac{3}{20}$

Simplifying Ratios with Units

We can simplify the ratio $30\,\text{cm}:80\,\text{cm}$ by first removing the units (because they are the same) and then dividing both sides by 10.

$30\,\text{cm}:80\,\text{cm}$ is the same as $3:8$

The ratio $5\,\text{cm}:2\,\text{m}$ can be simplified by making the units the same and then removing them.

$$5\,\text{cm}:2\,\text{m} \;=\; 5\,\text{cm}:200\,\text{cm}$$
$$=\; 5:200 \;=\; 1:40$$

(by dividing both sides by 5)

Exercise 6.5

Simplify the following ratios:

1. $400\,\text{cm}:100\,\text{cm}$ 2. $500\,\text{cm}:1\,\text{m}$

3. $60\,\text{g}:2\,\text{kg}$ 4. £1.50 : 50p

5. $5\,\text{kg}:250\,\text{g}$ 6. $400\,\text{mm}:4\,\text{m}$

7. $2\frac{1}{2}\,\text{km}:500\,\text{m}$ 8. $3\,\text{cm}:9\,\text{mm}$

9. £8.00 : 25p   10. $3\,\text{kg}:90\,\text{g}$

Proportional Parts

The line AB (Fig. 6.1), whose length is 15 cm, has been divided into two parts in the ratio $2:3$. The line has been divided into its **proportional parts** and, as can be seen from the diagram, the line has been divided into a total of 5 parts. The length AC contains 2 of these parts and the length BC contains 3 of them. Each part is $15\,\text{cm} \div 5 = 3\,\text{cm}$ long. Hence

$$AC = 2 \times 3\,\text{cm} = 6\,\text{cm} \ \text{ and}$$
$$BC = 3 \times 3\,\text{cm} = 9\,\text{cm}.$$

Fig. 6.1

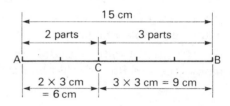

The problem of dividing the line into two parts in the ratio $2:3$ could be tackled in this way:

Total number of parts $= 2 + 3 = 5$

Length of each part $= 15\,\text{cm} \div 5 = 3\,\text{cm}$

Length of AC $= 2 \times 3\,\text{cm} = 6\,\text{cm}$

Length of BC $= 3 \times 3\,\text{cm} = 9\,\text{cm}$

Example 4

Divide £240 in the ratio $5:4:3$.

Total number of parts $= 5 + 4 + 3 = 12$

Amount of each part $=$ £240 $\div 12 =$ £20

Amount of first part $=$ £20 $\times 5 =$ £100

Amount of second part $=$ £20 $\times 4 =$ £80

Amount of third part $=$ £20 $\times 3 =$ £60

Example 5

Two lengths are in the ratio 8:5. If the first length is 120 metres, what is the second length?

> The first length is represented by 8 parts. Each part = 120 metres ÷ 8 = 15 metres. The second length is represented by 5 parts. Therefore the second length is 5 × 15 metres = 75 metres.

Alternatively the problem may be tackled by using fractions:

$$\text{The second length} = \tfrac{5}{8} \text{ of the first length}$$
$$= \tfrac{5}{8} \times 120 \text{ metres}$$
$$= 75 \text{ metres}$$

Exercise 6.6

1. Divide £800 in the ratio 5:3.

2. Divide 160 kg in the ratio 7:3.

3. Divide 120 m in the ratio 2:3:5.

4. A line 1.68 metres long is to be divided into three parts in the ratio 2:7:11. Find, in millimetres, the length of each part.

5. A sum of money is divided into two parts in the ratio 5:7. If the smaller amount is £200, find the larger amount.

6. An alloy consists of copper, zinc and tin in the ratio 2:3:5. Find the amount of each metal in 75 kg of the alloy.

7. A sum of money is shared into three parts in the ratio 2:4:5. If the largest share is £40, what is the total amount of money shared?

8. Four friends contribute sums of money to a charitable organization in the ratio 2:4:5:7. If the largest amount contributed is £1.40, calculate the total amount contributed by the four people.

Direct Proportion

Two quantities are said to **vary directly,** or be in **direct proportion,** if they increase or decrease at the same rate.

Thus the quantity of petrol used and the distance travelled by a motor car are in direct proportion.

Again, if we buy potatoes at 20 pence for 2 kilograms then we expect to pay 40 pence for 4 kilograms and 10 pence for 1 kilogram. If we double the amount bought we double the cost; if we halve the amount bought we halve the cost. The amount of potatoes bought and their cost are in direct proportion.

Example 6

If 7 pens cost 56p, how much will 5 pens cost?

> *Method 1* (the unitary method)
>
> 7 pens cost 56p
>
> 1 pen costs 56p ÷ 7 = 8p
>
> 5 pens cost 5 × 8p = 40p
>
> *Method 2* (the fractional method)
>
> 7 pens cost 56p
>
> 5 pens cost $\tfrac{5}{7} \times \tfrac{56}{1} = 40$p

You should decide for yourself which method, unitary or fractional, to use when solving a problem.

Exercise 6.7

1. If 74 exercise books cost £11.84, how much do 53 cost?

2. If 40 articles cost £35, how much do 55 articles cost?

3. Eggs cost 70 pence for 10. How much will 25 eggs cost?

4. A car travels 205 km on 20 litres of petrol. How much petrol will be needed for a journey of 369 km?

5. If 9 metres of stair carpet cost £21, how much will 96 metres cost?

6. A train travels 200 km in 4 hours. If it travels at the same rate, how long will it take to complete a journey of 350 km?

7. A machine makes 15 articles in $\frac{1}{2}$ hour. How many articles will it make in 3 hours?

8. 4 books weigh 800 grams. What is the weight, in kilograms, of 20 similar books?

Inverse Proportion

If an increase (or decrease) in one quantity produces a decrease (or increase) in a second quantity in the same ratio, the two quantities are in inverse proportion.

Example 7

Five men building a wall take 20 days to complete it. How long would it take four men to complete it?

   5 men take 20 days

   1 man takes 5×20 days $= 100$ days

   (1 man takes longer so multiply by 5)

      4 men take $100 \div 4 = 25$ days

   (4 men take less time so divide by 4)

   Alternatively: 4 men will take longer than 5 men so the time ratio is $5:4$. Hence

      time taken by 5 men $= \frac{5}{4} \times \frac{20}{1}$

                        $= 25$ days

Exercise 6.8

1. 20 men working in a factory produce 3000 articles in 12 working days. How long would it take 15 men to produce the articles?

2. A farmer employs 12 men to harvest his potato crop. They take 9 days to do the job. If he had employed only 8 men, how long would it have taken them?

3. A bag contains sweets. When divided amongst 8 children each child receives 9 sweets. If the sweets are divided amongst 12 children, how many sweets would each child receive?

4. 4 people can clean an office in 6 hours. How many people would be needed to clean the office in 4 hours?

5. A farmer employs 15 women to pick his fruit. It takes them 3 days. How long would it take 10 women to pick the fruit?

6. Seven ladies take 30 minutes to make 21 toys. How long would it take 20 women to make the toys?

7. 8 women can do a piece of work in 60 hours. If the work is to be completed in 20 hours, how many women are needed?

Exercise 6.9

1. A car does 8 km per litre of petrol. How far will it go on $5\frac{1}{2}$ litres?

2. 7 men take 1 hour to make 21 model cars. How long would it take 21 men?

3. 10 men dig up a row of potatoes in 40 minutes. How long would it take 5 men?

4. A car travels 80 miles in 2 hours. If it keeps the same speed, how far will it travel in (a) 1 hour, (b) 5 hours?

5. If 100 g of sweets costs 44p, find the cost of 300 g.

6. 8 people take 5 hours to pick a row of raspberries. How long would it take 4 people to do the work?

7. It took 4 men 2 hours to dig a hole in a road. How long would 12 men take?

8. 15 kg of potatoes costs £3.00. Calculate the cost of (a) 5 kg, (b) 40 kg.

Measures of Rate

If a car travels 8 km on 1 litre of fuel we say that its fuel consumption is 8 km per litre of petrol. This is the rate at which the car consumes fuel.

The flow of water from a pipe or a tap is usually measured in gallons per minute or litres per minute. This is the rate of flow of the water.

When a car has a speed of 40 miles per hour this gives the rate at which it travels, that is 40 miles in one hour.

Example 8

(a) A car has a fuel consumption of 7 km per litre. How much fuel will be needed for a journey of 105 km?

$$\text{Fuel required} = \frac{\text{Length of journey}}{\text{Fuel consumption}}$$

$$= \frac{105}{7} \text{ litres}$$

$$= 15 \text{ litres}$$

(b) The flow of water from a pipe is 20 litres per minute. How long will it take to fill a container with a capacity of 70 litres?

$$\text{Time taken} = \frac{\text{Capacity of container}}{\text{Rate of flow}}$$

$$= \frac{70}{20} \text{ minutes}$$

$$= 3.5 \text{ minutes}$$

We see from this example that if we know the rate of flow we can work out the time taken to fill a container of known capacity. We can also work out the amount of water delivered in a given time. If butter, for instance, is sold at £1.60 per kg then this is the rate at which butter is sold. We can, knowing the cost per kilogram, easily work out the cost of any amount of butter.

Exercise 6.10

1. Potatoes are sold at £3.30 per 55 kg bag.
   (a) Calculate the cost per kilogram.
   (b) How much will 12 kg cost?

2. The flow of water from a tap is measured as 60 litres in 5 minutes.
   (a) Work out the rate of flow in litres per minute.
   (b) Calculate the time needed to fill a container with a capacity of 45 litres.
   (c) How much water will flow in 7 minutes if this rate of flow is maintained?

3. A car travels a distance of 60 miles on 2 gallons of petrol.
   (a) Calculate the fuel consumption of the car in miles per gallon.
   (b) How far would the car be expected to travel on 5 gallons of fuel?
   (c) How many gallons of fuel would be needed for a journey of 120 miles?

4. A car travels at an average speed of 50 km/h.

   (a) How far will the car travel in 3 hours if this speed is maintained?

   (b) How long will it take the car to travel 175 km?

5. The density of aluminium is 2700 kg per cubic metre. How much will 5 cubic metres of aluminium weigh?

6. The density of nylon is 1.16 grams per cubic centimetre. How many cubic centimetres of nylon weigh 8.12 grams?

7. A schoolboy found that he walked 100 metres in 50 seconds.

   (a) Calculate his speed in metres per second.

   (b) If he continues to walk at this speed, how long will it take him to walk a further 350 metres?

Rate of Exchange

Every country has its own monetary system. If there is to be trade and travel between any two countries there must be a rate at which the money of one country can be converted into money of the other country. This rate is called the **rate of exchange**.

Foreign Exchange Rates at July 1992

Country	Rate of exchange
Belgium	58.10 francs = £1
France	9.49 francs = £1
Germany	2.83 marks = £1
Greece	342 drachmas = £1
Italy	2145 lire = £1
Spain	177 pesetas = £1
United States	$1.86 = £1

Example 9

(a) If 177 Spanish pesetas = £1, find to the nearest penny the value in British money of 5000 pesetas.

Method 1 (the unitary method)

$$177 \text{ pesetas} = £1$$

$$1 \text{ peseta} = £\frac{1}{177}$$

$$5000 \text{ pesetas} = £\frac{1}{177} \times \frac{5000}{1}$$

$$= £\frac{5000}{177}$$

$$= £28.25$$

Method 2 (the fractional method)

$$5000 \text{ pesetas} = £\frac{5000}{177}$$

$$= £28.25$$

(b) A tourist changes traveller's cheques for £50 into French francs at 9.49 francs to £1. How many francs does he get?

$$£50 = 50 \times 9.49 \text{ francs}$$

$$= 474.50 \text{ francs}$$

Exercise 6.11

Use a calculator and where necessary state the answers correct to 2 decimal places.

Using the exchange rates given above find:

1. The number of German marks equivalent to £15.

2. The number of Spanish pesetas equivalent to £25.

3. The number of United States dollars equivalent to £8.

4. The number of pounds equivalent to 225 dollars.

5. The number of Belgian francs equivalent to £98.50.

6. The number of pounds equivalent to 8960 lire.

7. A transistor set costs £52.60 in London. An American visitor wishes to buy a set but wants to pay in dollars. How much will he pay in dollars?

8. A tourist changes traveller's cheques for £100 into Greek currency at 350 drachmas to £1. He spends 30 000 drachmas and changes the remainder back into British money at £1 = 320 drachmas. How much will the tourist get for his drachmas?

Map Drawing and Model Scales

When drawings or models of large objects are to be made they are usually made to a scale, for example $1 \text{ cm} = 10 \text{ m}$. This means that an actual distance of 10 m on the object will be represented by 1 cm on the drawing or model.

Example 10

(a) A house is drawn to a scale of $1 \text{ cm} = 2 \text{ m}$. On the drawing the kitchen has a length of 4 cm. What is the actual length of the kitchen?

$$\text{Actual length of kitchen} = 4 \times 2 \text{ m}$$
$$= 8 \text{ m}$$

(b) A model of a train is made $\frac{1}{72}$nd full size. If the actual length of the locomotive is 108 ft, how long will the model be?

$$\text{Length of model} = \frac{1}{72} \times \frac{108}{1} \text{ ft}$$
$$= 1.5 \text{ ft}$$

On maps and drawings scales are often expressed as a ratio, for instance $1 : 1000$. This means that a distance of 1000 m on the ground would be represented by 1 m on the map or drawing.

Example 11

(a) A road map is drawn to a scale of 2 km to 1 cm. Express this scale as a ratio.

$$2 \text{ km} = 2 \times 1000 \text{ m}$$
$$= 2 \times 1000 \times 100 \text{ cm}$$
$$= 200\,000 \text{ cm}$$

Since the map scale is 2 km to 1 cm, 1 cm on the map represents 200 000 cm on the ground. Therefore the map scale expressed as a ratio is $1 : 200\,000$.

(b) A map is drawn to a scale of $1 : 1\,000\,000$. How many miles is represented by 1 inch?

1 inch on the map represents 1 000 000 inches on the ground.

$$1\,000\,000 \text{ inches} = \frac{1\,000\,000}{12} \text{ feet}$$
$$= \frac{1\,000\,000}{12 \times 5280} \text{ miles}$$
$$= 15.8 \text{ miles}$$

So the map scale is 15.8 miles to 1 inch (approximately 16 miles to the inch).

Exercise 6.12

1. A drawing of a house is made to a scale of $1 \text{ cm} = 2 \text{ m}$. The length of the lounge measures 4 cm on the drawing. What is the actual length of the lounge?

2. The scale of a map is $1 : 20\,000$. What distance, in metres, does 8 cm on the map represent?

3. The model of a car is 50 times smaller than the real car.
   (a) What length on the real car does 4 cm on the model represent?
   (b) What length on the model would represent 3 m on the real car? Give your answer in centimetres.

4. The scale of a map is $1 : 100\,000$. The actual distance between two towns shown on the map is 25 km. What is the distance between the two towns, in centimetres, on the map?

5. A model aeroplane is built 72 times smaller than the real aeroplane. What is the real measurement represented by (a) 3 cm, (b) 0.5 cm?

6. On a map 1 cm represents 5 km.

    (a) What distance on the ground is represented by 0.6 cm on the map?

    (b) What length on the map represents 7.5 km on the ground?

7. The distance between two islands is measured as 8.1 cm on a map drawn to a scale of 1 : 5 000 000. What is the actual distance between the two islands?

8. A model car is made to a scale of 1 : 25. If the actual length of the car is 12 ft, how long is the model?

Miscellaneous Exercise 6

1. If 5 pencils cost 45p, work out the cost of 12 pencils.

2. A car does 7 km per litre of petrol. How far will it go on $4\frac{1}{2}$ litres?

3. If a man walks 3 km in 40 minutes, how many kilometres will he have walked in 60 minutes if he keeps going at the same rate?

4. At a wedding reception for 120 people it was estimated that one bottle of wine would be sufficient for 8 people.

    (a) How many bottles would supply the 120 guests?

    (b) At a cost of £2.90 per bottle, what would be the total bill for wine?

5. Simplify the ratio 175 : 200.

6. Simplify the ratio 2 km : 500 m.

7. Express the ratio 8 : 64 as a fraction in its lowest terms.

8. Is 100 g of cornflour at 70p a better buy than 160 g at £1.20?

9. In a school the ratio of the number of pupils to the number of teachers is 18 : 1. If the number of pupils is 540, how many teachers are there?

10. A model of an aeroplane is made to a scale of 1 : 78. If the wing span of the model is 35 cm find, in metres, the wing span of the actual aeroplane.

11. £979 is to be divided into three parts in the ratio 6 : 3 : 2. Calculate the value of the smallest part.

12. A car goes 12 km per litre of petrol. How many whole litres of petrol will be needed to be sure of completing a journey of 100 km?

13. A councillor gets a car allowance of 17.4 pence per mile. How much will he be paid for a journey of 20 miles?

14. A vending machine needs 20 litres of orangeade to fill 50 cups. How many litres are needed to fill 60 cups?

15. Change £150 into dollars when the exchange rate is $1.85 to £1.

16. If two men can paint a fence in 6 hours, how long will it take three men to paint it?

17. 2 dratt equal 1 assam and 1 dratt equals 4 yeda. How many yeda equal 1 assam?

18. Taking £1 as 9.50 francs, express 5.50 francs as pence correct to the nearest penny.

19. A car which travels 10 km on a litre of petrol requires $22\frac{1}{2}$ litres for a journey. Find the number of litres which would be required for the same journey by a car which has a fuel consumption of 4 km per litre.

20. The annual rent of a field amounts to £93.50. The rent is shared by two farmers in the ratio 15 : 7. Find the difference between the amounts of their shares.

21. Twelve bottles of claret cost £48.96. How many bottles can be purchased for £61.20?

22. When a petrol tank is $\frac{7}{8}$ full it contains $31\frac{1}{2}$ litres of petrol. How much petrol does the tank hold when full?

23. A car runs 4 km on a litre of petrol which costs 47 pence. Find the cost of petrol for a journey of 392 km.

Mental Test 6

Try to answer the following questions without writing anything down except the answer.

1. An alloy is made by mixing copper and zinc in the ratio $5:2$. If 4 kg of zinc is used, how much copper is used?

2. Put the ratio $6:15$ in its simplest terms.

3. Express the ratio $15:20$ as a fraction in its lowest terms.

4. Simplify the ratio £2:50p.

5. Divide £200 in the ratio $2:3$.

6. If 8 rubbers cost 72p, how much do 2 rubbers cost?

7. 2 men digging a ditch take 4 days to complete it. How long would 8 men take?

8. A car has a fuel consumption of 10 miles per gallon. How many gallons are needed for a journey of 50 miles?

9. If the exchange rate is 10 francs = £1, how many francs are equivalent to £50.

10. A map is drawn to a scale of 1 cm : 5 km. A road is 20 km long on the ground. What length, in centimetres, will represent it on the map?

Percentages

Introduction

When comparing fractions it is often convenient to express them with a denominator of 100. Thus

$$\tfrac{1}{2} = \tfrac{50}{100} \quad \text{and} \quad \tfrac{2}{5} = \tfrac{40}{100}$$

Fractions with a denominator of 100 are called **percentages**. Thus

$$\tfrac{1}{4} = \tfrac{25}{100} = 25 \text{ per cent}$$
$$\tfrac{3}{10} = \tfrac{30}{100} = 30 \text{ per cent}$$

The symbol % is usually used instead of the words per cent. Thus

$$\tfrac{3}{20} = \tfrac{15}{100} = 15\%$$

Changing Fractions and Decimals into Percentages

To convert a fraction or a decimal into a percentage multiply it by 100.

Example 1

(a) Convert $\tfrac{17}{20}$ into a percentage.

$$\frac{17}{20} = \frac{17}{\cancel{20}_{1}} \times \frac{\cancel{100}^{5}}{1}\%$$

$$= \frac{17 \times 5}{1 \times 1}\%$$

$$= 85\%$$

(b) Convert 0.3 into a percentage.

$$0.3 = \frac{3}{10}$$

$$= \frac{3}{\cancel{10}_{1}} \times \frac{\cancel{100}^{10}}{1}\%$$

$$= \frac{3 \times 10}{1 \times 1}\%$$

$$= 30\%$$

Alternatively, multiplying the decimal number by 100

$$0.3 = 0.3 \times 100\% = 30\%$$

(c) Convert 0.56 into a percentage.

$$0.56 = 0.56 \times 100\% = 56\%$$

Exercise 7.1

Convert the following fractions into percentages:

1. $\tfrac{7}{10}$ 2. $\tfrac{4}{5}$ 3. $\tfrac{11}{20}$ 4. $\tfrac{9}{25}$

5. $\tfrac{31}{50}$ 6. $\tfrac{1}{4}$ 7. $\tfrac{9}{10}$ 8. $\tfrac{1}{5}$

Change the following decimal numbers into percentages:

9. 0.2 10. 0.34 11. 0.73

12. 0.68 13. 0.813 14. 0.927

15. 0.333

Not all percentages are whole numbers. For instance

$$\frac{3}{8} = \frac{3}{\cancel{8}_2} \times \frac{\cancel{100}^{25}}{1}\%$$

$$= \frac{3 \times 25}{2}\%$$

$$= \frac{75}{2}\%$$

$$= 37\frac{1}{2}\%$$

Also
$$\frac{1}{3} = \frac{1}{3} \times \frac{100}{1}\%$$

$$= \frac{100}{3}\%$$

$$= 33\frac{1}{3}\%$$

Banks, business houses and building societies frequently state their interest rates as mixed numbers.

For instance $12\frac{1}{2}\%$ and $7\frac{1}{4}\%$.

Changing Percentages into Fractions and Decimals

To convert a percentage into a fraction or a decimal divide by 100.

Example 2

(a) Convert 45% into a fraction in its lowest terms.

$$45\% = \frac{45}{100}$$

$$= \frac{9}{20}$$

(b) Convert $7\frac{1}{2}\%$ into a fraction in its lowest terms.

$$7\frac{1}{2}\% = \frac{15}{2} \div 100$$

$$= \frac{\cancel{15}^3}{2} \times \frac{1}{\cancel{100}_{20}}$$

$$= \frac{3 \times 1}{2 \times 20}$$

$$= \frac{3}{40}$$

(c) Convert 3.9% into a decimal number.
$$3.9\% = 3.9 \div 100 = 0.039$$

Note that all we have done is to move the decimal point 2 places to the left.

Exercise 7.2

Convert the following percentages into fractions in their lowest terms:

1. 32%   2. 24%   3. 30%
4. 45%   5. 6%   6. $37\frac{1}{2}\%$
7. $66\frac{2}{3}\%$ 8. $7\frac{1}{4}\%$ 9. $8\frac{1}{3}\%$
10. $62\frac{1}{2}\%$

Convert the following percentages into decimals:

11. 78%   12. 31%   13. 48.2%
14. 2.5%   15. 1.25%   16. 3.95%
17. 20.1%   18. 1.96%

The following table gives corresponding fractions, decimals and percentages. Copy the table and write in the figures which should be placed in each of the spaces filled with a question mark. Put fractions into their lowest terms.

	Fraction	Decimal	Percentage
	$\frac{1}{4}$	0.25	25
19.	$\frac{11}{20}$	?	?
20.	$\frac{17}{50}$	?	?
21.	$\frac{7}{8}$	?	?
22.	?	0.76	?
23.	?	0.08	?
24.	?	0.375	?
25.	?	?	15
26.	?	?	27
27.	?	?	45
28.	?	?	$5\frac{1}{4}$

Percentage of a Quantity

To find the percentage of a quantity we must first express the percentage as a fraction or a decimal.

Example 3

(a) What is 10% of 40?

    Method 1 (the fractional method)

$$10\% \text{ of } 40 = \frac{10}{100} \text{ of } 40$$

$$= \frac{10}{100} \times \frac{40}{1}$$

$$= 4$$

    Method 2 (the decimal method)

    Since $10\% = 10 \div 100 = 0.1$

$$10\% \text{ of } 40 = 0.1 \times 40 = 4$$

If a calculator is to be used the decimal method is to be preferred.

(b) What is $7\frac{1}{2}\%$ of £50?

    Since $7\frac{1}{2}\% = 7.5 \div 100 = 0.075$

$$7\frac{1}{2}\% \text{ of } £50 = £50 \times 0.075 = £3.75$$

(c) 22% of a certain length is 55 cm. What is the complete length?

    Method 1 (the unitary method)

    22% of the length $= 55$ cm

     1% of the length $= 55 \div 22$ cm

$$= 2.5 \text{ cm}$$

The complete length will be 100%, hence

  100% of the length $= 100 \times 2.5$ cm

$$= 250 \text{ cm}$$

Therefore the complete length is 250 cm.

    Method 2 (the fractional method)

22% of the complete length $= 55$ cm

$$\text{Complete length} = \frac{55}{1} \times \frac{100}{22} \text{ cm}$$

$$= 250 \text{ cm}$$

(d) What percentage is 37 of 264? Give the answer correct to 5 significant figures.

$$\text{Percentage} = \frac{37}{264} \times \frac{100}{1}$$

$$= \frac{37 \times 100}{264}$$

$$= 14.015\%$$
$$\text{(correct to 5 s.f.)}$$

Exercise 7.3

1. What is:
   (a) 20% of 50?   (b) 30% of 80?
   (c) 5% of 120?   (d) 12% of 20?
   (e) 20.3% of 105?   (f) 3.7% of 68?

2. What percentage is:
   (a) 25 of 200?   (b) 30 of 150?
   (c) 24 of 150?   (d) 29 of 178?
   (e) 15 of 33?

Where necessary give the percentage correct to 5 significant figures.

3. In a test a girl scores 36 marks out of 60.

    (a) What is her percentage mark?

    (b) The percentage needed to pass the test is 45%. How many marks are needed to pass?

4. If 20% of a length is 23 cm, what is the complete length?

5. Given that 13.3 cm is 15% of a certain length, what is the complete length?

6. What is

    (a) 9% of £80

    (b) 12% of £110

    (c) 75% of £250

7. 27% of a consignment of fruit is bad. If the consignment weighs 800 kg, how much fruit is good?

8. In a certain county the average number of children eating lunches at school was 29 336, which represented 74% of the total number of children attending school. Calculate the total number of children attending school in the county.

Percentage Change

An increase of 5% in a number means that the number has been increased by $\frac{5}{100}$ of itself. Thus if the number is represented by 100, the increase is 5 and the new number is 105. The ratio of the new number to the old number is 105:100.

Example 4

An increase of 10% in salaries makes the wage bill for a factory £55 000.

(a) What was the wage bill before the increase?

(b) What was the amount of the increase?

(a) If 100% represents the wage bill before the increase then 110% represents the wage bill after the increase.

$$\text{Wage bill before the increase} = \frac{100}{110} \times £55\,000$$

$$= £50\,000$$

(b) The amount of the increase
$$= 10\% \text{ of } £50\,000$$
$$= 0.1 \times £50\,000$$
$$= £5000$$

Example 5

When a sum of money is decreased by 20% it becomes £320. What was the original sum?

If 100% represents the original sum then the sum after the decrease of 20% is represented by 80%. Therefore

$$\text{Original sum} = \frac{100}{80} \times £320$$

$$= £400$$

Exercise 7.4

1. The duty on an article is 20% of its value. If the price of the article after the duty has been paid is £960, find the price exclusive of tax.

2. When a sum of money is decreased by 10% it becomes £18. What was the original sum?

3. A man sells a car for £850 thus losing 15% of what he paid for it. How much did the car cost him?

4. During an epidemic 40% of the people in a certain place in Africa died and 1200 were left. How many people died?

5. The value of a machine depreciates by 15% of its value at the beginning of the year. If its value at the end of the year was £1360, what was its value at the beginning of the year?

6. 25% of a consignment of fruit was bad. If 1500 kg of the consignment was good, how much did the consignment weigh?

7. A man pays 20% of his salary in income tax. If his salary net of tax is £6400 per annum, what was his gross salary?

Miscellaneous Exercise 7

1. Change to fractions in their lowest terms
   (a) 48%   (b) 6%

2. Change to decimal numbers
   (a) 64%   (b) 7%

3. Below is a table of corresponding fractions, decimals and percentages. Write down the figures which should be placed in the spaces marked a, b, c, d, e and f.

Fraction	Decimal	Percentage
$\frac{1}{2}$	0.5	50
a	0.85	b
$\frac{7}{20}$	c	d
e	f	12

4. Find the value of
   (a) 9% of £20   (b) 38% of £70

5. A coat has been reduced by 15% in a sale. The old price was £80.
   (a) How much was the reduction in price?
   (b) How much will the customer pay for the coat?

6. In a class of 30 children, 40% are boys.
   (a) How many boys are there in the class?
   (b) What percentage of the class are girls?

7. Express $\frac{132}{150}$ as a percentage.

8. Find 5% of £260.

9. After prices have risen by 8% the new price of an article is £70.47. Calculate the original price.

10. Next year a man will receive a 12% wage increase and his weekly wage will then be £161.28. What is his present weekly wage?

11. The number of people working for a company at the end of 1984 was 1210. This was an increase of 10% on the number working for the company at the beginning of 1984. How many people worked for the company at the beginning of 1984?

12. The entry fee for an examination was £5.00 in 1984 and it rose to £6.40 in 1985. Express the increase in the fee as a percentage of the fee in 1984.

13. Calculate $7\frac{1}{2}$% of £160.

14. After 6% of a woman's wages have been deducted she receives a net amount of £98.70. Calculate the amount deducted.

15. A girl scores 66 marks out of 120 in her maths examination. What is her percentage mark? If the percentage required for a pass is 45%, how many marks must be obtained for a pass?

16. 8% of a sum of money is equal to £9.60. Find:
    (a) 1% of the sum of money
    (b) the sum of money
    (c) 92% of the sum of money

Mental Test 7

Try to answer the following questions without writing anything down except the answer.

1. Convert $\frac{3}{10}$ into a percentage.

2. Convert 0.4 into a percentage.

3. Convert 70% into a fraction in its lowest terms.

4. Convert 24% into a decimal number.

5. What is 10% of 50?

6. What percentage is 8 of 50?

7. Calculate 25% of £80.

8. 50% of a length is 20 cm. What is the whole length?

9. What is 7% of 200?

10. Change $66\frac{2}{3}\%$ into a fraction.

Wages and Salaries

Introduction

Everyone who works for an employer receives a wage or salary in return for his or her labour. However, the payments may be made in several different ways. Wages are usually paid weekly and salaries monthly.

Payment by the Hour

Many people are paid a certain amount of money for each hour that they work. They usually work a fixed number of hours known as the **basic week**. It is this basic week which determines the hourly (or basic) rate of pay. The basic week and the basic rate of pay are often fixed by negotiation between the employer and the trade union which represents the workers.

Example 1

A man works a basic week of 35 hours and his weekly wage is £280. Find the hourly rate of payment.

$$\text{Hourly rate of pay} = \frac{\text{Weekly wage}}{\text{Basic week}}$$

$$= \frac{£280}{35}$$

$$= £8.00$$

Example 2

A man works a basic week of 38 hours and his hourly rate of pay is £7. Calculate his weekly wage.

$$\text{Weekly wage} = 38 \text{ hours} \times £7 \text{ per hour}$$

$$= £266$$

1. A woman works a basic week of 35 hours and her basic rate is £5 per hour. Calculate her weekly wage.

2. A man works a basic week of 40 hours for which he is paid £200. Calculate his hourly rate.

3. Calculate the weekly wage for a 35-hour week if the basic rate is

   (a) £4   (b) £5.50   (c) £8

4. The basic week is 42 hours and the weekly wage is £294. What is the basic rate?

5. If the basic rate is £5.84 and the basic week is 35 hours, what is the weekly wage?

Overtime

Hourly paid workers are usually paid extra for working more hours than the basic week requires. These extra hours of work are called **overtime**. Overtime is usually paid at time-and-a-quarter ($1\frac{1}{4}$ times the basic rate), time-and-a-half ($1\frac{1}{2}$ times the basic rate) or double time (twice the basic rate).

Example 3

A girl is paid a basic rate of £4 per hour. Calculate the hourly rate of overtime when this is paid at (a) time-and-a-quarter, (b) time-and-a-half,   (c) double time.

(a) Overtime rate at time-and-a-quarter

$$= 1\frac{1}{4} \times £4$$

$$= 1.25 \times £4$$

$$= £5.00 \text{ per hour}$$

(b) Overtime rate at time-and-a-half

$$= 1\frac{1}{2} \times £4$$

$$= 1.5 \times £4$$

$$= £6.00 \text{ per hour}$$

(c) Overtime rate at double time

$$= 2 \times £4$$

$$= £8 \text{ per hour}$$

Example 4

Peter Taylor works a 40-hour week for which he is paid £280. He works 5 hours overtime at time-and-a-half. What is his total wage for the week?

$$\text{Basic rate of pay} = \frac{\text{Weekly wage}}{\text{Basic week}}$$

$$= \frac{£280}{40}$$

$$= £7.00 \text{ per hour}$$

Overtime rate at time-and-a-half

$$= 1\frac{1}{2} \times £7.00$$

$$= 1.5 \times £7.00$$

$$= £10.50 \text{ per hour}$$

Payment for overtime

$$= 5 \times £10.50$$

$$= £52.50$$

Total wage for the week

$$= £280 + £52.50$$

$$= £332.50$$

Exercise 8.2

1. A man is paid a basic rate of £5 per hour. What is his overtime rate at (a) time-and-a-quarter, (b) time-and-a-half,   (c) double time?

2. Find the hourly overtime rate at time-and-a-half when the basic rate is

    (a) £3   (b) £4   (c) £7

3. A shop assistant works a basic week of 46 hours for which she is paid £121.44. During a certain week she works 4 hours overtime for which she is paid time-and-a-half. How much does she earn in overtime?

4. Michael Evans is paid £100.80 for a basic week of 40 hours. During a certain week he works 12 hours overtime which is paid at time-and-a-quarter. Calculate his total wage for the week.

5. A woman's basic hourly rate is £3.76. She works a basic week of 35 hours from Monday to Friday. On Sunday she works 7 hours for which she is paid double time.

    (a) What is her overtime rate?

    (b) How much money does she earn in overtime?

    (c) What is her total wage for the week?

Piecework

Some workers are paid a fixed amount for each article or piece of work that they make. This is called piecework. Often, if they can make more than a certain number of articles they are paid a bonus.

Example 5

A man is paid 9p for every handle he fixes to an electric iron up to 200 per day. For each handle over 200 that he fixes he is paid a bonus of 2p. If he fixes 250 handles in a day, calculate how much he earns.

$$\text{Money earned on first } 200 = 200 \times 9p$$
$$= £18.00$$
$$\text{Money earned on next } 50 = 50 \times (9 + 2)p$$
$$= 50 \times 11p$$
$$= £5.50$$
$$\text{Total amount earned} = £18.00 + £5.50$$
$$= £23.50$$

Exercise 8.3

1. A woman is paid 8p for each bag of sweets she packs up to a limit of 350. After that she earns a bonus of 2p per bag. Work out how much she earns if she packs 450 bags in a day.

2. Nancy Jones is paid 9p for each article she makes up to a limit of 200 per day. For each article over 200 that she completes she is paid a bonus of 3p. Calculate how much she earns in a day in which she completes 320 articles.

3. A man is paid 12p for each spot weld that he makes up to 300 per day. After that he is paid a bonus of 2p. If he makes 450 spot welds in a day, calculate how much he earns in the day.

4. For each article up to a limit of 150, a pieceworker is paid 16p. After that she is paid 20p per article. If she completes 220 articles in a day, how much will she earn?

5. Workers in a shirt factory are paid 25p for each shirt they complete up to a maximum of 25. After that they are paid 90p for each completed shirt. If a worker completes 35 shirts in a day, how much will she earn?

Commission

Shop assistants, salesmen and representatives are often paid commission on top of their basic wage. This is calculated as a small percentage of the value of the goods that they have sold.

Example 6

A salesman is paid a commission of 2% on the value of the goods which he has sold. Calculate his commission if he sells goods to the value of £5000.

$$\text{Commission} = 2\% \text{ of } £5000$$
$$= £5000 \times 0.02$$
$$= £100$$

Example 7

A shop assistant is paid a basic wage of £80. In addition she is paid 3% commission on the goods which she sells. During a certain week she sells goods to the value of £2000. What is her earnings during this week?

$$\text{Commission} = 3\% \text{ of } £2000$$
$$= £2000 \times 0.03$$
$$= £60$$
$$\text{Total earnings} = £80 + £60$$
$$= £140$$

Exercise 8.4

1. Find the commission at 2% on goods sold to the value of

    (a) £100   (b) £300

    (c) £2000   (d) £5000

2. A car salesman sells a car for £3500. He is paid commission of 2% on the sale. How much commission does he earn?

3. A representative sells farm machinery for £8000. He is paid commission of 3% on his sales. How much commission is he paid?

4. Furniture worth £900 is sold by a salesman who, in addition to his weekly wage of £70 per week, is paid commission of 3% on his sales. How much will he earn this week?

5. A sales assistant is paid a basic wage of £68 per week. In addition she is paid a commission of 4% on her sales. Work out her total wage for a week in which she sold £2500 of goods.

Salaries

People like teachers, civil servants, secretaries and company managers are paid a fixed amount each year. The money is usually paid monthly and they are not paid overtime or commission.

Example 8

A secretary is paid £14 400 per annum. How much is she paid per month?

Monthly salary $= £14\,400 \div 12 = £1200$

Exercise 8.5

The following are the salaries of five people. How much is each paid monthly?

1. £5320   2. £6000   3. £8400

4. £6096   5. £7440

Deductions from Earnings

The wages and salaries which we have discussed previously are not the take-home pay or salary of the workers. A number of deductions are made first. The most important of these are

(1) Income tax

(2) National insurance

(3) Pension fund payments.

Before deductions the wage (or salary) is known as the **gross wage** (or **salary**).

After the deductions have been made the wage (or salary) is known as the **net wage** (or **salary**). It is also called the **take-home pay**.

National Insurance

Both employees and employers pay National Insurance. Employees pay a contribution based on a certain percentage of their gross pay or salary above a certain amount (in 1991–2 the figures were 9% per annum and £54 per week). Men over 65, women over 60 and employees earning less than £54 per week do not pay National Insurance.

The rates for National Insurance are fixed by the Chancellor of the Exchequer and they may vary from year to year. National Insurance contributions pay for such things as hospitals, doctors, sick pay and unemployment benefit.

Example 9

A man earns £12 000 per annum. His deductions for national insurance are 9% of his gross salary. How much does he pay in national insurance?

$$\begin{aligned} \text{Amount of national insurance} &= 9\% \text{ of } £12\,000 \\ &= 0.09 \times £12\,000 \\ &= £1080 \end{aligned}$$

Income Tax

Taxes are levied by the Chancellor of the Exchequer to produce money to pay for such items as the armed services, the Civil Service and motorways. The largest producer of revenue is income tax.

Every person who earns more than a certain amount has to pay income tax. Tax is not paid on the entire income. Certain allowances are made such as a personal allowance, a married couple's allowance, etc.

The taxable income is the gross income minus allowances. At the time of going to press, the basic rate of tax is 20% payable on the first £2000 of taxable income, and 25% payable on the next £21 700 of taxable income. A higher rate of 40% is payable on the remainder of taxable income.

Example 10

Eric earns £15 000 per annum and his wife Freda earns £9000 per annum. How much do they each pay in income tax per annum?

Eric:	£	£
Total income (earnings)		15 000
Less personal allowances	3445	11 555
Less married couple's allowance	1720	9 835
Taxable income		9 835

Tax payable = 20% of £2000 + 25% of (£9835 − £2000)

= £400 + £1958.75

= £2358.75

Freda:	£	£
Total income (earnings)		9000
Less personal allowances	3445	5555
Taxable income		5555

Tax payable = 20% of £2000 + 25% of (£5555 − £2000)

= £400 + £888.75

= £1288.75

Example 11

A single person has a total income of £56 000. Calculate the amount of income tax payable when the rates are as given above.

	£	£
Total income		56 000
Less personal allowances	3445	52 555
Taxable income		52 555

Tax payable = 20% of £2000 + 25% of £21 700 + 40% of (£52 555 − £21 700 − £2000)

= £400 + £5425 + £11 542

= £17 367

PAYE

Most people pay tax by a method known as **pay-as-you-earn** or PAYE for short. The tax is deducted from their wages or salaries before they receive it. The taxpayer and his employer receive a notice of coding which gives the employee's allowances (based upon tax forms which he or she has previously completed) and sets a code number. The employer then knows from tax tables the amount of tax to deduct from the wages of an employee.

Pension Fund

Many firms operate their own private pension scheme to provide a pension in addition to that provided by the state. Usually, both employer and employee contribute, the employee's share being of the order of 5 or 6% of gross annual salary.

The amount of pension received depends upon the length of service and earnings at the time of retirement.

Example 12

An employee earns a salary of £8000 per annum. His pension fund payments amount to 6% of his annual salary. How much are his pension fund payments?

$$\text{Pension fund payments} = 6\% \text{ of } £8000$$
$$= 0.06 \times £8000$$
$$= £480$$

Example 13

A woman earns £9000 per annum. She pays £540 in national insurance contributions, £1950 in income tax and £456 in pension fund payments. Calculate her monthly take-home pay.

$$
\begin{array}{ll}
\text{The deductions are} & £540 \\
 & £1950 \\
 & \underline{£456} \\
 & £2946
\end{array}
$$

$$\text{Annual salary after deductions} = £9000$$
$$- £2946$$
$$= £6054$$
$$\text{Monthly take-home pay} = £6054 \div 12$$
$$= £504.50$$

Exercise 8.6

1. A person earns £8500 per annum. He pays 9% of his salary in national health contributions. How much does he pay?

2. Calculate the amount paid into a pension fund at a rate of 5% of the annual wage if this is £7000.

3. A woman earns £95 per week. Her deductions are: national insurance £3.66, income tax £19.36 and pension fund payments £5.70. Work out her take home pay.

4. A man pays tax on £7200. If tax is paid at 25%, calculate the amount of income tax paid.

5. When income tax is levied at 20% a man pays £240 per annum in income tax. What is his taxable income?

6. A man earns £7500 per year. £1400 is deducted for income tax, £675 for national insurance and £450 for his pension fund.

    (a) Find his yearly take-home pay after deductions.

    (b) What is his monthly take-home pay to the nearest penny?

7. A single man has a personal allowance of £3445 and no other allowances. If he earns £8000 per annum, calculate

    (a) his taxable income

    (b) the amount of tax payable when this is levied at 20% for the first £2000 and then 25%.

8. A married man earns £21 000 per annum. His tax-free allowances to be set against his income are a personal allowance of £3445 and a married man's allowance of £1720. Calculate

    (a) his taxable income

    (b) the amount of tax payable when this is levied at 20% for the first £2000 of taxable income and 25% for the remainder.

9. After the first £2000, on which tax is paid at 20%, the next £21 700 of a man's taxable income is taxed at 25%. The remainder of his taxable income, if any, is taxed at 40%. Calculate

    (a) the amount of tax payable on a taxable income of £32 000

    (b) the taxable income on which the tax payable is £9125.

10. Tom Green earns a salary of £26 000 per annum and he also has an investment income of £7500 per annum. His wife Pat has a total income of £9600 per annum. How much do they pay in total in income tax when Tom's allowances are a personal allowance of £3445 and a married person's allowance of £1720 whilst Pat has a personal allowance of £3445? The basic rate of taxation is 20% on the first £2000 of taxable income, 25% on the next £21 700 and 40% on the remainder.

Miscellaneous Exercise 8

1. Margaret worked for 40 hours. She is paid £2.90 per hour. What was her wage?

2. A man's taxable income is £3000. He pays income tax at 25%. How much tax did he pay?

3. A woman's total income is £8000 per annum. Her tax free allowances amount to £3250. How much is her taxable income?

4. A man is paid £200 for a 40-hour week. Calculate the man's basic hourly rate of pay.

5. A girl is paid £2.50 per hour. Overtime is paid at time-and-a-half. What is her overtime rate?

6. A salesman is paid commission of 3% on his weekly sales. During one week he sold goods to the value of £8000. How much commission did he earn?

7. A woman is paid 12p for every article she makes. During one week she made 750 articles. How much money did she make?

8. A trainee earned £78 for a week's work. He paid £9.36 in income tax. What percentage of his weekly wage was paid in income tax?

9. A man's income is £7000 per annum. He is allowed £3000 in tax-free allowances.
(a) Calculate his taxable income.
(b) If tax was levied at 30%, find the amount of tax payable in a year.

10. A man is paid £240 for a 40-hour week. Overtime is paid at time-and-a-half.
(a) Calculate the man's basic hourly rate of pay.
(b) Calculate his overtime rate of pay per hour.
(c) If he works 7 hours overtime during a certain week, work out his total wages for the week.

11. A salesman is paid a basic wage of £1100 per month. On top of this he is paid commission of $2\frac{1}{2}\%$ on goods that he sells. During one month he sells goods to the value of £20 000. Calculate his total wages for the month.

12. A woman is paid 8p for every article that she makes up to a limit of 150 per day. For each article made over this limit she is paid a bonus of 2p per article. In one day she made 220 articles. How much did she earn that day?

13. A woman earns £8000 per year. Deductions from her yearly pay are £720 for national insurance, £1850 for income tax and £450 for pension fund payments.
(a) What are her total deductions for the year?
(b) What is her net annual pay?
(c) What is her monthly take-home pay?

14. A man has a taxable income of £43 000. For up to £2000 of taxable income, tax is levied at 20%. After this the next £21 700 is taxed at 25% and the remainder, if any, is taxed at 40%. Calculate the amount of tax payable per annum.

15. A woman is paid £3.50 per hour for a 40-hour week.

   (a) Calculate her gross weekly wage.

   (b) If she pays national insurance at 3.85% of her gross salary find how much she pays in national insurance contributions per week.

   (c) If, in addition, she pays £18.30 per week in income tax, find her net wage for the week.

16. A man works a basic week of 40 hours at £3 per hour. He then earns overtime at time-and-a-half. Calculate his gross wage for a 44-hour week.

Mental Test 8

Try to answer the following questions without writing anything down except the answer.

1. A man's basic wage is £120 for a 40-hour week. What is his basic hourly rate of pay?

2. A man earns £4 per hour. How much will he earn for a 40-hour week?

3. A girl earns £2 per hour and overtime is paid at time-and-a-half. What is her overtime rate?

4. A jobbing gardener is paid £3 per hour. How much will he earn for 5 hours work?

5. A woman is paid 20p for every article she completes. In one day she completed 50 articles. How much did she earn?

6. A man has a taxable income of £200. If tax was levied at 30%, how much tax would be payable?

7. National insurance is paid at 9% of gross salary. A woman earns £5000 per year. How much does she pay in national insurance contributions?

8. Commission is paid at 5% to an agent who sold goods to the value of £500. How much commission did he earn?

9. A man earns £200 per week. His deductions are £20 for national insurance and £25 for income tax. What is his take-home pay for the week?

10. A woman employee is paid an annual salary of £6000 per year. What is her monthly salary?

Interest

Introduction

If you invest money with a bank or a building society then interest is paid to you for lending them the money. On the other hand, if you borrow money from the bank or building society, then you will be charged interest.

Example 1

I invest £1500 for 1 year at 12% interest. How much interest will I get at the end of the year?

$$\text{Amount of interest} = 12\% \text{ of } £1500$$
$$= 0.12 \times £1500$$
$$= £180$$

Simple Interest

With simple interest the amount of interest earned is the same every year. If the money is invested for 2 years the amount of interest will be doubled; for 3 years the amount of interest will be multiplied by 3; for 6 months (i.e. $\frac{1}{2}$ year) the amount of interest will be halved.

Example 2

A man borrows £800 for 4 years. If the rate of simple interest is 15%, find the amount of interest that he pays.

$$\text{Amount of interest for 1 year} = 15\% \text{ of } £800$$
$$= 0.15 \times £800$$
$$= £120$$
$$\text{Amount of interest for 4 years} = 4 \times £120$$
$$= £480$$

Find the interest for 1 year on

1. £500 at 5% per annum.
2. £1100 at 10% per annum.
3. £5000 at 12% per annum.
4. £20 000 at 8% per annum.
5. £750 at 11% per annum.

Find the amount of simple interest on

6. £600 at 10% for 3 years.
7. £1200 at 8% for 5 years.
8. £600 at 12% for 6 months.
9. £850 at 7% for 2 years.
10. £900 at 15% for 7 years.

The Simple Interest Formula

From what has been said we can see that the simple interest can be calculated from

$$I = P \times R \times T$$

where I is the amount of simple interest, P is the principal (i.e. the amount of money invested or borrowed), R is the rate of interest as a decimal and T the time in years.

Since the rate R can be expressed as a fraction with a denominator of 100

$$I = \frac{P \times R \times T}{100}$$

Example 3

A person borrows £4500 for a period of 4 years at 20% simple interest. Calculate the amount of interest that the person must pay.

We are given that $P = 4500$, $R = 20$ and $T = 4$.

$$I = \frac{P \times R \times T}{100}$$

$$= \frac{4500 \times 20 \times 4}{100}$$

$$= £3600$$

The simple interest formula can be transposed to give

$$T = \frac{100 \times I}{P \times R}; \quad R = \frac{100 \times I}{P \times T}; \quad P = \frac{100 \times I}{R \times T}$$

The form to be used depends upon what we have to find.

Example 4

(a) £1500 is invested at 15% simple interest per annum. How long will it take for the investment to reach £2850?

$$\text{Interest} = £2850 - £1500$$

$$= £1350$$

We now have $I = 1350$, $P = 1500$ and $R = 15$. We have to find T.

$$T = \frac{100 \times I}{P \times R}$$

$$= \frac{100 \times 1350}{1500 \times 15}$$

$$= 6$$

Hence the investment will reach £2850 after 6 years.

(b) What principal is needed to earn interest of £96 if the money is invested for 5 years at 6% per annum?

We are given that $I = 96$, $R = 6$ and $T = 5$ and we have to find P.

$$P = \frac{100 \times I}{R \times T}$$

$$= \frac{100 \times 96}{6 \times 5}$$

$$= 320$$

Hence a principal of £320 is required.

In practice, for periods of more than 1 year, simple interest is virtually unknown.

Exercise 9.2

Find the simple interest on

1. £1400 for 4 years at 5% per annum.

2. £2000 for 3 years at 10% per annum.

3. £500 for 5 years at 8% per annum.

4. £1500 for 6 months at 12% per annum.

5. £800 for 4 years at 24% per annum.

Find the length of time for

6. £1000 to be the interest on £5000 invested at 5% per annum.

7. £480 to be the interest on £2000 invested at 8% per annum.

8. £1200 to be the interest on £3000 invested at 10% per annum.

Find the rate per cent per annum simple interest for

9. £420 to be the interest on £1200 invested for 5 years.

10. £72 to be the interest on £200 invested for 3 years.

11. £1200 to be the interest on £3000 invested for 4 years.

Find the principal required for

12. The simple interest to be £600 on money invested for 3 years at 5% per annum.

13. The simple interest to be £40 on money invested for 2 years at 10% per annum.

14. The simple interest to be £432 on money invested for 4 years at 9% per annum.

15. £500 is invested for 7 years at 10% per annum simple interest. How much will the investment be worth after this period?

16. £1000 is invested for 3 years and after this time the investment is worth £1240. Calculate the rate per cent per annum simple interest.

17. £2000 was invested at 12% per annum simple interest. If the investment is now worth £2960, find how long ago the money was invested.

Compound Interest

Compound interest is different from simple interest in that the interest which is added to the principal also attracts interest. Thus after 2 years there will be more interest than after 1 year because there is more capital to attract interest.

Example 5

£5000 is invested for 2 years at 15% compound interest. Calculate the value of the investment after this period.

	£
Principal	5000
1st year's interest	750
Value of the investment after 1 year	5750
2nd year's interest	802.50
Value of the investment at the end of 2 years	6612.50

The value of an investment is often called the amount. Thus in Example 5 we would say that the amount after 2 years was £6612.50.

Compound interest tables are available which give the amount to which a principal of £1 grows.

Years	5%	6%	7%	8%	9%
1	1.050	1.060	1.070	1.080	1.090
2	1.103	1.124	1.145	1.166	1.188
3	1.158	1.191	1.225	1.260	1.295
4	1.216	1.262	1.311	1.360	1.412
5	1.276	1.338	1.403	1.469	1.539
6	1.340	1.419	1.501	1.587	1.677
7	1.407	1.504	1.606	1.714	1.828
8	1.477	1.594	1.718	1.851	1.993
9	1.551	1.689	1.838	1.999	2.172
10	1.629	1.791	1.967	2.159	2.367

Years	10%	11%	12%	13%	14%
1	1.100	1.110	1.120	1.130	1.140
2	1.210	1.232	1.254	1.277	1.300
3	1.331	1.368	1.405	1.443	1.482
4	1.464	1.518	1.574	1.630	1.689
5	1.611	1.685	1.762	1.842	1.925
6	1.772	1.870	1.974	2.082	2.195
7	1.949	2.076	2.211	2.353	2.502
8	2.144	2.304	2.476	2.658	2.853
9	2.358	2.558	2.773	3.004	3.252
10	2.594	2.839	3.106	3.395	3.707

Example 6

Using the compound interest table, find the compound interest earned by £800 invested for 7 years at 12% per annum.

From the table, £1 becomes £2.211. To find to what £800 grows we multiply £2.211 by 800.

Amount = £2.211 × 800

= £1768.80

Interest = Amount − Principal

= £1768.80 − £800

= £968.80

Exercise 9.3

Calculate the compound interest earned on

1. £300 invested for 2 years at 5% per annum.

2. £1000 invested for 3 years at 10% per annum.

3. £500 invested for 2 years at 15% per annum.

Calculate the amount when

4. £800 is invested for 3 years at 9% per annum.

5. £2000 is invested for 2 years at 12% per annum.

6. £20 000 is invested for 2 years at 15% per annum.

7. A woman borrows £1500 for 2 years at 15% per annum compound interest. How much will the woman have to pay back?

8. An investor can invest money at 10% simple interest or 8% compound interest per annum. If she invests £8000 for 3 years, calculate which is the most profitable investment and find the difference between the two.

Use the compound interest table on page 80 to calculate the amount when

9. £1500 is invested for 7 years at 9% per annum.

10. £2500 is invested for 9 years at 11% per annum.

11. £500 is invested for 6 years at 7% per annum.

Use the compound interest table on page 80 to find the amount of compound interest earned on

12. £700 invested for 5 years at 8% per annum.

13. £3000 invested for 8 years at 10% per annum.

14. £4500 invested for 8 years at 14% per annum.

Depreciation

When the value of an article decreases with age this reduction in its value is called **depreciation**. Examples of articles which depreciate in value are machinery, motor vehicles and typewriters.

Problems on depreciation are similar to those on compound interest except that with compound interest the interest is **added** to the amount of the previous year, whilst for depreciation the interest is **subtracted** from the amount of the previous year.

Example 7

A small business buys a computer costing £2000. The rate of depreciation is 20% per annum. What is its value at the end of 3 years?

Cost of computer	= £2000
1st year's depreciation	= 20% of £2000
	= £400
Value at end of 1st year	= £2000 − £400
	= £1600
2nd year's depreciation	= 20% of £1600
	= £320
Value at end of 2nd year	= £1600 − £320
	= £1280
3rd year's depreciation	= 20% of £1280
	= £256
Value at end of 3rd year	= £1280 − £256
	= £1024

Hence the computer is reckoned to be worth £1024 at the end of 3 years.

Exercise 9.4

1. A firm buys a machine for £1600. If its value depreciates at 10% per annum, calculate its value at the end of 3 years.

2. The rate of depreciation of office machinery bought for £60 000 is 15% per annum. How much will the machinery be worth at the end of 5 years?

3. A lorry costs £25 000. It depreciates in value at 20% per annum. Find its value at the end of 7 years.

4. A machine costs £40 000 when new. Its value depreciates at 25% per annum. Work out its value at the end of 3 years.

Miscellaneous Exercise 9

1. What is the interest for 1 year on £800 at 10% per annum?

2. What is the interest for six months on £500 at 8% per annum?

3. £200 is invested for 2 years at 12% per annum simple interest.

    (a) What is the interest at the end of the first year?

    (b) How much is the interest at the end of the two years?

4. A woman invests £200 for 2 years at 10% compound interest.

    (a) How much interest does the money earn in the first year?

    (b) At the end of the second year, how much interest has been earned?

    (c) How much money will she get at the end of the second year?

5. What sum of money must be invested to produce simple interest of £100 at the end of 1 year. The rate is 20% per annum.

6. Calculate the simple interest on £1600 invested for 1 year at 5% per annum.

7. A man invests £5000 at 8% compound interest per annum. Calculate the value of his investment at the end of 2 years.

8. A woman has £3400 to invest. She can buy savings bonds which pay 9% per annum simple interest or she can put the money into a building society account paying 8% compound interest per annum.

    (a) Calculate the value of her investment at the end of 3 years if she buys savings bonds.

    (b) If she puts the money into a building society account, calculate the amount standing in her account at the end of 3 years.

    (c) Which is the better investment and what is the difference in the amount of interest earned?

9. A man earned £12 000 during a certain year. He invested 15% of his earnings in a savings account paying 10% per annum compound interest. Work out the value of this investment at the end of two years.

10. A sum of money was invested at 12% per annum simple interest. After 1 year the amount of interest was £600. How much money was invested?

11. A machine which costs £5500 depreciates at a rate of 15% per annum. How much will the machine be worth at the end of 2 years?

Mental Test 9

Try to answer the following questions without writing anything down except the answer.

1. Find the interest on £200 invested at 5% for 1 year.

2. What is the simple interest on £100 invested for 2 years at 10% per annum?

3. How much does £1000 become when it is invested at 20% per annum for 1 year?

4. Use the compound interest table on page 80 to find the amount accruing when £100 is invested for 5 years at 8% compound interest per annum.

5. A machine depreciates at 10% per annum. If it cost £1000 when new, how much is it worth at the end of 1 year?

Money in Business and the Community

Profit and Loss

When a dealer buys or sells goods, the cost price is the price at which he buys the goods and the selling price is the price at which he sells the goods. The profit is the difference between the selling price and the cost price. That is

$$\text{Profit} = \text{Selling price} - \text{Cost price}$$

The profit per cent is usually calculated on the cost price, and

$$\text{Profit \%} = \frac{\text{Selling price} - \text{Cost price}}{\text{Cost price}} \times 100$$

The profit per cent is called the **mark-up**.

Example 1

A shopkeeper buys an article for £5.00 and sells it for £6.00. What is his profit per cent?

Cost price = £5   and   Selling price = £6

$$\text{Profit \%} = \frac{6-5}{5} \times 100$$

$$= \frac{1}{5} \times \frac{100}{1}$$

$$= 20\%$$

It is often easier for a retailer to calculate the profit as a percentage of the selling price, because the bill shows the amount of the takings and it is then easy to calculate the profit as a percentage of the takings. This is called the **margin**.

$$\text{Margin} = \frac{\text{Profit}}{\text{Selling price}} \times 100$$

If a loss is made the cost price is greater than the selling price. That is

$$\text{Loss} = \text{Cost price} - \text{Selling price}$$

As with profit per cent, the loss per cent is usually calculated on the cost price. Thus

$$\text{Loss \%} = \frac{\text{Cost price} - \text{Selling price}}{\text{Cost price}} \times 100$$

Example 2

A man buys a car for £3200 and sells it for £2400. Calculate his percentage loss.

Cost price = £3200   and

Selling price = £2400

$$\text{Loss \%} = \frac{3200 - 2400}{3200} \times 100$$

$$= 25\%$$

Example 3

A dealer buys 20 similar articles for £5. He sells them for 30p each. What is his profit per cent?

Cost per article = £5 ÷ 20

$$= £0.25$$

$$= 25\text{p}$$

$$\text{Profit \%} = \frac{30 - 25}{25} \times 100$$

$$= 20\%$$

Exercise 10.1

1. A shopkeeper buys an article for 40p and sells it for 50p. What is the profit per cent?

2. An article is bought for £5 and sold for £4. What is the loss per cent?

3. A greengrocer buys a box of 150 grapefruit for £12. He sells them for 12p each. What is his profit per cent?

4. A second-hand car is bought for £3100 and sold for £775. What is the percentage loss?

5. A dealer buys 200 pencils for £50 and sells them for 30p each. Calculate his profit per cent.

6. A retailer buys 30 pens at 8p each. Three are damaged and unsaleable but he sells the others at 10p each. What profit per cent does he make?

7. A motor bike is bought for £168 and sold for £126.
   (a) What is the loss?
   (b) What is the loss per cent?

8. A dealer buys a chair for £30 and sells it for £40.
   (a) Calculate his profit.
   (b) What is the profit per cent?

The selling price (sometimes called the marked price) may be calculated when the cost price and the profit per cent are given. This calculation is of great importance to people such as shopkeepers and dealers who sell things.

Example 4

A dealer buys a table for £20 and wants to make a profit of 30%. What should his selling price be?

If we let the cost price be 100, then since the profit is 30% the selling price is 130. That is

$$\text{Selling price} = \frac{130}{100} \times \frac{20}{1}$$

$$= £26$$

Example 5

A lady buys a typewriter for £200 and when she sells it she shows a loss of 40%. How much did she sell the typewriter for?

If the cost price is 100, then since the loss is 40% the selling price is 60.

$$\text{Selling price} = \frac{60}{100} \times \frac{200}{1}$$

$$= £120$$

When the selling price and the profit per cent are known, the cost price may be calculated by the method shown in Example 6.

Example 6

By selling a wardrobe for £375 a trader makes a profit of 25%. For how much did he buy the wardrobe (i.e. what was the cost price of the wardrobe)?

If the cost price is 100 then, since the profit is 25% the selling price is 125. Therefore

$$\text{Cost price} = \frac{100}{125} \times \frac{375}{1}$$

$$= £300$$

Exercise 10.2

1. A dealer buys an article for £30. He wishes to make a profit of 20%. How much should he sell the article for?

2. A dealer buys an armchair for £200. If his profit is to be 40%, for how much should he sell the armchair?

3. A bicycle is sold for £125 and the trader makes a profit of 25%. What was the cost price of the bicycle?

4. A car was sold for £21 000 and the garage that sold the car made a profit of 40%. Calculate the cost price of the car.

5. A chair was marked for sale at £260 thereby making a profit of 30% for the shopkeeper. How much did the shop-keeper pay for the chair?

6. A shopkeeper buys an article for £14.40. His marked selling price is 20% more than the cost price. What is his marked selling price?

7. Calculate the selling price of potatoes per kilogram if they are bought for £4 per 50-kg bag and the profit per cent is $12\frac{1}{2}$%.

Discount

Sometimes a dealer will deduct a percentage of the selling price if the customer is prepared to pay cash. This is called **discount**.

Example 7

A piece of furniture is offered for sale at £270. A customer is offered a discount of 10% for cash. How much does the customer actually pay?

    Method 1

      Discount $=$ 10% of £270

               $=$ £27

      Amount actually paid $=$ £270 $-$ £27

                        $=$ £243

    Method 2

    Since a discount of 10% is given, the customer pays only 90% (100% $-$ 10%) of the cost.

      Amount actually paid $=$ 90% of £270

                   $=$ 0.9 $\times$ £270

                   $=$ £243

Exercise 10.3

1. An armchair is marked for sale at £200. The trader offers a discount of 10% for cash. How much will a cash-paying customer actually pay for the armchair?

2. During a sale a clothing shop offers a $12\frac{1}{2}$% discount for cash. If the price of a suit is £64, how much will it sell for during the sale?

3. A furniture store offers a 7% discount for cash. How much discount will be allowed on furniture whose marked price is £1100?

4. A shop offers a discount of 5p in the pound (i.e. 5%) for cash. How much discount will the shop allow on a washing machine whose selling price is £170?

5. A grocer offers a discount of $2\frac{1}{2}$% to his customers provided they pay their grocery bill within one week. If a bill of £36.25 is paid within one week, how much discount will the grocer allow?

Value Added Tax

Value added tax or VAT for short is a tax on goods and services which are purchased. Some services and goods bear no tax, for instance food, books and insurance. The rate varies from time to time but at the moment it is levied at 17.5%.

Example 8

A man buys a lawnmower which is priced at £120 plus VAT. If the rate of tax is 17.5%, how much will he actually pay for the mower?

      VAT $=$ 17.5% of £120

          $=$ 0.175 $\times$ £120

          $=$ £21

  Total cost of mower $=$ £120 $+$ £21

                     $=$ £141

Example 9

A housewife buys a table for £138, the price includes VAT at 17.5%. What is the price of the table exclusive of VAT?

Let 100 be the price exclusive of VAT. Then

Price inclusive of VAT at 17.5% is 117.5

$$\text{Price exclusive of VAT} = \frac{100}{117.5} \times \frac{138}{1}$$

$$= £117.45$$

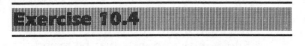

Exercise 10.4

1. A woman buys a washing machine whose price exclusive of VAT is £240. If VAT is charged at 17.5%, how much will the woman pay for the machine?

2. A chair is priced at £54 inclusive of VAT which is charged at 17.5%. What is its price exclusive of VAT?

3. A woman buys a refrigerator for £320 but on top of this VAT at 17.5% is charged. How much does she pay altogether?

4. A telephone bill is £40 excluding VAT. How much is the bill when VAT at 17.5% is added on?

5. A radio costs £138 including VAT at 17.5%. What is the cost without VAT?

6. A carpet costs £230 including VAT. How much does it cost without VAT which is charged at 17.5%?

7. Calculate the amount of VAT which is charged on a garage bill of £160 which includes VAT at 17.5%.

Rates

Every property in a town or city is given a **rateable value** which is fixed by the local district valuer. This rateable value depends upon the size, condition and location of the property.

The rates of the town or city are levied at so much in the pound of rateable value, for instance, 90p in the pound. The money raised by the rates is used to pay for such things as libraries, police, education, etc.

Yearly rates = Rateable value × Rate in £1

Example 10

The rateable value of a house is £240. If the rate is £1.25 in the pound, how much must the householder pay in rates?

Yearly rates = 240 × £1.25

= £300

Example 11

A householder pays rates of £270 on property which has a rateable value of £300. What is the rate in the pound?

For a rateable value of £300 rates payable are £270. For a rateable value of £1 rates payable are

$$£\frac{270}{300} = £0.90$$

$$= 90p$$

Hence the rate is 90p in the pound.

Example 12

A local authority needs to raise £9 020 000 to cover its expenditure for the current year. If the rateable value of all the property in the area is £8 200 000, what rate in the pound should be levied?

$$\text{Rate chargeable in £1} = \frac{9\,020\,000}{8\,200\,000}$$

$$= 1.10$$

Hence a rate of £1.10 in £1 should be levied.

Most local authorities state on their rate demand the product of a penny rate. This is the amount that would be raised if the rate levied was 1p in the pound, that is £0.01 in £1.

Example 13

The rateable value of a town is £9 350 000. What is the product of a penny rate?

Product of a penny rate $= £0.01 \times 9\,350\,000$

$= £93\,500$

Example 14

The cost of highways and bridges in a town is equivalent to a rate of 9.28p in the pound. The rateable value of all the property in the town is £15 400 000. Calculate the amount of money available for spending on highways and bridges.

Amount available $= £0.0928 \times 15\,400\,000$

$= £1\,429\,120$

Exercise 10.5

1. The rateable value of a house is £180. Calculate the yearly rates payable by the householder when the rate is £1.20 in the pound.

2. A householder pays £180 in rates when the rate is levied at £1.50 in the pound. What is the rateable value of the house?

3. A house is assessed at a rateable value of £450. The owner pays £405 in rates for the year. What is the rate in the pound?

4. What rate should a local authority charge if they need to raise £1 000 000 from a total rateable value of £4 000 000?

5. Calculate the total income from the rates of a town of rateable value £4 300 000 when the rates are £1.08 in the pound.

6. A town having a rateable value of £7 720 000 needs to raise £704 000 from the rates. What local rate should be levied?

7. The rateable value of the property in a city is £8 500 000. What is the product of a penny rate?

8. The rateable value of all the property in a city is £8 796 000.
   (a) Calculate the product of a penny rate.
   (b) The total expenses for running the city are £4 837 800. Calculate the rate that should be levied.

9. The total rateable value of the property in a locality is £850 000. Calculate the total cost of the public library if a rate of 4.6p in the pound must be levied for this purpose.

10. The expenditure of a town is £9 000 000 and its rates are 87p in the pound. The cost of its libraries is £300 000. What rate in the pound is needed for the up-keep of the libraries?

Insurance

Our future is something which is far from certain. We could become too ill to work or we could be badly injured or even killed in an accident. Our house could be burgled or burnt down. We might be involved in a car accident and be liable for injuries and damage. How do we take care of such eventualities? The answer is to take out insurance policies. The insurance company collects premiums from thousands of people who wish to insure themselves. It invests this money to earn interest and this money is then available to pay to people who claim for their loss.

Example 15

A householder values his house and its contents at £35 000. His insurance company charges a premium of £3 per £1000 insured. How much is the annual premium?

$$\text{Annual premium} = \frac{3 \times 35\,000}{1000}$$

$$= £105$$

Car Insurance

By law a vehicle must be insured, and the owner of a vehicle can be prosecuted for not having third-party insurance. That is, a policy must be taken out in case someone is injured or damage is caused by an accident which is the fault of the policy holder.

Third-party insurance covers only the other person; it does not cover the policy holder who will have to pay himself for any damage to his car. A fully comprehensive policy is needed to cover damage to the policy holder's car plus damage to other people and their property.

The size of the premium depends upon
(1) The value of the policy holder's vehicle.
(2) The engine size.
(3) The area in which the policy holder lives.
(4) Special risks (sports cars and young drivers).
(5) The use to which the vehicle is put (taxi, private, goods, etc.).

If a driver makes no claims during a year he gets a bonus (called a no-claims bonus) which means that he or she will pay a smaller premium next year.

Example 16

The insurance on a car is £180 but the owner is allowed a $33\frac{1}{3}\%$ no-claims bonus. How much is the insurance premium for the year?

$$\text{No-claims bonus} = 33\tfrac{1}{3}\% \text{ of } £180$$

$$= \tfrac{1}{3} \text{ of } £180$$

$$= £60$$

$$\text{Insurance premium} = £180 - £60$$

$$= £120$$

Cost of Running a Car

When calculating the cost of running a car, the costs of tax, insurance, petrol, maintenance and depreciation should all be taken into account. Every motor vehicle requires a Road Fund Licence which has to be paid when the vehicle is first registered and renewed periodically. It is also a legal requirement that every motor vehicle must be insured to protect a third party who may be injured in an accident. The vehicle will need maintaining and its value will depreciate with age.

Example 17

A car is bought for £3600 and used for one year. It is then sold for £2600. During this year it did 20 000 km averaging 12 km per litre of petrol which cost 45p per litre. Insurance cost £87, tax £100, repairs and maintenance £350. What is the total cost for a year's motoring and what is the cost per kilometre?

Depreciation	= £1000
Insurance	= £87
Tax	= £100
Repairs and maintenance	= £350
Cost of petrol	$= £\dfrac{20\,000}{12} \times \dfrac{45}{100}$
	= £750
Total cost for the year	= £2287
Cost per kilometre	$= \dfrac{£2287 \times 100\text{p}}{20\,000}$
	= 11.44p

Life Assurance

With this type of assurance a sum of money, depending upon the size of the premium etc., is paid to the dependents (wife or husband and children) of the policy holder upon his or her death. The size of the premium depends upon

(1) The age of the person (the younger the policy holder is, the less he pays because there is less risk of him or her dying suddenly).

(2) The amount of money the policy holder wants the dependents to receive (the greater the amount the larger the premium).

Example 18

A man aged 35 years wishes to assure his life for £7500. The insurance company quotes an annual premium of £13.90 per £1000 assured. Calculate the amount of the monthly premium.

$$\text{Annual premium} = £13.90 \times \frac{7500}{1000}$$

$$= £13.90 \times 7.5$$

$$= £104.25$$

$$\text{Monthly premium} = £104.25 \div 12$$

$$= £8.69$$

Endowment Assurance

This is very similar to life assurance but the person can decide for how long he or she is going to pay the premiums. At the end of the chosen period a certain sum of money will be paid to the policy holder. If, however, the policy holder dies before the end of the chosen period, the assured sum of money will be paid to his or her dependents. Some endowment and life policies are 'with profits', which means that the sum assured may increase over the period of the policy depending upon the profits made by the insurance company.

Example 19

A man aged 30 years wishes to take out an endowment 'with profits' assurance policy. He is quoted a price of £48.50 per £1000 assured over a period of 20 years. Calculate the monthly premiums if he wishes to assure himself for £6000.

$$\text{Annual premiums} = £48.50 \times \frac{6000}{1000}$$

$$= £48.50 \times 6$$

$$= £291.00$$

$$\text{Monthly premiums} = £291.00 \div 12$$

$$= £24.25$$

Exercise 10.6

1. A householder wishes to insure his house for £25 000. His insurance company charges a premium of 15p per £100 insured. Calculate the householder's annual premium.

2. An insurance company offers the following rates to customers: buildings £1.25 per £1000 insured; contents 25p per £100 insured. Calculate the annual premium paid by a householder if his house is valued at £27 000 and the contents at £6500.

3. The insurance on a car is £120 but a 20% no-claims bonus is allowed. How much is the yearly premium?

4. If the insurance premium is £20 on £5000, how much will the premium be on £8000?

5. A man aged 25 years wishes to assure his life for £7000. He is quoted a rate of £9.00 per £1000 assured per annum. How much in premiums will he pay monthly?

6. A car owner is quoted an annual premium of £90 but she is allowed a no-claims bonus of $33\frac{1}{3}$%. Calculate the amount of her annual premium.

7. A person aged 36 years is quoted a rate of £13.90 per £1000 assured per annum. He wishes to assure his life for £8000 but he wishes to pay monthly. The company states that for monthly payments the premiums will be increased by 3%. How much per month will the policy holder actually pay?

8. For a monthly premium of £50 a woman aged 40 is guaranteed £6624 at the end of 10 years. However, the company states that the maturity value of the policy is likely to be £11 241.

    (a) Work out how much the woman will pay in premiums over the 10-year period.

    (b) Calculate how much profit she is likely to make on the policy.

9. A man drives 15 000 km in a year. If he averages 10 km per litre of petrol costing 40p per litre how much does petrol cost him?

10. A second-hand car is bought for £900. It is run for a year and then sold for £640. The cost of insurance is £63.20, tax £100, repairs £76 and maintenance £16. It is driven 12 000 km in the year and it averages 11 km per litre of petrol costing 42p per litre. What is the cost of a year's motoring and how much is the cost per kilometre?

11. A new car is bought for £3000 and it is sold two years later for £1900. During this time it travelled 42 000 km at an average petrol consumption of 14 km per litre. If petrol costs 40p per litre and other expenses were £210, calculate the cost per kilometre of running the car.

Miscellaneous Exercise 10

1. The cost price of a table is £120 and it is sold for £200.
    (a) How much is the profit?
    (b) What is the profit per cent?

2. A motor cycle is bought for £300 and sold for £150.
    (a) How much loss is made?
    (b) What is the loss per cent?

3. A dealer offers a discount of 25% on all the clothes in his shop. A coat is priced at £80. How much will a customer pay when he buys the coat?

4. Value added tax is charged at 17.5%. An electric fire is priced at £50 exclusive of tax.
    (a) How much tax will be charged?
    (b) How much will a customer pay for the fire?

5. A house has a rateable value of £400. Rates are levied at £1.25 in the pound. How much will the householder pay in rates?

6. A man wants to insure his house for £30 000. The insurance company quotes him £3 per £1000 insured. How much will the man pay in premiums per annum?

7. The insurance on a car is £120 but there is a 30% no-claims bonus. How much is the premium for the year?

8. A woman pays a 10% deposit on a house which she bought for £30 000. How much mortgage does she need?

9. The rateable value of Mr Richards' house is £280. He pays rates of 93p in the pound. How much does Mr Richards pay in rates for the year?

10. A car owner is quoted a premium of £80 per annum but he is allowed a no-claims bonus of 40%. How much does he actually pay?

11. Given that the value added tax on goods supplied is 17.5% of their value, calculate the value added tax payable on goods valued at £90.

12. By selling a chair for £31.05 a shopkeeper makes a profit of 15% on his cost price. Calculate his profit.

13. By selling an article for £18 a dealer made a profit of 44% on the cost price. At what price must it be sold to make a profit of 40%?

14. A woman sells her car for £2430 and as a result loses 10% of the price she paid for it. What price did she pay for the car?

15. A city treasurer estimated that the city would need a rate of 82.5p in the pound in order to provide revenue of £68 887 500 and that the cost to the rates of Family and Community Service would be 8.6p in the pound. Calculate **(a)** the product of a penny rate, **(b)** the amount of money to be raised for Family and Community Service.

16. Find the rateable value of a house on which the amount paid in rates was £192 when the rate was £1.60 in the pound.

17. The price of a coat was £68.25 but a 5% discount for cash is allowed. How much does a cash-paying customer actually pay for the coat?

18. A man aged 45 years takes out a 'with profits' insurance policy for £5000. He is charged premiums of 45p per £100 assured. If he pays the premiums monthly, how much will he pay?

Mental Test 10

Try to answer the following questions without writing anything down except the answer.

1. The cost price of an article is £8 and the selling price is £10. What is the profit?

2. The cost of a car is £3000 and it is sold for £2500. How much loss is made?

3. If the profit is £10 and the cost price is £50, what is the percentage profit?

4. The cost price of an article is £100 and the selling price is £120. What is the profit per cent?

5. A dealer offers a discount of 20% on all the goods in his shop. An article is priced at £200. How much will a customer pay for it?

6. Value added tax is charged at 15%. How much tax is payable on goods priced at £50 exclusive of tax?

7. A house has a rateable value of £300. If the rates are levied at £1.50 in the pound, how much will the householder pay in rates?

8. All the property in a small town has a rateable value of £500 000. What is the product of a penny rate?

9. A man wishes to insure the contents of his house for £5000. The insurance company quotes him £2 per £1000 insured. How much will the man pay in premiums per year?

10. A woman insures her car for £200 but a no-claims bonus of 20% is given. How much will she pay in insurance premiums?

Household Finance

Rent

Rent is a charge for accommodation and it is usually paid weekly or monthly to the landlord who owns the property. The biggest owners of rented property are local authorities.

The landlord may include the rates in his charge for accommodation but if he does not it is the tenant's responsibility to see that these are paid.

Example 1

A landlord charges a rent of £17.50 per week and the rates are £101.40 per annum. What weekly inclusive amount should the landlord charge the tenant?

$$\text{Rates per week} = £101.40 \div 52$$
$$= £1.95$$

$$\text{Inclusive charge per week} = £17.50 + £1.95$$
$$= £19.45$$

Mortgages

If someone buying a house or flat cannot pay outright, he or she arranges a loan called a mortgage from a building society. The society generally requires a deposit of 5 or 10% of the purchase price of the property, although sometimes 100% mortgages are available.

The balance of the loan plus interest is paid back over a number of years — perhaps 20 or 25. The interest rates charged by the building society vary from time to time.

Example 2

(a) A man wishes to buy a house costing £32 000. He pays a 10% deposit to the building society. How much mortgage will he require?

Method 1

$$\text{Deposit} = 10\% \text{ of } £32\,000$$
$$= £3200$$

$$\text{Mortgage required} = £32\,000 - £3200$$
$$= £28\,800$$

Method 2

Since a deposit of 10% is paid the mortgage required is 90% of the purchase price.

$$\text{Mortgage required} = 90\% \text{ of } £32\,000$$
$$= £28\,800$$

(b) A building society quotes the repayments on a mortgage as being £12.69 per month per £1000 borrowed. What will be the monthly repayments on a mortgage of £25 000?

$$\text{Monthly repayments} = \frac{25\,000}{1000} \times £12.69$$
$$= £317.25$$

Sometimes a combined mortgage and endowment assurance can be arranged. An endowment policy (see Chapter 10) is taken out for the value of the loan. Interest is paid on the full amount of the loan for the whole period of the endowment, after which the money received from the insurance policy is used to pay off the loan.

Example 3

A man wishes to borrow £20 000 to buy a house. The loan is to be covered by an insurance policy for a 20-year period which costs £3.88 per £1000 assured per month. The building society charges interest on the loan at 13%. What will be the total monthly payments?

Annual loan interest $= 13\%$ of £20 000

$$= £2600$$

Monthly payments of interest $= £2600 \div 12$

$$= £216.67$$

Endowment payments per month $= \dfrac{20\,000}{1000} \times £3.88$

$$= £77.60$$

Total monthly payments $= £216.67 + £77.60$

$$= £294.27$$

Exercise 11.1

1. A woman wishes to buy a house costing £28 000. She pays a 10% deposit. How much mortgage does she require?

2. A house is rented for £44.16 per week. Its rateable value is £174 and rates are levied at £1.08 in the pound. What will be the weekly charge for rent and rates?

3. A flat is rented for £39.50 per week and the rates are £194.40 per annum. Calculate the weekly charge for rent and rates.

4. A man borrows £24 000 from a building society in order to buy a house. The society charges £12.20 per month per £1000 borrowed. How much are the monthly repayments?

5. A person borrows £16 000 from a building society to buy a house. The loan is covered by an insurance policy for the ten-year term of the loan, the premiums being £7.95 per £1000 assured. The building society charges interest at 10% per annum. Work out the total monthly outgoings of the borrower.

6. Calculate the monthly cost of owning a house, given that the expenditure is as follows: mortgage repayments £220 per month; insurances £330 per year; rates £120 per half year.

7. A woman took out an £8000 mortgage when the building society rate was $8\frac{1}{2}\%$ per annum. If the rate is now 11% per annum, calculate the increase in her monthly payments to the society.

Credit Sale

When we purchase goods and pay for them by instalments we are said to have purchased them on **credit**. Usually the purchaser pays a deposit. The remainder of the purchase price (called the balance) plus interest is repaid in a number of instalments.

Example 4

A woman buys some furniture for £280. She pays a deposit of 25%, and interest at 20% is paid on the outstanding balance. She pays back the balance plus interest in twelve monthly instalments. Calculate the amount of each instalment.

$$\text{Deposit} = 25\% \text{ of } £280$$
$$= £70$$
$$\text{Balance} = £280 - £70$$
$$= £210$$
$$\text{Interest} = 20\% \text{ of } £210$$
$$= £42$$
$$\text{Total to be repaid} = £210 + £42$$
$$= £252$$
$$\text{Monthly instalments} = £252 \div 12$$
$$= £21$$

Exercise 11.2

1. If the credit sale payments on a record player are £5 per month, how much will be paid at the end of 2 years?

2. A woman buys a washing machine for £150. She pays a deposit of £30 and is charged 10% interest on the balance.

   (a) How much is the balance?

   (b) Calculate the interest payable.

   (c) How much is the interest plus the balance?

   (d) If the balance plus interest is payable in ten equal monthly instalments, what is the amount of each instalment?

3. A man buys a television set for £320 and pays a deposit of £50. He is to pay the balance plus interest in twelve monthly instalments. If interest is charged at 10% on the balance for the full period of the loan, calculate the amount of each instalment.

4. A woman buys a suite of furniture for £640. A deposit of 20% is paid and interest at 12% per annum is charged on the balance for the full period of the loan. The balance is to be paid in four quarterly payments.

   (a) Work out the amount of the deposit.

   (b) What is the amount of the balance?

   (c) Calculate the interest payable.

   (d) Work out the total amount to be repaid.

   (e) Calculate the amount of each instalment.

5. The cash price of a table is £200. To buy it on credit a deposit of £50 is payable plus ten equal payments of £21.

   (a) Work out the amount paid for the furniture if it is bought on credit.

   (b) Find the difference between the cash price and the credit sale price.

Bank Loans

Many people take out personal loans from a bank. The bank will calculate the interest for the whole period of the loan and the loan plus interest is usually repaid in equal monthly periods.

Example 5

A woman borrows £300 from a bank. The bank charges 18% interest for the whole period of the loan. If the repayments are in twelve equal monthly instalments, calculate the amount of each payment.

$$\text{Interest} = 18\% \text{ of } £300$$
$$= £54$$
$$\text{Total amount to be repaid} = £300 + £54$$
$$= £354$$
$$\text{Amount of each instalment} = £354 \div 12$$
$$= £29.50$$

Exercise 11.3

1. A man borrows £500 from a bank and interest amounts to £100. He repays the loan plus interest in ten equal monthly instalments. What is the amount of each instalment?

2. A bank lends a woman £1200. They charge interest at 20%. She pays the loan plus interest back in ten equal instalments How much does she pay each time?

3. An advertisement in a newspaper states that if you borrow £1000 over a period of 10 years, the repayments will be £20.70 per month.

   (a) A man borrows £4500 over 10 years. How much will be his monthly repayments?

   (b) How much will the man repay altogether over the 10-year period?

4. Over 5 years a man who borrows £7500 repays the loan at £210.07 per month. How much does the finance house charge per £1000 borrowed per month?

5. A man borrows £2000 from a bank and repays £2400 over a period of one year. What rate of interest does the bank charge?

Gas Bills

Gas is charged according to the number of kilowatt-hours (kWh) used. To find the cost of gas multiply the number of kilowatt-hours by the charge per kilowatt-hour. In addition a standing charge may also be made.

Example 6

A customer uses 8700 kWh during one quarter. She is charged 1.566p per kWh plus a standing charge of £8.65. Work out the amount of the customer's gas bill.

$$\text{Cost of gas used} = (8700 \times 1.566) \text{ pence}$$
$$= £136.24$$
$$\text{Gas bill} = \text{Cost of gas used} + \text{Standing charge}$$
$$= £136.24 + £8.65$$
$$= £144.89$$

Exercise 11.4

1. Copy and complete the following table:

Charge per kWh	Number of kWh used	Cost of gas used in pence
1.256	3263	
1.492	2531	
2.063	7235	
2.412	12 500	

2. A customer uses 2700 kWh of gas in a certain quarter. If the cost per kWh is 1.497 pence and the standing charge is £9.93, work out the amount of the gas bill.

3. The present reading of a gas meter is equivalent to 28 976 kWh and the previous reading was 22 579 kWh.

   (a) How many kWh of gas have been used?

   (b) If gas is charged at 1.653 pence per kWh and the standing charge is £8.75, calculate the amount of the gas bill.

4. A housewife used 9600 kWh of gas during a certain quarter. The housewife received a bill for £158.40 which included a standing charge of £8.35. Work out the cost of gas in pence per kilowatt-hour.

5. A man received a gas bill for £145.10 which included a standing charge of £9.25. If gas was charged at 1.627 pence per kWh, find the number of kWh that he used.

Electricity Bills

Electricity is charged according to the number of units used (1 unit = 1 kilowatt hour). There is also a fixed charge which is added to the bill.

Example 7

A user of electricity uses 1765 units in a quarter. If the standing charge is £8.62 and the price per unit is 7.72 pence, find the amount of the quarterly electricity bill.

1765 units at 7.72p per unit

$$= 1765 \times £0.0772$$

$$= £136.26$$

Quarterly electricity bill $= £136.26 + £8.62$

$$= £144.88$$

Reading Electricity Meters

When reading an electricity meter remember that adjacent dials revolve in opposite directions. The five dials are read from left to right. The method of reading an electricity meter is shown in Fig. 11.1.

(1) Always note down the number the pointer has passed. In Fig. 11.1, on the dial labelled 10 000 the pointer is between 4 and 5 so write down 4.

(2) If the pointer is directly over a figure write down that figure and underline it. On the dial marked 1000 in the diagram, the reading is 5 so write 5.

(3) The reading of the dials in the diagram gives 45 9<u>2</u>8.

(4) Now look at the figures underlined. If one of these is followed by a 9 reduce the underlined figure by 1. Thus the corrected reading is 44 928.

(5) When you have worked out the meter reading, subtract the previous reading to find the number of units of electricity used.

Fig. 11.1

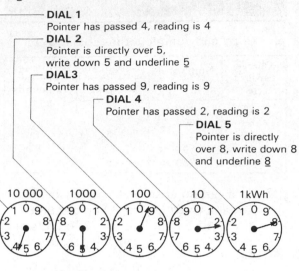

DIAL 1
Pointer has passed 4, reading is 4
DIAL 2
Pointer is directly over 5, write down 5 and underline <u>5</u>
DIAL3
Pointer has passed 9, reading is 9
DIAL 4
Pointer has passed 2, reading is 2
DIAL 5
Pointer is directly over 8, write down 8 and underline <u>8</u>

Example 8

Fig. 11.2 shows the dials of an electricity meter.

(a) Read the meter.

(b) If the previous meter reading was 53 432, find the number of units used.

(c) If the charge per unit is 5.93p and the standing charge is £9.80 per quarter, calculate the amount of the electricity bill.

Fig. 11.2

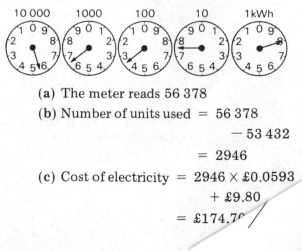

(a) The meter reads 56 378

(b) Number of units used $= 56\,378$

$$-\,53\,432$$

$$= 2946$$

(c) Cost of electricity $= 2946 \times £0.0593$

$$+\,£9.80$$

$$= £174.7^{\circ}$$

$$=$$

Exercise 11.5

1. Copy and complete the table below:

Charge per unit (p)	Number of units used	Cost of electricity used	
		In pence	In pounds
7.71	100		
2.63	600		
9.93	450		
8.17	800		
12.95	750		

2. In a certain area electricity is charged for at a fixed rate of 9.80p per unit used. A householder uses 780 units in a quarter. How much will his electricity bill be?

3. A householder uses 2500 units of electricity in a quarter. Each unit used costs 8.17p and there is a standing charge of £12.40 per quarter. Calculate the amount the householder will pay for his electricity.

4. The meter reading on 15 Feb was 009 870 and on 17 May it was 009 986. How many units of electricity have been used? If each unit costs 7.72p and there is a standing charge of £8.70, find how much the electricity bill will be.

5. An economy tariff charges for electricity as follows: day rate 8.17p per unit; night rate 2.63p per unit; standing charge £12.40. Work out the amount of the electricity bill if 1325 units are used during the day and 932 units during the night.

6. Copy the set of dials shown in Fig. 11.3 and put in arrows to show readings of
   (a) 81 372   (b) 56 325   (c) 73 454

Fig. 11.3

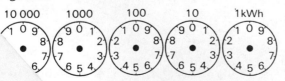

7. Read the dials shown in Fig. 11.4.

Fig. 11.4

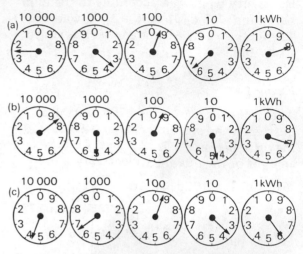

8. A householder receives an electricity bill for £131.50. If electricity is charged for at 10.24p per unit and the standing charge is £8.62, find the number of units used by the householder.

Telephone Bills

The amount of a telephone bill depends upon the number of calls made and the charge for rental of the telephone itself. The bills are paid quarterly and VAT at 17.5% is charged.

Example 9

A householder dialled 500 calls in a certain quarter. Each call was charged for at 4.20 pence. The quarterly rental charge is £7.60. If VAT is charged at 17.5%, find the total amount of the householder's telephone bill.

$$500 \text{ calls at } 4.20\text{p} = £21.00$$

$$\text{Cost less VAT} = £21.00 + £7.60$$
$$= £28.60$$

$$\text{VAT at } 17.5\% = 17.5\% \text{ of } £28.60$$
$$= £5.01$$

$$\text{Total cost} = £28.60 + £5.01$$
$$= £33.61$$

Exercise 11.6

Work out the total cost of the following telephone bills:

1. Standing charge of £14.27.
    100 units at 4.8p per unit.

2. Standing charge of £16.15.
    180 units at 4.8p per unit.

3. Standing charge of £17.80.
    150 units at 4.3p per unit.

4. A telephone bill is made up of the following charges: rental £16.15 per quarter, 603 dialled units at 4.8p each and VAT at 17.5%

    (a) Find the cost of the dialled units.

    (b) Calculate the total bill not including VAT.

    (c) Work out the total amount of the bill including VAT.

5. The telephone meter was read on 17th Feb as 007 963 and on 15th May as 009 215.

    (a) How many units have been used during this period?

    (b) If each unit is charged for at 4.9p, calculate the cost of the units used.

    (c) The standing charge is £17.29. What is the total bill excluding VAT?

    (d) VAT is charged at 17.5% on the total bill. How much is VAT?

    (e) Work out the total telephone bill including VAT.

Miscellaneous Exercise 11

1. Mrs Evans insured her jewellery for £2400. The premium is 25p per £100 of cover. What is her total premium?

2. A small computer is priced at £300 plus VAT at 17.5%.

    (a) How much is the VAT?

    (b) What is the total cost of the computer including VAT?

    (c) Tindalls offer the computer with £30 off its total price. Lyons offer the computer with 20% off its total price. Find the difference between the two offers.

    (d) Thompsons offer the computer on credit as follows: Deposit £90 plus 12 equal monthly payments of £18.75. What is the credit sale price of the computer?

3. (a) Fig. 11.5 shows the positions of the dials of an electricity meter at the beginning of a quarter. Write down the meter reading.

Fig. 11.5

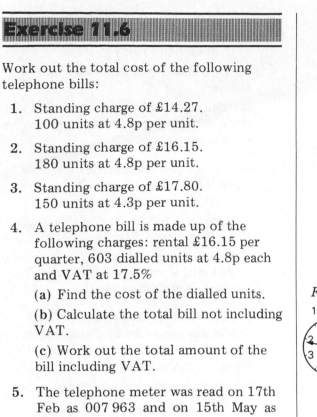

    (b) If during this quarter 760 units of electricity are used, write down the new meter reading.

    (c) Copy the dials shown in Fig. 11.6 and show the new meter reading on the dials.

Fig. 11.6

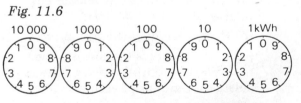

    (d) If electricity is charged at 7.75 pence per unit plus a standing charge of £8.50, calculate the total cost of the 760 units in the quarter.

4. Copy and complete the following telephone bill:

Quarterly rental charge	£9.75
Dialled units 600 at 4.7p per unit	
Total cost (exclusive of VAT)	
VAT at 17.5%	
Total payable	

5. A small firm uses 2500 units of electricity at night and 3500 units during the day per quarter. They have the option of choosing the small supplies tariff or the small supplies economy tariff.

   (a) The small supplies tariff makes a standing charge of £8.62 per quarter plus 10.11 pence for each unit used up to 1000 units supplied per quarter and 7.72 pence for each additional unit supplied per quarter. If the firm opts for this tariff, work out the cost of electricity per quarter.

   (b) The economy tariff makes a standing charge of £12.40 per quarter plus 2.63 pence per unit for each unit supplied at night and 10.24 pence for each of the first 1000 units supplied in the day. For each additional unit supplied in the day, 8.17 pence per unit is charged. Calculate the cost of electricity using this tariff.

   (c) Which tariff should be used and what is the difference in cost between the two tariffs?

6. A woman borrows £2500 from a bank which charges interest at 15% for the entire period of the loan.

   (a) How much does the woman pay in interest?

   (b) How much does she pay in total?

   (c) If the loan plus interest is to be repaid in 12 monthly instalments, calculate the amount of each instalment.

7. A man purchases a house using a combined life insurance and mortgage.

   (a) He borrows £18 000 from the building society which charges an interest rate of 13% per annum. Calculate the amount of interest he has to pay per annum.

   (b) The insurance premium is £2.95 per month per £1000 of cover. What is the total premium per month?

   (c) Calculate the total monthly payments the man must make?

8. A woman wants to buy a house for £30 000. She pays a deposit of 10% to the building society.

   (a) How much mortgage does she require?

   (b) The monthly repayments are £11.60 per £1000 borrowed. How much are her monthly repayments?

9. A man wants to borrow £960 from a bank. The bank agrees to the loan at 12% per annum simple interest.

   (a) How much does he pay in interest if the loan lasts for two years?

   (b) The loan plus interest is to be repaid in 24 equal instalments. Work out the amount of each instalment.

10. A suite of furniture is priced at £700. Credit sale terms are available which are: deposit 25% of cash price and 18 equal monthly instalments of £36.75.

    (a) Work out the amount of the deposit.

    (b) Calculate the total credit sale price.

    (c) Find the difference between the credit sale price and the cash price.

Mental Test 11

Try to answer the following questions without writing anything down except the answer.

1. A landlord charges an annual rent of £480. What is the monthly rent?

2. A building society charges a person £20 per month per £1000 borrowed. If the person borrows £10 000 what are the monthly repayments?

3. A woman buys furniture priced at £800. She pays a deposit of 10%. What is her outstanding balance?

4. A young man buys a motor cycle on credit. He pays a deposit of £50 and 20 monthly instalments of £30 each. How much does he pay in total for his motor bike?

5. A man borrows £500 from his bank which charges 20% interest for the whole period of the loan, which he pays off in 10 equal monthly instalments. How much is each instalment?

6. A housewife uses 3000 kWh of gas per quarter which is charged for at 2p per kWh. How much is her quarterly gas bill?

7. A householder uses 2000 units of electricity which costs 5p per unit. How much is his electricity bill?

12 Time, Distance and Speed

Measurement of Time

The units of time are:

$$60 \text{ seconds (s)} = 1 \text{ minute (min)}$$
$$60 \text{ minutes (min)} = 1 \text{ hour (h)}$$
$$24 \text{ hours (h)} = 1 \text{ day (d)}$$

Example 1

(a) Change 720 s into minutes.

$$720 \text{ s} = 720 \div 60 \text{ min}$$
$$= 12 \text{ min}$$

(b) Change 8 h into minutes.

$$8 \text{ h} = 8 \times 60 \text{ min}$$
$$= 480 \text{ min}$$

(c) How many seconds are there in 1 hour?

$$1 \text{ h} = 60 \text{ min}$$
$$= 60 \times 60 \text{ s}$$
$$= 3600 \text{ s}$$

The Clock

There are two ways of showing the time:

(1) With a 12-hour clock (Fig. 12.1), in which there are two periods each of 12 hours duration during each day.

The period between midnight and noon is called a.m. whilst the period between noon and midnight is called p.m. Thus 8.45 a.m. is a time in the morning whilst 8.45 p.m. is a time in the evening.

Fig. 12.1

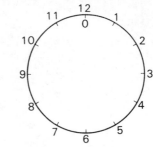

Example 2

Find the length of time between 2.15 a.m. and 8.30 p.m.

2.15 a.m. to 12 noon is	9 h 45 min
12 noon to 8.30 p.m. is	8 h 30 min +
Total length of time is	18 h 15 min

(2) With a 24-hour clock (Fig. 12.2) in which there is one period of 24 hours. This clock is used for railway and airline timetables. Times between midnight and noon are given times of 0000 hours to 1200 hours whilst times between noon and midnight are given times of 1200 hours to 2400 hours.

Fig. 12.2

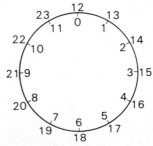

3.30 a.m. is written 0330 hours whilst 3.30 p.m. is written 1530 hours and this avoids confusion between the two times.

Note that 24-hour-clock times are written with four figures. 1528 hours should be read as 'fifteen twenty-eight hours' and 1300 hours as 'thirteen hundred hours'.

Example 3

Find the length of time between 0425 hours and 1812 hours.

1812 hours is	18 h 12 min
0425 hours is	4 h 25 min −
Total length of time is	13 h 47 min

Exercise 12.1

1. How many minutes are there in
   (a) 4 hours   (b) $3\frac{1}{2}$ hours

2. How many seconds are there in
   (a) 9 min   (b) $2\frac{1}{4}$ min

3. Change 350 min into hours and minutes.

Find the length of time, in hours and minutes, between

4. 2.39 a.m. and 8.46 p.m.

5. 2.15 a.m. and 7.30 p.m.

6. 0.36 a.m. and 1.15 p.m.

7. 0049 hours and 1236 hours.

8. 0242 hours and 1448 hours.

9. 0836 hours and 2115 hours.

10. 0425 hours and 1712 hours.

Timetables

Timetables are usually written using the 24-hour clock. The timetable at the foot of the page is a railway timetable for the Paddington to Cheltenham service.

Example 4

If I catch the 0935 from Paddington, at what time do I arrive at Stonehouse and how long will the journey take?

According to the timetable the 0935 from Paddington arrives at Stonehouse at 1121 hours.

1121 hours is	11 h 21 min
0935 hours is	9 h 35 min −
Time taken is	1 h 46 min

Exercise 12.2

Using the Paddington to Cheltenham railway timetable answer the following questions:

1. I can catch either the 0638 or the 0730 train from Swindon. Which is the faster train to Cheltenham? What is the difference in the times taken for the journey?

2. What time does the 0725 from Paddington arrive at Gloucester? How long does the journey take?

3. Which is the faster train, the 0835 from Paddington to Swindon or the 0935 from Paddington to Swindon? How much shorter is the journey?

MONDAYS TO SATURDAYS

London, Paddington . . . dep	—	—	0725	0835	0935	1040	1100	1200	1405	1430	1535	1625	1650	1742
Swindon . . . . dep	0638	0730	0848	0942	1045	1134	1200	1305	1503	1542	1642	1738	1748	1835
Kemble. . . . . dep	0653	0745	0903	0957	1100	1148	1215	1320	1518	1557	1657	1755	1803	1849
Stroud . . . . dep	0709	0801	0919	1013	1116	1202	1231	1336	1534	1613	1713	1810	1819	1903
Stonehouse . . . . dep	0714	0806	0924	1018	1121	■	1236	1341	1539	1618	1718	1815	1824	1908
Gloucester . . . . arr	0727	0819	0937	1032	1135	1217	1249	1354	1552	1631	1731	1828	1838	1920
Cheltenham . . . . arr	0756	0833	1012	1046	1215	1231	1310	1415	1607	1718	1744	1845	1851	1935

4. I want to board the 1405 train from Paddington at Kemble. At what time is it scheduled to arrive? At what time does this train arrive at Cheltenham? How long is the journey from Kemble to Cheltenham?

5. Which is the fastest train from Stroud to Cheltenham? How long is the shortest journey?

Average Speed

The road from Campton to Burley is 50 miles long. It takes me two hours to drive from Campton to Burley. Obviously I do not drive at the same speed all the time. Sometimes I drive slowly and sometimes I have to stop at crossroads and traffic lights.

However, the time it takes me is exactly the same as if I were driving along a straight road at 25 miles per hour, i.e. 50 miles ÷ 2 hours. So we say that my average speed is 25 miles per hour.

$$\text{Average speed} = \frac{\text{Total distance travelled}}{\text{Total time taken}}$$

The unit of speed depends upon the unit of distance and the unit of time. If the distance is measured in kilometres and the time in hours then the average speed will be measured in kilometres per hour (km/h). If the distance is measured in miles and the time in hours then the average speed will be measured in miles per hour (mile/h). If the distance is measured in metres and the time in seconds then the average speed will be measured in metres per second (m/s).

Note carefully that

$$\text{Distance travelled} = \text{Average speed} \times \text{Time taken}$$

$$\text{Time taken} = \frac{\text{Distance travelled}}{\text{Average speed}}$$

Example 5

(a) A car travels a total distance of 400 km in 5 hours. What is the average speed?

$$\text{Average speed} = \frac{\text{Distance travelled}}{\text{Time taken}}$$

$$= \frac{400}{5} \text{ km/h}$$

$$= 80 \text{ km/h}$$

(b) A lorry travels at 35 miles per hour for 3 hours. How far does it travel?

$$\text{Distance travelled} = \text{Average speed} \times \text{Time taken}$$

$$= 35 \times 3 \text{ miles}$$

$$= 105 \text{ miles}$$

(c) A train travels 120 km at 40 km/h. How long does the journey take?

$$\text{Time taken} = \frac{\text{Distance travelled}}{\text{Average speed}}$$

$$= \frac{120}{40} \text{ hours}$$

$$= 3 \text{ hours}$$

Remember that average speed is defined as *total distance travelled* divided by *total time taken*.

Example 6

(a) A car travels 60 km at an average speed of 30 km/h and 30 km at an average speed of 40 km/h. Calculate its average speed for the entire journey.

Time taken to travel 60 km at 30 km/h

$$= \frac{60}{30} \text{ hours}$$

$$= 2 \text{ hours}$$

Time taken to travel 30 km at 40 km/h

$$= \frac{30}{40} \text{ hours}$$

$$= 0.75 \text{ hours}$$

Total time taken $= (2 + 0.75)$ hours

$\qquad = 2.75$ hours

Total distance travelled $= (60 + 30)$ km

$\qquad = 90$ km

Average speed $= \dfrac{90}{2.75}$ km/h

$\qquad = 32.7$ km/h

(b) A train travels between two towns 270 km apart in 9 hours. If on the return journey the average speed is reduced by 3 km/h, calculate the time taken for the return journey.

Average speed on outward journey

$$= \frac{270}{9} \text{ km/h}$$

$$= 30 \text{ km/h}$$

Average speed on return journey

$$= (30 - 3) \text{ km/h}$$

$$= 27 \text{ km/h}$$

Time taken for return journey

$$= \frac{270}{27} \text{ hours}$$

$$= 10 \text{ hours}$$

Exercise 12.3

Find the time taken to travel

1. 10 miles at an average speed of 5 miles per hour.

2. 48 km at an average speed of 12 km/h.

3. 60 m at an average speed of 15 m/s.

Find the distance travelled if

4. Average speed is 20 miles/h and time is 3 h.

5. Average speed is 40 km/h and time is $3\frac{1}{2}$ h.

6. Average speed is 60 m/s and time is 8 s.

Find the average speed if

7. Distance is 30 miles and time is 2 hours.

8. Distance is 140 km and time is $3\frac{1}{2}$ hours.

9. Distance is 80 m and time is 5 seconds.

10. An aeroplane took $\frac{3}{4}$ hour to travel from one airport to another. If its average speed was 400 km/h, how far apart are the airports?

11. A boy walks 3 km at a speed of 6 km/h. He then cycles 6 km at a speed of 12 km/h. What is his average speed for the entire journey?

12. A car travels 136 miles at an average speed of 32 miles/h. On the return journey the speed is increased to 39 miles/h. Calculate the average speed for the complete journey.

13. For the first $1\frac{1}{2}$ hours of a 91 km journey the average speed was 30 km/h. If the average speed for the remainder of the journey was 23 km/h, calculate the average speed for the entire journey.

14. A motorist travelling at a steady speed of 90 km/h covers a section of motorway in 25 minutes. After a speed limit is imposed he finds that, when travelling at the maximum speed allowed, he takes 5 minutes longer than before to cover the same section. Calculate the speed limit.

15. In winter a train travels between two towns 264 km apart at an average speed of 72 km/h. In summer the journey takes 22 minutes less than in winter. Find the average speed in summer.

Miscellaneous Exercise 12

1. (a) A clock at a bus station shows the departure of an evening bus to be 8.15. How would this time be written in a timetable which uses the 24-hour-clock system?

    (b) The bus arrived at its destination at 2045 hours. For how long was the bus travelling?

2. An aeroplane left town A at 1535 hours taking 2 h 40 min for its flight to town B. At what time did the aeroplane reach town B?

3. A car leaves Baxton at 1325 and reaches Flighton at 1555 hours. The distance between the two towns is 125 km.

    (a) How long did the journey take?

    (b) What was the average speed of the car?

4. A boat leaves a harbour at 0235 hours and arrives at its destination at 0405 hours.

    (a) How long did the journey take?

    (b) If the distance travelled was 12 miles, calculate the average speed of the boat?

5. The distance between Cardiff and London is 150 miles. A car leaves Cardiff at 9.35 a.m. and arrives in London 5 hours later.

    (a) What time was it when the car arrived in London?

    (b) What was the average speed of the car?

6. The distance between two towns is 160 km.

    (a) How long will it take a motorist to make the journey if his average speed is 40 km/h?

    (b) If the motorist completes the return journey in $2\frac{1}{2}$ hours, what was his average speed?

7. A car leaves Bristol at 1055 hours and arrives in Hereford $1\frac{1}{4}$ hours later. The distance from Bristol to Hereford is 50 miles.

    (a) What time did it arrive in Hereford?

    (b) What was its average speed?

8. On a train journey of 117 km, the average speed for the first 27 km was 45 km/h.

    (a) Find the time taken for this part of the journey.

    (b) For the remainder of the journey the average speed was $37\frac{1}{2}$ km/h. Find the time taken for this part of the journey.

    (c) What was the total time for the whole journey?

    (d) Calculate the uniform speed at which the train would have to travel in order to cover the whole distance in the same time.

9. A motorway journey takes 3 hours at an average speed of 120 km/h.

    (a) How long, in kilometres, was the journey?

    (b) How long, in hours, will the journey take if the average speed is decreased to 80 km/h?

10. A greyhound runs 400 metres in 50 seconds.

    (a) Calculate its average speed in metres per second.

    (b) Work out its average speed in kilometres per hour.

Mental Test 12

Try to answer the following questions without writing anything down except the answer.

1. Change 240 seconds into minutes.
2. Change 4 minutes into seconds.
3. Change 180 minutes into hours.
4. Change 5 hours into minutes.
5. How many days in 120 hours?
6. Find the length of time between 9.30 a.m. and 11.45 a.m.
7. What is the length of time between 1520 hours and 1840 hours?
8. A train leaves Paddington at 1430 hours and arrives in Cheltenham at 1738 hours. How long does the journey take?
9. A motorist travels 160 km in 4 hours. What is his average speed?
10. A train travels 200 miles in 4 hours. Calculate its average speed.
11. A car travels for 3 hours at an average speed of 50 km/h. How far has it travelled?
12. A train travels 80 km at an average speed of 40 km/h. How long does the journey take?

13. An aeroplane flies at an average speed of 500 miles per hour. It flies for 3 hours. How far has it flown?
14. A car takes 5 hours to travel 240 km. What is its average speed?
15. A train travels 150 miles at an average speed of 50 miles per hour. How long does the journey take?

Use the bus timetable and route map shown below to answer the following questions:

16. How long does the bus take to travel from Stroud to Sapperton?
17. What time must I arrive at Chalford, Marle Hill to make sure of arriving in Stroud by the 1239 bus?
18. What is the number of the bus service from Stroud to Eastcombe?
19. Work out the time for the bus to travel between Bourne Bridge and Daglingworth.
20. Is there a bus from Cirencester to Stroud which leaves Cirencester at 10 minutes to 3 in the afternoon?

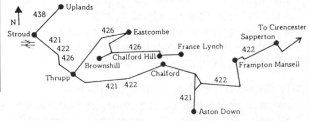

STROUD · CIRENCESTER via Chalford   Service 422

Service 422 via Bowbridge, Thrupp, Brimscombe, Chalford, Aston Down Turn, White Horse Inn, Frampton Mansell, Sapperton, Park Corner, Duntisbourne Rouse Turn, Daglingworth, Stratton.

Mondays to Fridays*

STROUD, Bus Station	0855	1055	1255	CIRENCESTER, Bus Station	0950	1150	1450
Thrupp, Brewery Lane	0903	1103	1303	Daglingworth	1000	1200	1500
Bourne Bridge	0907	1107	1307	Sapperton	1010	1210	1510
Chalford, Marle Hill.	0912	1112	1312	White Horse Inn	1017	1217	1517
White Horse Inn.	0917	1117	1317	Chalford, Marle Hill.	1022	1222	1522
Sapperton	0924	1124	1324	Bourne Bridge	1027	1227	1527
Daglingworth	0934	1134	1334	Thrupp, Brewery Lane	1031	1231	1531
CIRENCESTER, Bus Station	0944	1144	1344	STROUD, Bus Station	1039	1239	1539

CODE
* — Not Bank Holiday Mondays, Good Friday or Boxing Day.

Coursework 1

Coursework involves a range of activities which may or may not be directly related to the main core topics. Assignments can involve mathematical investigations, practical work, experimentation, historical research, statistical surveys, problem solving or any combination of these.

The assignments in this volume will help you to practise the skills you need in order to prepare your own folder of coursework studies. Some of the assignments extend topics covered in other chapters of the book with the aim of strengthening your understanding in these areas. Others explore new avenues of mathematical research.

Assessment of your mathematics course is based on five Attainment Targets. Coursework is assessed according to the requirements of Attainment Target 1 (Ma1), but it can also make contact with the other four Attainment Targets (Ma2 to Ma5). Coursework assignments are labelled to indicate which additional Attainment Targets are explored in each case.

A detailed written report on each coursework activity will be required, incorporating all relevant deductions, methods employed and sources of reference where necessary. Diagrams, tables, graphs and models should accompany the written analysis where appropriate. Assignments which involve repeated trials, statistical studies, etc. can often be explored more thoroughly using computers.

Your coursework will be assessed under three main headings:

(1) *Applications:* You must demonstrate that you can use mathematics to investigate an assignment by breaking it down into a series of simpler tasks. Your approach must be methodical and your calculations accurate.

(2) *Communication:* The presentation of your assignment should be orderly and thorough. You must explain your conclusions clearly through written work, diagrams, etc. and, where appropriate, through discussion.

(3) *Reasoning:* You must be able to prove that your conclusions are valid and logical as well as show that you fully understand their meaning.

Assignment 1 $\diagdown$ Ma 2

There is something interesting about various pairs of digits. For instance, let us take the four digits 1, 2, 4 and 8. We can form the two numbers 21 and 48.

By multiplication,

$$21 \times 48 = 1008$$

Reverse the digits in each number to give 12 and 84.

By multiplication,

$$12 \times 84 = 1008$$

Similar results are obtained with the digits 2, 3, 4 and 6:

$$24 \times 63 = 1512$$
$$42 \times 36 = 1512$$

Here are some clues to assist you in solving the problem:

There are other pairs you could use with the number 12.

There are pairs you could use with the digits 1 and 3.

No more clues!

Investigate the problem to find the total of fourteen sets of four digits that will produce a similar result.

Assignment 2   Ma 2

(a) Investigate factors of numbers. What can you say if the total number of factors is odd? Examine numbers which are the product of two primes. What types of number have only three factors?

(b) One method for locating the prime factors of a number is to box it off and perform successive divisions by prime numbers until we reach 1:

$$
\begin{array}{r|r}
2 & 300 \\ \hline
2 & 150 \\ \hline
3 & 75 \\ \hline
5 & 25 \\ \hline
5 & 5 \\ \hline
  & 1
\end{array}
\qquad
\begin{aligned}
300 &= 2 \times 2 \times 3 \times 5 \times 5 \\
    &= 2^2 \times 3 \times 5^2
\end{aligned}
$$

Another method is to plant 'factor trees' on which the prime factors grow like fruit. Here is a tree which can be used to find the prime factors of 300:

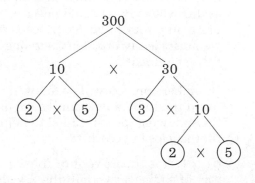

Study how this works. How many other trees can you construct for 300? How do the trees compare with one another?

Draw trees for a range of other numbers. Do odd numbers produce different branch patterns from even numbers? Is the number of branches significant? Investigate squares and numbers which have only one prime factor such as 27 or 128. Try numbers where each prime factor only appears once in the prime factor product. Draw conclusions from your discoveries.

What is special about numbers which only produce one tree? Find the smallest number which will produce a tree with eight branches. Can you predict the next number which will produce eight branches? Which will be the first number to generate twelve branches? Work out a general rule.

Some numbers can produce a symmetrical tree where the branches are in balance on either side. Other numbers cannot. What conditions must be met for a symmetrical tree to be possible? What further conditions would be necessary for a number to produce more than one symmetrical tree?

Does the tree method reveal more information than the process of successive division? Which method do you prefer? Give reasons.

(c) Demonstrate how factor trees can be used to determine the HCF and LCM of two or more numbers. Will this method work for all numbers?

(d) A rectangular number is one which can be represented as a rectangle of dots. Here are two ways of representing the number 24:

```
.  .  .  .  .      .  .  .  .  .  .
.  .  .  .  .      .  .  .  .  .  .
.  .  .  .  .      .  .  .  .  .  .
.  .  .  .  .
```

What is the connection between rectangular numbers and factors? What type of number cannot be rectangular?

Assignment 3   Ma 2, 4

The pieces in a set of dominoes consist of rectangles made up of two squares. Each domino has two numbers from 1 to 6 and there are 28 in the set. This investigation concerns a set of trominoes. A tromino will be three squares forming a right angle. Only the numbers 1 to 4 will be used in the three squares. There are no blanks. Here are two examples:

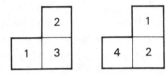

It should be noted that some formations will be mirror images of each other. These are classified as different trominoes. On the other hand, other trominoes will be simply rotated versions of one another and not actually different:

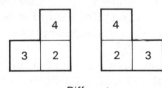

Different
(reflection)

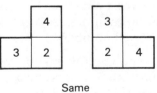

Same
(rotation)

Investigate the set of trominoes. How many are there in the set?

Invent different sets of rules which could be used to play games with the set of trominoes.

Decide on a suitable box in which the trominoes could be sold as a game.

Assignment 4   Ma 2

This assignment is concerned with building numbers from four fours. Firstly, the ground rules:

(i) For each number you build, you must use only four fours. No other digit is allowed and you must not use more or less than four fours.

(ii) The fours may be combined using any of the normal operations of addition, subtraction, multiplication or division.

(iii) You may not 'glue' the fours together to form larger numbers such as 44 or 444.

(a) Some numbers are easy to build, especially even numbers. For instance, here are two ways of making the number 2:

$$\frac{4 \times 4}{4 + 4} = 2; \qquad \frac{4}{4} + \frac{4}{4} = 2$$

Odd numbers are more difficult to construct, and, not surprisingly, prime numbers present the greatest challenge of all. See how many numbers up to 100 you can construct using only four fours.

(b) You will certainly have many gaps in your list at the moment and must have realised that the present rules will not allow you to build the more awkward numbers such as 53 or 79. In fact, you have probably concluded that numbers as difficult as these could never be made using only four fours. To proceed further, we must clearly broaden the scope of the experiment.

From here on you will be allowed to 'bend the rules' a little and make use of the following special 'fours' to help you fill the gaps in your list:

(i) $\sqrt{4}$: 'the square root of four', a neat way of sliding a two into the calculations cunningly disguised as a four!

(ii) .4: 'point four' – not to be written as '0.4' in this exercise, because zeros are not allowed.

(iii) .$\dot{4}$: 'point four recurring' (which is $\frac{4}{9}$ as a fraction).

(iv) $\sqrt{.\dot{4}}$: 'the square root of point four recurring'. This one comes in handy when you are struggling with the more obstinate numbers.

(v) 4!: 'four factorial' – extremely useful for building the larger numbers. It represents the product $(4 \times 3 \times 2 \times 1)$ or 24. Scientific calculators have a factorial key which is usually labelled $x!$ or $n!$. Look for it on your calculator and try entering 4!.

By including these extra 'fours' in your calculations, it is now possible to construct every number from 1 to 100 using four fours. Before returning to your number building, see what happens when you pair up some of these 'fours'. For instance, what results do you get from $(4 \div .4)$ or $(4! \div .\dot{4})$?

To prove the versatility of the new rules, here is a way of making that awkward prime 79 using only four fours. Check it out for yourself:

$$\frac{4! - \sqrt{4}}{.4} + 4! = 79$$

Now you are on your own. The toughest number to construct is 73 and you deserve a prize if you can manage it!

(c) How many ways can you make the number '1' using four fours? There are many more ways than you might imagine at first.

(d) What is the largest number you can construct with four fours using any of the various types of 'four' we have met in this investigation?

(e) Can you make all the numbers from 1 to 18 using only three threes?

Assignment 5   Ma 2

Time and motion studies in industry are carried out to ensure that the firm is making the best use of its workforce. To eliminate waste of time and effort, the movement of workers is studied and regulated. What kind of movement occurs at your centre when the timetable demands a change of lesson?

This is a specimen timetable for one day:

Tutorial	Room 10	Lunch	
History	Room 25	English	Room 30
Mathematics	Room 16	PE	Room 5
Break		Geography	Room 28
Practical	Room 7	Session ends	

A drawing can be constructed to show the movements through the building which a student following the timetable must make going from lesson to lesson. The drawing assumes there is a ground floor and two other floors in the building (see page 112).

The scale is 1 millimetre to 1 metre for all horizontal movements on a single floor and 1 millimetre to 1 step for the use of stairs between floors.

The numbers show the positions of the rooms in the building. If a centre's rooms were all on the ground floor, the drawing could be modified to show movements left and right instead of horizontally and vertically.

Investigate parts of the timetable at your own centre. Indicate the sections of a timetable which in your opinion contain excessive movement and then make enquiries to discover whether such movement is actually necessary.

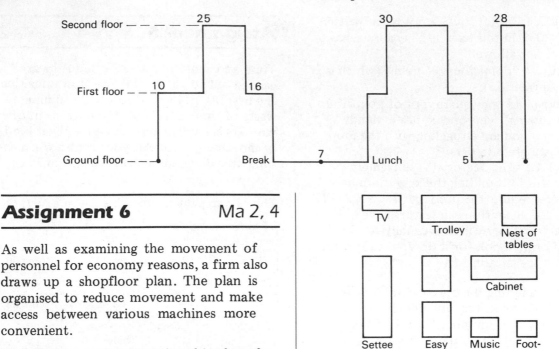

Assignment 6   Ma 2, 4

As well as examining the movement of personnel for economy reasons, a firm also draws up a shopfloor plan. The plan is organised to reduce movement and make access between various machines more convenient.

As a simple exercise, examine this plan of a lounge in which certain items of furniture are to be positioned. Make a drawing of the plan twice the size of the one shown in the diagram and produce paper cut-outs of the furniture, also on twice the scale shown in the drawings.

Use the cut-outs to establish a final arrangement of furniture in the room and present a detailed report explaining the reasons for your choice of position for each item of furniture.

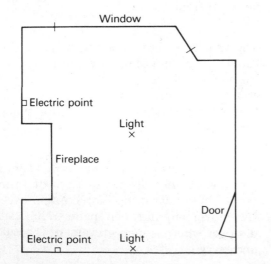

Now investigate one of these situations using the same approach:

(a) A kitchen contains a cooker, refrigerator, freezer, cupboards for cooking utensils, crockery and some food items, work-tops, sink, draining boards and waste disposal units. The area of the floor is a rectangle measuring about 5 metres by 3 metres. Consideration should also be given to the building of an extension to take a washing machine, drier, airing space and ironing facilities.

(b) A workshop measures 10 metres by 6 metres and sixteen students will be working at benches using wood and metal at various times. The two activities may be taking place simultaneously on some occasions. There is a separate outside store for supplies of wood and metal, but workbenches, machines, tools and all other facilities available to the students are contained within the workshop area. An extension for light crafts is under consideration.

Assignment 7   Ma 2

The following is a printout of a single month extracted from a calendar:

Sun		5	12	19	26
Mon		6	13	20	27
Tue		7	14	21	28
Wed	1	8	15	22	29
Thu	2	9	16	23	
Fri	3	10	17	24	
Sat	4	11	18	25	

What month is represented here? Given that this month appeared in a calendar printed after 1960, work out which year it must have come from. How many times has this month appeared exactly like this since 1899? When will it look like this again?

Investigate number patterns within this month. Isolate different shaped arrays of numbers from the date grid and identify relationships and rules which govern each number arrangement. Here are some suggestions:

Square

2	9	16
3	10	17
4	11	18

Triangle

		27
	21	28
15	22	29

T-shape

13	20	27
	21	
	22	

Cross

13		27
	21	
15		29

Select other months from a calendar and see whether the relationships you have discovered are maintained on other date grids.

Assignment 8   Ma 2

The digits 0 to 9 are to be arranged in the form of a triangle. Before that is done, one of them is cast out and placed at the centre of the triangle so that it takes no further part in the problem. In this example, the number 0 is cast out to the centre:

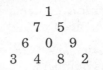

The 0 is in the centre and can be ignored. The other nine numbers form the outer edges of the triangle so that there are four numbers arranged along each side. The total on each side of the triangle is 17.

In the next example, the number 6 has been cast out and placed in the centre and is not used in this problem.

The total on each side this time is 14.

Cast out each number in turn and find as many magic triangles as you can. There is more than one for each number that is placed in the centre, so there are plenty for you to discover.

Assignment 9   Ma 2

Any four numbers are chosen to start this investigation.

$$18 \quad 41 \quad 5 \quad 76$$

A second row is established from the differences between numbers next to each other and the difference between the two end numbers:

i.e. $41 - 18 \quad 41 - 5 \quad 76 - 5 \quad 76 - 18$

The second row therefore becomes:

$$23 \quad 36 \quad 71 \quad 58$$

The process is continued until something significant occurs.

Investigate the sequence using any four numbers.

Does a mixture of small and very large numbers affect the final position in any way?

Does the inclusion of decimal quantities make any difference?

Investigate particular groups such as odd, even and prime numbers.

What happens with groups of three numbers and groups of five numbers?

Is there anything significant about any other particular set?

Are any patterns of behaviour worth noting?

Assignment 10   Ma 2

A local jeweller had two tricky problems with chains that had suffered some damage.

(a) The first customer brought in five lengths of chain, each of which had four links. She wanted the links to join together to make a continuous loop with no break.

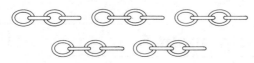

'I charge £1 to break a link and £1 to join it together again,' said the jeweller, 'so that will cost you £10.'

'Oh no,' replied the young lady, 'let me explain how you can do it for £8, charging the same rates.'

What was her explanation?

(b) A dowager duchess came in with a more expensive chain which had snapped into nine lengths. There were two with 5 links, three with 6 links and four separate lengths of 3 links, 4 links, 7 links and 8 links.

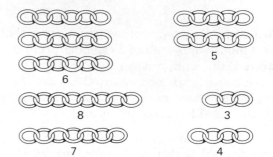

She also wanted all the links joined into a continuous loop. The rates of £1 for a break and £1 for a make were explained to her. Expecting no argument, the jeweller said, 'Ah, yes. I can break each link in the eight-link portion. I shall then use the eight links to join the remaining eight lengths in a continuous chain. That will cost you £16.'

'My man,' said the duchess, 'I shall expect you to work out a method which will only cost me £14.'

What was that method?

13

Basic Algebra

Introduction

In algebra, letters are used in place of numbers. Let us take two numbers and call them a and b. The sum of the two numbers is $a + b$. The difference of the two numbers is $a - b$. The product of the two numbers is $a \times b$ which is usually written ab (i.e. the multiplication sign is usually omitted in algebra).

The first number divided by the second number is $a \div b$ or $\dfrac{a}{b}$.

Example 1

(a) Translate into symbols six times a number x.

Six times a number x is $6 \times x = 6x$

(b) Translate into symbols 5 times a number y minus four.

Five times a number y minus four is

$$5 \times y - 4 = 5y - 4$$

Exercise 13.1

Translate each of the following into algebraic symbols:

1. Seven times a number a.

2. Five times a number b minus seven.

3. The sum of five times a number x and three times a number y.

4. The product of the three numbers x, y and z.

5. Eight times the product of the two numbers m and n.

6. Twice the number p divided by three times the number q.

7. Nine times a number r plus three times a number s.

8. The product of two numbers x and y divided by a third number p.

Substitution

Substitution is the process of assigning numbers to symbols in an algebraic expression.

Example 2

Find the value of $3x + 5y - 2z$ when $x = 6$, $y = 7$ and $z = 4$.

$$3x + 5y - 2z = 3 \times 6 + 5 \times 7 - 2 \times 4$$
$$= 18 + 35 - 8$$
$$= 45$$

(Remember that we must multiply before adding or subtracting.)

A Complete GCSE Mathematics General Course

If $x = 3$ and $y = 6$, find numerical values for the following algebraic expressions:

1. $x + 3y$
2. $3x - 4y$
3. $5x - 2y - 3x + 4y$
4. $x - 5$
5. $2x + 7$
6. $2x \div y$
7. $\dfrac{3y}{2x}$
8. $\dfrac{3y + 6}{4x}$

If $a = 3$, $b = 6$ and $c = -7$, find numerical values for the following expressions:

9. $b + 3$
10. $a - 3$
11. $12 - c$
12. $5a$
13. $3c$
14. bc
15. abc
16. $5ab$
17. $\dfrac{12}{a}$
18. $\dfrac{ab}{9}$
19. $a + 3b - 4c$
20. $a - 3b + 8c$
21. $\dfrac{c}{a}$
22. $\dfrac{b}{c}$

Powers

The quantity $a \times a \times a$ is usually written as a^3. The number 3 which shows the number of as to be multiplied together is called the **index** (plural: indices) and we say that a has been raised to the third power.

Example 3

(a) Find the value of y^6 when $y = 2$.

$$y^6 = 2^6$$
$$= 2 \times 2 \times 2 \times 2 \times 2 \times 2$$
$$= 64$$

(b) Find the value of a^3 when $a = -2$.

$$a^3 = (-2)^3$$
$$= (-2) \times (-2) \times (-2)$$
$$= -8$$

When dealing with expressions like $5ab^4$, note that it is only b which is raised to the power of 4. Thus

$$5ab^4 = 5 \times a \times b \times b \times b \times b$$

Example 4

Find the numerical value of $5b^2c^3$ when $b = 4$ and $c = 2$.

$$5b^2c^3 = 5 \times 4^2 \times 2^3$$
$$= 5 \times 4 \times 4 \times 2 \times 2 \times 2$$
$$= 640$$

If $p = 2$, $q = -3$ and $r = 5$, find numerical values for the following algebraic expressions:

1. p^3 >
2. q^4
3. r^2
4. pq^4
5. $3p^4$
6. $4p^2q$
7. p^3r^2
8. pqr^3
9. $4pq^2r$
10. $p^2 + q^2$
11. $q^2 - r^2$
12. $p^4 - q^3$

Addition and Subtraction of Algebraic Terms

Like terms are numerical multiples of the same algebraic quantity. Thus $5x$, $8x$ and $-3x$ are three like terms.

The **numerical coefficient** is the number which appears before the symbols of a term. Thus in the term $8x$, the numerical coefficient is 8. If a term has no number in front of the symbols, for instance ab, the numerical coefficient is 1.

Like terms are added or subtracted by adding or subtracting their numerical coefficients.

Example 5

(a) $3x + 5x - 2x = (3 + 5 - 2)x = 6x$

(b) $3y - 5y = (3 - 5)y = -2y$

(c) $9b + b + 3b = (9 + 1 + 3)b = 13b$

Only like terms can be added or subtracted. $7a - 3b + c$ is an expression which contains three unlike terms. Hence it cannot be simplified further.

It is quite possible to have several sets of like terms in an expression, and each set may then be simplified by adding and/or subtracting.

Example 6

Simplify $8p + 2q - 5r + 6r - 4p + q - 3p + 5r$.

$$8p + 2q - 5r + 6r - 4p + q - 3p + 5r$$
$$= (8 - 4 - 3)p + (2 + 1)q$$
$$+ (-5 + 6 + 5)r$$
$$= 1p + 3q + 6r$$
$$= p + 3q + 6r$$

Note that it is usual in algebra to write, for instance, $1p$ as p, the 1 being understood.

Exercise 13.4

Simplify each of the following by adding or subtracting like terms:

1. $3x + 5x$ 2. $7y - 3y$

3. $8p - 11p$ 4. $q - 4q$

5. $3x - 5x$ 6. $8p + 5p - 3p$

7. $-3x - 4x + x$ 8. $5m + 8m - 15m$

9. $5x - 4y + 2c + 7x + 5c - 2y - 6y + 8c + x + 3y$

10. $a - 3b + 2c - 5a - 6c + 8b - 7b + 9a - c$

Multiplication of Algebraic Terms

The rules for the multiplication of algebraic terms are the same as those for directed numbers. Thus

$$(+x) \times (+y) = +xy = xy$$
$$(-x) \times (+y) = -xy$$
$$(+x) \times (-y) = -xy$$
$$(-x) \times (-y) = +xy = xy$$

Example 7

(a) $2x \times (-3y) = 2 \times (-3) \times x \times y$
$$= (-6) \times x \times y$$
$$= -6xy$$

(b) $5a \times 2b \times (-3c) = -(5 \times 2 \times 3) \times a \times b \times c$
$$= -30abc$$

Note that in writing algebraic expressions such as those in Example 7, we have used the order: numbers first then letters in alphabetical order. It is not wrong to write terms in any other order but it looks better if we write the numbers first and then the letters in alphabetical order. Thus we would write $5xyz$ rather than $x5zy$.

Note also that it is bad practice to have two signs following one another, for instance $2x \times -3y$. A bracket should be used to avoid this. Thus $2x \times -3y$ would be written $2x \times (-3y)$.

When multiplying expressions containing the same symbols, indices are used.

Example 8

(a) $3m \times 5m = 3 \times m \times 5 \times m$
$$= 15m^2$$

(b) $3mn \times 2n^2 = 3 \times m \times n \times 2 \times n \times n$
$$= 6mn^3$$

Exercise 13.5

Simplify each of the following by making them into single terms:

1. $2y \times 3p$
2. $3m \times (-2p)$
3. $(-4a) \times 5b$
4. $3a \times 2b \times 5c$
5. $(-2x) \times (-3y) \times 2z$
6. $7m \times (-2n) \times p$
7. $3a \times 2b \times (-c)$
8. $(-3) \times (-5y) \times 2z \times (-4q)$
9. $2m \times (-3m)$
10. $2p \times 3q \times (-q) \times 4p$

Brackets

As in arithmetic, brackets are used for convenience in grouping terms together. When 'removing' a bracket each term within the bracket is multiplied by the term outside the bracket.

Example 9

(a) $5(2m + 3p) = (5 \times 2m) + (5 \times 3p)$

$$= 10m + 15p$$

(b) $-2(3a - 2b) = (-2 \times 3a)$

$$+ [-2 \times (-2b)]$$

$$= -6a + 4b$$

When a bracket has a minus sign outside it, it means that each term within the bracket has to be multiplied by -1 when the bracket is removed.

$$-(p + q) = (-1 \times p) + (-1 \times q)$$

$$= -p - q$$

Note that all the signs inside the bracket are changed when the bracket is removed.

Example 10

(a) $-(4m - 3n) = -4m + 3n$

(b) $-5(a - 2b) = -5a + 10b$

When simplifying expressions containing brackets remove the brackets first and then add and/or subtract like terms.

Example 11

Simplify $5(3a + 2b) - 3(2a - 6b)$.

$$5(3a + 2b) - 3(2a - 6b) = 15a + 10b$$

$$-6a + 18b$$

$$= 9a + 28b$$

Exercise 13.6

Remove the brackets and simplify:

1. $3(a + 2b)$
2. $5(2x - 3y)$
3. $5(4p - 3q)$
4. $-(p - q)$
5. $-2(3x - 4y)$
6. $-5(4m - 3n)$
7. $3(a - 2b) + 5(2a - 3b)$
8. $-(p + 2q) - (3p + 5q)$
9. $2(3m + 5n) - 3(2m - 5n)$
10. $5(2x - 3y) - 3(x - 2y)$

Division of Algebraic Terms

When dividing algebraic expressions, cancellation between numerator (top part) and denominator (bottom part) is often possible. Note that cancelling is equivalent to dividing top and bottom by the same quantity.

Example 12

(a) $\dfrac{pq}{p} = \dfrac{\cancel{p} \times q}{\cancel{p}} = q$

(b) $\dfrac{3x^2y^3z}{xyz} = \dfrac{3 \times \cancel{x} \times x \times \cancel{y} \times y \times y \times \cancel{z}}{\cancel{x} \times \cancel{y} \times \cancel{z}}$

$$= 3xy^2$$

Exercise 13.7

Simplify the following:

1. $12x \div 6$
2. $(-3a) \div 3b$
3. $(-5x) \div (-5y)$
4. $4a \div 2b$
5. $4ab \div 2a$
6. $12x^2yz^2 \div 4xz^2$
7. $(-12a^2b) \div 6a$
8. $8a^2bc^2 \div 4ac^2$
9. $7a^2b^2 \div ab$
10. $2x^3y^2z^4 \div x^2yz^2$

Factorising

The expression $3x + 3y$ has the number 3 common to both terms.

$$3x + 3y = 3 \times (x + y) = 3(x + y)$$

3 and $(x + y)$ are said to be the **factors of** $3x + 3y$.

Example 13

(a) $2x + 6 = 2(x + 3)$

Note that $2x$ and 6 have a common factor of 2. This common factor is placed outside the bracket. To find the terms inside the bracket, divide each of the terms making up the original expression by the common factor. Thus

$$\frac{2x}{2} = x \quad \text{and} \quad \frac{6}{2} = 3$$

(b) $px - px^2 = px(1 - x)$

Note that the terms px and px^2 have a common factor of px and that
$$\frac{px}{px} = 1 \quad \text{and} \quad \frac{px^2}{px} = x$$

(c) $m(x - y) - n(x - y) = (x - y)(m - n)$

Note that the terms $m(x - y)$ and $n(x - y)$ have a common factor of $(x - y)$ and that
$$\frac{m(x - y)}{x - y} = m \quad \text{and} \quad \frac{n(x - y)}{x - y} = n.$$

Exercise 13.8

Factorise the following:

1. $3x + 6$
2. $4y - 8$
3. $2p - 4$
4. $5m + 10$
5. $2x + 2y$
6. $3p - 3q$
7. $ax^2 + bx$
8. $mx - mx^2$
9. $x(a + b) + y(a + b)$
10. $a(x - y) - b(x - y)$

Multiplying Algebraic Fractions

As with arithmetic fractions, to multiply algebraic fractions we first multiply the numerators and then multiply the denominators. We then cancel common factors.

Example 14

$$\frac{3ab}{5c^2} \times \frac{cd}{7a^2} = \frac{3ab \times cd}{5c^2 \times 7a^2}$$

$$= \frac{3 \times \cancel{a} \times b \times \cancel{c} \times d}{5 \times \cancel{c} \times c \times 7 \times \cancel{a} \times a}$$

$$= \frac{3 \times b \times d}{5 \times c \times 7 \times a}$$

$$= \frac{3bd}{35ac}$$

Dividing Algebraic Fractions

As with arithmetic fractions, when dividing algebraic fractions we invert the second fraction and then proceed as in multiplication.

Example 15

$$\frac{5pq^2}{8m^2n} \div \frac{3pq}{7mn} = \frac{5pq^2}{8m^2n} \times \frac{7mn}{3pq}$$

$$= \frac{5pq^2 \times 7mn}{8m^2n \times 3pq}$$

$$= \frac{5 \times \cancel{p} \times \cancel{q} \times q \times 7 \times \cancel{m}n \times \cancel{n}}{8 \times \cancel{m}n \times m \times \cancel{n} \times 3 \times \cancel{p} \times \cancel{q}}$$

$$= \frac{5 \times q \times 7}{8 \times m \times 3}$$

$$= \frac{35q}{24m}$$

Exercise 13.9

Simplify each of the following:

1. $\dfrac{6a}{b^2} \times \dfrac{b}{3a^2}$ 2. $\dfrac{9x^2}{6y^2} \times \dfrac{y^3}{x^3}$

3. $\dfrac{6pq}{4rs} \times \dfrac{8s^2}{3p}$ 4. $\dfrac{6ab}{c} \times \dfrac{ad}{2b} \times \dfrac{8cd^2}{4bc}$

5. $\dfrac{ab^2}{bc^2} \div \dfrac{a^2}{bc^3}$ 6. $\dfrac{3xy^2}{pq} \div \dfrac{6y^3}{p^2}$

7. $\dfrac{xy^2}{a^2b} \div \dfrac{x^2y}{ab^2}$ 8. $\dfrac{m^2n^3}{p^2q} \div \dfrac{m^3n^4}{p^3q^2}$

Addition and Subtraction of Algebraic Fractions

The procedure for adding and subtracting algebraic fractions is the same as for arithmetic fractions:

(1) Find the LCM of the denominators.

(2) Express each fraction with this common denominator.

(3) Add or subtract the numerators of these equivalent fractions.

Example 16

Express $\dfrac{a}{3} + \dfrac{a}{4} + \dfrac{a}{5}$ as a single fraction.

The LCM of 3, 4 and 5 is 60.

$$\frac{a}{3} + \frac{a}{4} + \frac{a}{5} = \frac{20a}{60} + \frac{15a}{60} + \frac{12a}{60}$$

$$= \frac{20a + 15a + 12a}{60}$$

$$= \frac{47a}{60}$$

Example 17

Simplify $\dfrac{x}{2} - \dfrac{x-1}{3}$

The LCM of 2 and 3 is 6.

$$\frac{x}{2} - \frac{x-1}{3} = \frac{3x}{6} - \frac{2(x-1)}{6}$$

$$= \frac{3x - 2(x-1)}{6}$$

$$= \frac{3x - 2x + 2}{6}$$

$$= \frac{x+2}{6}$$

Exercise 13.10

Simplify each of the following by expressing them as single fractions in their lowest terms:

1. $\dfrac{a}{4} + \dfrac{a}{5}$ 2. $\dfrac{p}{2} - \dfrac{q}{3}$

3. $\dfrac{3a}{4} - \dfrac{a}{5}$ 4. $\dfrac{x}{2} + \dfrac{x}{3} + \dfrac{x}{4}$

5. $\dfrac{3}{2x} + \dfrac{2}{3x}$ 6. $\dfrac{m}{x} - \dfrac{3m}{2x} + \dfrac{5m}{3x}$

7. $\dfrac{4x}{3y} - \dfrac{2x}{5y}$ 8. $1 + \dfrac{x+2}{3}$

9. $\dfrac{x}{3} + \dfrac{2x-1}{4}$ 10. $5m - \dfrac{m-2}{2}$

11. $\dfrac{x-3}{4} - \dfrac{x}{3}$ 12. $\dfrac{4}{x} - \dfrac{5}{2x} + \dfrac{3}{4x}$

13. $\dfrac{x}{2} - \dfrac{x-3}{3}$

14. $\dfrac{2a+3b}{3} - \dfrac{a-2b}{2}$

15. $\dfrac{5-x}{5} - \dfrac{x-3}{2}$

Miscellaneous Exercise 13

1. What is the value of $x^2 + x + 6$ when $x = 4$?

2. Express $\dfrac{a-b}{a+b}$ as a fraction in its lowest terms given that $a = 5$ and $b = 3$.

3. If $a = 0.5$ and $b = 37$, what is the value of $\dfrac{b}{a}$?

4. If $p = 3xy$, what is the value of p when $x = 2$ and $y = 5$?

5. An approximate way of converting degrees Celsius into degrees Fahrenheit is $F = 2C + 30$, where F is the temperature in degrees Fahrenheit and C is the temperature in degrees Celsius. Convert a temperature of $12°C$ into degrees Fahrenheit.

6. If $a = 3$ and $b = 6$, find the value of $5a^2 b$.

7. Work out the value of $a + 2b$ when $a = 1$ and $b = \frac{1}{2}$.

8. Make into single terms:
   (a) $8a - 2a$ (b) $8a \times 2a$
   (c) $8a \div 2a$

9. Remove the brackets and simplify:
   (a) $3(x + y) + 2(x + 2y)$
   (b) $3(x + y) - (x - 2y)$
   (c) $(2x - y) - (2x + y)$

10. Find the value of $9a^2 + ab - c^2$ when $a = 1$, $b = 2$ and $c = -3$.

11. Simplify the expressions
    (a) $5a \times 2b \times 3a$
    (b) $8x - y - 3x - 5y$

12. Given that $x = 6$, $y = 4$ and $z = -2$, work out the values of the following expressions:
    (a) $x - y + z$ (b) $x - (y - z)$ (c) $\dfrac{x^2}{3y^2}$

13. If $x = 3$ and $y = -5$, find the value of
    (a) $(x + y)^2$ (b) $x^2 + 3xy + y^2$

14. Express as a single fraction in its simplest form $\dfrac{2x}{3} - \dfrac{x}{6}$

15. Express $\dfrac{x}{2} - \dfrac{x-4}{3}$ as a single fraction, simplifying your answer as far as possible.

16. Simplify $\dfrac{3a^2 b^3}{2cd} \times \dfrac{c^3 d^2}{ab}$

17. Simplify $\dfrac{6ab}{5cd} \div \dfrac{4a^2}{7bd}$

Mental Test 13

Try to answer the following questions without writing anything down except the answer.

1. Make into a single term $8x + 3x$.

2. Make into a single term $a + 3a$.

3. Subtract $5b$ from $8b$.

4. Make into a single term $5p - p$.

5. Simplify $2a \times 3b$.

6. Make into a single term $p \times p$.

7. Simplify $3m \times 4m$.

8. Divide $5a^2$ by a.

9. Divide $4a^3$ by a^2.

10. Make into a single term $6p - 2p$.

11. Remove the brackets from $3(p + q)$.

12. Remove the brackets from $-(x - y)$.

13. Remove the brackets from $-2(m + n)$.

14. Simplify $\dfrac{a}{b} \times \dfrac{p}{q}$.

15. Make into a single term $5x - 3x - 4x$.

14 Indices

Introduction

We have seen that a quick way to write $2 \times 2 \times 2$ is 2^3. The figure 3 which shows the number of twos to be multiplied together is called the **index** (plural: indices). 2^3 is said to be 2 raised to the power of 3. The number 2 is called the **base**.

a^4 is a raised to the power of 4; a is the base.

Multiplication

$$a^2 \times a^3 = a \times a \times a \times a \times a = a^5$$

We can see that in the case of multiplication, it is quicker to **add** the indices. Thus

$$a^2 \times a^3 = a^{2+3} = a^5$$

Example 1

(a) $3^4 \times 3^5 = 3^{4+5} = 3^9$

(b) $p^2 \times p^4 \times p^6 = p^{2+4+6} = p^{12}$

(c) $5y \times 2y^2 \times 3y^4 = 5 \times 2 \times 3 \times y^{1+2+4}$
$$= 30y^7$$

Division

$$a^5 \div a^2 = \frac{a^5}{a^2}$$
$$= \frac{a \times a \times a \times \cancel{a} \times \cancel{a}}{\cancel{a} \times \cancel{a}}$$
$$= a^3$$

We can see that in the case of division, it is quicker to **subtract** the indices. Thus

$$a^5 \div a^2 = a^{5-2}$$
$$= a^3$$

Example 2

(a) $p^9 \div p^4 = p^{9-4} = p^5$

(b) $\dfrac{y^3 \times y^5 \times y^4}{y^2 \times y^7} = \dfrac{y^{3+5+4}}{y^{2+7}}$
$$= \frac{y^{12}}{y^9}$$
$$= y^{12-9}$$
$$= y^3$$

Exercise 14.1

Make each of the following into a single term:

1. $5^2 \times 5^4$

2. $a^3 \times a^4$

3. $p^2 \times p^3 \times p^4$

4. $y^3 \times y^5 \times y^7 \times y^8$

5. $2 \times 2^3 \times 2^4$

6. $2a \times 3a^2 \times 4a^3$

7. $3y^2 \times y \times 5y^4$

8. $6p^4 \times p^2 \times 3p$

Exercise 14.2

Simplify each of the following:

1. $3^6 \div 3^4$

2. $p^7 \div p^5$

3. $d^5 \div d^4$

4. $(q^5 \times q^3) \div q^4$

5. $\dfrac{m^2 \times m^7}{m^5 \times m^3}$

6. $\dfrac{t^4 \times t^9}{t^3 \times t^5}$

7. $\dfrac{am^3}{am^2}$

8. $\dfrac{3p^2 \times 8p^4}{6p^3}$

Raising the Power of a Quantity

$$(a^3)^2 = a^3 \times a^3$$
$$= a^{3+3}$$
$$= a^6$$

We can see that when raising the power of a quantity it is quicker to *multiply* the indices so that

$$(a^3)^2 = a^{3 \times 2}$$
$$= a^6$$

Example 3

(a) $(p^4)^3 = p^{4 \times 3} = p^{12}$

(b) $(2m^3)^5 = 2^{1 \times 5} \times m^{3 \times 5} = 2^5 \times m^{15}$

(c) $\left(\dfrac{x^2}{y^3}\right)^4 = \dfrac{x^{2 \times 4}}{y^{3 \times 4}} = \dfrac{x^8}{y^{12}}$

Exercise 14.3

Simplify each of the following:

1. $(y^2)^4$ 2. $(7^2)^3$

3. $(5c^2)^3$ 4. $(5ab^2c^3)^5$

5. $\left(\dfrac{3a^2}{2b^3}\right)^4$ 6. $(m^3)^5$

7. $(3b)^4$ 8. $(2x^2y)^3$

9. $(\frac{3}{4})^4$ 10. $(\frac{1}{2})^5$

Negative Indices

$$a^3 \div a^5 = \frac{a^3}{a^5}$$

$$= \frac{\cancel{a} \times \cancel{a} \times \cancel{a}}{\cancel{a} \times \cancel{a} \times \cancel{a} \times a \times a}$$

$$= \frac{1}{a \times a}$$

$$= \frac{1}{a^2}$$

Using the rule for division,

$$a^3 \div a^5 = a^{3-5}$$
$$= a^{-2}$$

so that

$$\frac{1}{a^2} = a^{-2}$$

The reciprocal of a number is 1/number. Hence the reciprocal of 5 is 1/5 and the reciprocal of a^2 is $1/a^2$.

Therefore a negative index indicates the **reciprocal** of a quantity.

Example 4

(a) $5^{-4} = \dfrac{1}{5^4}$

(b) Express 5^{-1} as a fraction.

$$5^{-1} = \frac{1}{5^1} = \frac{1}{5}$$

(c) $7p^{-3} = 7 \times p^{-3} = \dfrac{7}{1} \times \dfrac{1}{p^3} = \dfrac{7}{p^3}$

Zero Index

$$a^3 \div a^3 = \frac{a^3}{a^3}$$

$$= \frac{\cancel{a} \times \cancel{a} \times \cancel{a}}{\cancel{a} \times \cancel{a} \times \cancel{a}}$$

$$= 1$$

Using the rule for division,

$$a^3 \div a^3 = a^{3-3}$$
$$= a^0$$

Hence $a^0 = 1$.

No matter what numbers or symbols we use we find that any quantity raised to the power of 0 equals 1. Thus

$$p^0 = 1, \quad 963^0 = 1 \quad \text{and} \quad 83^0 = 1$$

Example 5

Find the value of (a) 7×3^0, (b) $3x^0$.

(a) $7 \times 3^0 = 7 \times 1 = 7$

(b) $3x^0 = 3 \times x^0 = 3 \times 1 = 3$

Powers and Roots

The root of a number can be written as a fractional index. The square root of $5 = \sqrt{5} = 5^{1/2}$ because $5^{1/2} \times 5^{1/2} = 5^{1/2+1/2} = 5^1 = 5$. The cube root of $7 = \sqrt[3]{7} = 7^{1/3}$ because $7^{1/3} \times 7^{1/3} \times 7^{1/3} = 7^{1/3+1/3+1/3} = 7^1 = 7$.

In general the nth root of a number a is written $\sqrt[n]{a} = a^{\frac{1}{n}}$

Example 6

Find the value of (a) $\sqrt[6]{64}$, (b) $\sqrt[3]{27}$.

(a) $\sqrt[6]{64} = 64^{1/6} = (2^6)^{1/6} = 2^{6 \times 1/6} = 2^1$
$= 2$

(b) $\sqrt[3]{27} = (3^3)^{1/3} = 3^{3 \times 1/3} = 3^1 = 3$

Exercise 14.4

Write each of the following fractions with a negative index:

1. $\dfrac{1}{5}$ 　　 2. $\dfrac{1}{3^2}$ 　　 3. $\dfrac{1}{a^3}$

4. $\dfrac{1}{b^5}$ 　　 5. $\dfrac{2}{x^3}$ 　　 6. $\dfrac{5}{x}$

7. $\dfrac{1}{m^6}$ 　　 8. $\dfrac{8}{y^7}$

Write each of the following as a fraction:

9. 3^{-1} 　　 10. 5^{-2} 　　 11. a^{-3}

12. m^{-5} 　　 13. $3b^{-1}$ 　　 14. $7q^{-3}$

Find the value of each of the following, stating the answer as a fraction or a whole number:

15. 10^{-1} 　　 16. 3^{-2} 　　 17. 2^{-4}

18. 5^{-2} 　　 19. 3^{-3} 　　 20. 10^{-3}

21. b^0 　　 22. $3q^0$ 　　 23. $(\frac{1}{5})^0$

24. $\left(\dfrac{x}{y}\right)^0$

25. Express as powers of x
(a) $\sqrt[5]{x}$, (b) $\sqrt[4]{x}$, (c) $\sqrt[7]{x}$.

26. Find the values of
(a) $4^{1/2}$, (b) $8^{1/3}$, (c) $32^{1/5}$.

27. Find the value of $32^{1/5} \times 25^{1/2} \times 27^{1/3}$.

28. Find the value of $\sqrt{\dfrac{p}{q}}$ when
$p = 16^{1/2}$ and $q = 8^{1/3}$.

Place Value

The number 2478 is 2 thousand 4 hundred 7 tens and 8 units. Now 1 thousand is 10^3, 1 hundred is 10^2, 1 ten is 10^1 and 1 unit is 10^0. We can show our number in a table:

10^3	10^2	10^1	10^0
2	4	7	8

This may be written in expanded notation as:

$$2 \times 10^3 + 4 \times 10^2 + 7 \times 10^1 + 8 \times 10^0$$

Example 7

Write the number 8 632 917 in expanded notation.

The number 8 632 917 is 8 millions 6 hundred-thousands 3 ten-thousands 2 thousands 9 hundreds 1 ten and 7 units, i.e.

$8\,632\,917 = 8 \times 1\,000\,000 + 6 \times 100\,000$
$+ 3 \times 10\,000 + 2 \times 1000$
$+ 9 \times 100 + 1 \times 10$
$+ 7 \times 1$

Now,

$$1\,000\,000 = 10^6 \qquad 100\,000 = 10^5$$
$$10\,000 = 10^4 \qquad 1000 = 10^3$$
$$100 = 10^2 \qquad 10 = 10^1$$
$$1 = 10^0$$

Hence

$$8\,632\,917 = 8 \times 10^6 + 6 \times 10^5 + 3 \times 10^4$$
$$+ 2 \times 10^3 + 9 \times 10^2$$
$$+ 1 \times 10^1 + 7 \times 10^0$$

Exercise 14.5

Copy the following table:

10^6	10^5	10^4	10^3	10^2	10^1	10^0

By writing figures in the appropriate columns use your table to represent the following numbers:

1. 936
2. 7254
3. 8619
4. 18 342
5. 278 948
6. 563 740
7. 8 621 975
8. 7300
9. 83 507
10. 900 007

Write in expanded notation the numbers given in the table below:

	10^6	10^5	10^4	10^3	10^2	10^1	10^0
11.					7	5	3
12.		3	8	4	6	8	5
13.		1	7	3	2	7	8
14.		2	3	0	0	0	5
15.	3	9	0	7	6	0	0

Decimal Numbers

We can do the same thing with decimal numbers but in this case we use negative indices:

$$10^{-1} = \frac{1}{10} \qquad 10^{-2} = \frac{1}{100}$$
$$= 0.1 \qquad\qquad = 0.01$$
$$10^{-3} = \frac{1}{1000}$$
$$= 0.001 \quad \text{and so on}$$

We can show the number 0.3465 in a table:

10^{-1}	10^{-2}	10^{-3}	10^{-4}
3	4	6	5

This may be written in expanded notation as

$$3 \times 10^{-1} + 4 \times 10^{-2} + 6 \times 10^{-3} + 5 \times 10^{-4}$$

Now consider the number 9346.825. Using a table as before we have

10^3	10^2	10^1	10^0	10^{-1}	10^{-2}	10^{-3}
9	3	4	6	8	2	5

This may be written in expanded notation as

$$9 \times 10^3 + 3 \times 10^2 + 4 \times 10^1 + 6 \times 10^0$$
$$+ 8 \times 10^{-1} + 2 \times 10^{-2} + 5 \times 10^{-3}$$

We see that the decimal point separates the positive powers of 10 from the negative powers of 10.

Exercise 14.6

Copy the following table:

10^4	10^3	10^2	10^1	10^0	10^{-1}	10^{-2}	10^{-3}	10^{-4}

By writing figures in the various columns, use your table to represent the following numbers:

1. 0.52
2. 0.375
3. 0.9134
4. 56 327.58
5. 96.3
6. 187.25
7. 3059.273
8. 0.0635
9. 7300.2763
10. 83 507.0038

11. Write each of the above numbers in expanded notation.

Write in decimal notation the numbers given in the table below:

	10^6	10^5	10^4	10^3	10^2	10^1	10^0	10^{-1}	10^{-2}	10^{-3}
12.					2	7	5	3		
13.			8	4	6	0	5			
14.				8	0	0	0	7	3	
15.	9	0	7	6	0	8	3	2	6	
16.						6	2	7	5	4

Large Numbers in Standard Form

Any number can be expressed as a value between 1 and 10 multiplied by a power of 10. For instance

$$8372 = 8.372 \times 10^3$$

A number expressed in this way is said to be written in **standard form** or in **scientific notation**.

This way of writing numbers is very useful particularly if the numbers are very large or very small.

Note that $100 = 10^2$, $1000 = 10^3$ and $1\,000\,000 = 10^6$.

The power of 10 is found by counting the number of figures to the left of the decimal point and subtracting 1.

Example 8

(a) $563 = 5.63 \times 100 = 5.63 \times 10^2$

(b) $75\,532 = 7.5532 \times 10\,000$
$$= 7.5532 \times 10^4$$

(c) $25\,000\,000 = 2.5 \times 10\,000\,000$
$$= 2.5 \times 10^7$$

Exercise 14.7

Write the following numbers in standard form:

1. 827
2. 1734
3. 17 632
4. 893 762
5. 8 036 000

Write the following as ordinary numbers (i.e. not in standard form):

6. 3.2×10^3
7. 5×10^2
8. 1.87×10^6
9. 4.32×10^4
10. 8.762×10^5

Small Numbers in Standard Form

Now $0.1 = 10^{-1}$, $0.01 = 10^{-2}$ and $0.001 = 10^{-3}$.

The negative power of 10 is found by adding 1 to the number of zeros following the decimal point.

Example 9

(a) $0.043 = 4.3 \div 100 = 4.3 \div 10^2 = \dfrac{4.3}{10^2}$
$$= 4.3 \times 10^{-2}$$

(b) $0.0036 = 3.6 \div 1000 = 3.6 \div 10^3$
$$= \dfrac{3.6}{10^3} = 3.6 \times 10^{-3}$$

Exercise 14.8

Write the following numbers in standard form:

1. 0.3   2. 0.05   3. 0.0056

4. 0.000 007   5. 0.000 067

Write the following as ordinary numbers (i.e. not in standard form):

6. 2×10^{-3} 7. 5.67×10^{-1}

8. 3.2×10^{-2} 9. 5×10^{-6}

10. 4.1×10^{-4}

Adding and Subtracting Numbers in Standard Form

(1) If the numbers to be added or subtracted have the same power of 10, then the numbers may be added or subtracted directly.

Example 10

(a) $1.859 \times 10^2 + 2.387 \times 10^2 + 9.163 \times 10^2$

$$= (1.859 + 2.387 + 9.163) \times 10^2$$

$$= 13.409 \times 10^2$$

$$= 1.3409 \times 10^3$$

(b) $8.768 \times 10^{-3} - 4.381 \times 10^{-3}$

$$= (8.768 - 4.381) \times 10^{-3}$$

$$= 4.387 \times 10^{-3}$$

(2) If the numbers have different powers of 10, first convert them to decimal form and then add or subtract them.

Example 11

$$3.478 \times 10^3 + 4.826 \times 10^2 = 3478 + 482.6$$

$$= 3960.6$$

$$= 3.9606 \times 10^3$$

Multiplying and Dividing Numbers in Standard Form

By using the laws of indices, numbers expressed in standard form can easily be multiplied or divided.

Example 12

(a) $(8.463 \times 10^2) \times (1.768 \times 10^3)$

$$= (8.463 \times 1.768) \times (10^2 \times 10^3)$$

$$= 14.96 \times 10^5 \text{ (to 4 sig. fig.)}$$

$$= 1.496 \times 10^6$$

(b) $(3.258 \times 10^2) \div (7.197 \times 10^4)$

$$= (3.258 \div 7.197) \times (10^2 \div 10^4)$$

$$= 0.4527 \times 10^{-2} \text{ (to 4 sig. fig.)}$$

$$= 4.527 \times 10^{-3}$$

Exercise 14.9

State which of the following numbers is the larger and by how much:

1. 5.8×10^2 and 2.1×10^3

2. 9.4×10^3 and 3.95×10^4

3. 8.58×10^4 and 9.87×10^3

4. 2.1×10^{-2} and 5.4×10^{-3}

5. 8.73×10^{-3} and 1.26×10^{-1}

State the answers to the following in standard form:

6. $3.582 \times 10^3 + 8.907 \times 10^3$

7. $7.81 \times 10^{-2} + 1.88 \times 10^{-2} + 8.89 \times 10^{-2}$

8. $1.809 \times 10^2 - 1.705 \times 10^2$

9. $8.89 \times 10^{-3} - 8.85 \times 10^{-3}$

10. $1.78 \times 10^2 + 2.58 \times 10^3$

11. $5.987 \times 10^3 + 8.91 \times 10^2 + 7.635 \times 10^4$

12. $8.902 \times 10^{-2} - 7.652 \times 10^{-3}$

13. $1.832 \times 10^{-1} - 9.998 \times 10^{-2}$

14. $7.58 \times 10^2 \times 6 \times 10^3$
    (to 3 significant figures)

15. $6 \times 10^{-1} \times 2.58 \times 10^{-2}$
    (to 2 significant figures)

16. $5 \times 10^3 \times 2.11 \times 10^4 \times 4 \times 10^2$
    (to 3 significant figures)

17. $(2.68 \times 10^2) \div (8 \times 10^3)$
    (to 2 significant figures)

18. $(1.78 \times 10^{-1}) \div (3 \times 10^{-2})$
    (to 3 significant figures)

Miscellaneous Exercise 14

1. Rewrite the number 1 200 000 in standard form.

2. Find the value of $43 \times 3.16 \times 25$, giving your answer in standard form.

3. Write 5.01×10^{-3} as a decimal number not in standard form.

4. Write 0.000 893 in standard form.

5. If $p = 3 \times 10^3$ and $q = 2 \times 10^2$, work out the value of

   (a) $p + q$ (b) pq

   (c) $p - q$ (d) $p \div q$

   expressing the answer in standard form.

6. Find the value of
   (a) 3^{-2} (b) 3^0 (c) 10^{-1}

7. Which is greater, 8.73×10^2 or 1.2×10^3, and by how much?

8. Calculate $10^{20} \div 10^5$, stating the answer as a power of 10.

9. If $3^x = 81$, what is the value of x?

10. Find the value of $2^4 \times 2^2$.

11. Write 8×10^{-2} as a decimal number.

12. Find the value of n in each of the following:

    (a) $12^5 \times 12^n = 12^9$

    (b) $2^5 \div 2^4 = 2^n$

    (c) $37\,500 = 3.75 \times 10^n$

13. The radius, in centimetres, of the nucleus of an atom is

    0.000 000 000 000 3.

    Write this number in standard form.

14. Red light has a wavelength of 6.7×10^{-5} cm. Write this number in regular decimal notation.

15. Express 35.7×10^{-3} in the form $A \times 10^n$ where A is a number between 1 and 10 and n is an integer.

16. Evaluate (a) $(\tfrac{1}{2})^3$,   (b) $(3.5)^0$.

Mental Test 14

Try to answer the following questions without writing anything down except the answer.

1. Find the value of 2^3.

2. Write 3^{-1} as a fraction.

3. Remove the brackets from $(3a^2)^2$.

4. Simplify $a^2 \times a^4$.

5. Write $a^5 \div a^3$ as a single term.

6. Simplify $a^2 \times a^3 \times a^5$.

7. Remove the brackets from $\left(\dfrac{a^2}{b}\right)^3$.

8. Find the value of $2^2 \times 2^3$.

9. Write 8000 in standard form.

10. Write 0.003 in standard form.

Linear Equations, Formulae and Inequalities

Introduction

Fig. 15.1 shows a pair of scales which are in balance. That is, each scale pan contains exactly the same number of kilograms.

Therefore $x + 2 = 7$

This is an example of an equation.

Fig. 15.1

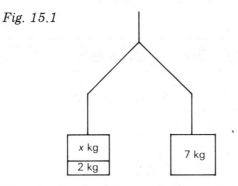

To solve the equation we have to find a value for x so that the scales remain in balance. We may keep the scales in balance by adding or subtracting the same amount from each pan. If we take 2 kg from the left-hand pan then we are left with x kg in this pan. To maintain balance we must take 2 kg from the right-hand pan, leaving 5 kg. That is

$$x + 2 - 2 = 7 - 2$$

$$x = 5$$

Let us take a second example as shown in Fig. 15.2. In the left-hand pan we have 4 packets of exactly the same weight, whilst in the right-hand pan there is 8 kg. The problem is to find the weight of each packet.

Fig. 15.2

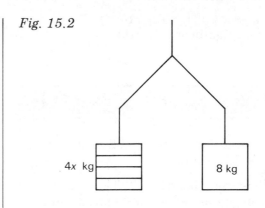

If we let each packet have a weight of x kilograms, there is $4x$ kg in the left-hand pan and hence we have the equation

$$4x = 8$$

We can maintain the balance of the scales by dividing or multiplying the quantities in each of the scales by the same amount. In our equation, if we divide each side by 4 we get

$$\frac{4x}{4} = \frac{8}{4}$$

Cancelling the fours on the left-hand side we get

$$x = 2$$

and therefore each packet has a weight of 2 kg.

From these two examples we can say:

(1) An equation expresses balance between two sets of quantities. That is, the equals sign means that one side of the equation is equal to the other side.

(2) If one side of the equation is changed in any way, the equation will still be true so long as the same operation is carried out on the other side. We can add the same amount to each side of the equation without destroying the balance. Also, we can multiply or divide each side of the equation by the same amount.

Example 1

(a) If $5x = 10$, by dividing both sides by 5,
$$x = 2$$

(b) If $x + 3 = 7$, by subtracting 3 from each side, $x = 4$

(c) If $x - 6 = 2$, by adding 6 to each side,
$$x = 8$$

(d) If $\dfrac{x}{2} = 3$, by multiplying each side by 2
$$x = 6$$

Exercise 15.1

Solve the following equations for x:

1. $3x = 6$
2. $7x = 14$
3. $8x = 16$
4. $2x = 12$
5. $x + 3 = 9$
6. $x + 3 = 7$
7. $x + 7 = 11$
8. $x + 5 = 8$
9. $x - 5 = 7$
10. $x - 3 = 6$
11. $x - 2 = 8$
12. $x - 4 = 8$
13. $\dfrac{x}{2} = 4$
14. $\dfrac{x}{3} = 5$
15. $\dfrac{x}{7} = 3$
16. $\dfrac{x}{4} = 6$
17. $\dfrac{x}{5} = 2$

Solving Harder Equations

Equations Requiring Addition, Subtraction and Division

Example 2

Solve for x the equation $x - 4 = 8$

Adding 4 to each side
$$x - 4 + 4 = 8 + 4$$
$$x = 12$$

The method shown in Example 2 is time-consuming. Provided you understand the method, you can use the quicker method which follows:

The operation of adding 4 to each side can be accomplished by taking 4 to the RHS (right-hand side) and changing its sign.

$$x - 4 = 8$$
$$x = 8 + 4$$
$$x = 12$$

The solution of an equation is that value of the unknown which, when substituted into the equation, makes the LHS equal to the RHS.

Check: LHS $= 12 - 4 = 8 =$ RHS

Since LHS = RHS the solution is correct.

Example 3

Solve $4x - 1 = 7$
$$4x = 7 + 1$$
$$4x =$$

Dividing both sid
$$x$$

Check: Substituting for x in the original equation:

$$LHS = 4 \times 2 - 1$$
$$= 8 - 1$$
$$= 7$$
$$= RHS$$

Since the LHS equals the RHS, the solution is correct. In many cases the given equation contains terms in the unknown and/or constant terms on both sides of the equation. We then take all the terms containing the unknown quantity to one side of the equation and all the other terms to the other side.

Example 4

Solve $5x + 8 = 3x - 2$

$$5x - 3x = -2 - 8$$
$$2x = -10$$
$$x = -5$$

Check:
$$LHS = 5 \times (-5) + 8$$
$$= -25 + 8$$
$$= -17$$
$$RHS = 3 \times (-5) - 2$$
$$= -15 - 2$$
$$= -17$$

Since the LHS equals the RHS the solution is correct.

Example 5

Solve $9 - 4x = 3x - 12$

Gather the terms containing x on the RHS to give a positive term in x. (Taking the terms in x to the LHS would give a negative term in x.)

$$9 - 4x = 3x - 12$$
$$9 + 12 = 3x + 4x$$
$$21 = 7x$$
$$x = 3$$

When an equation contains brackets we remove these first and then solve the equation by the methods previously shown.

Example 6

Solve $3(4x - 5) = 5 - 2(x + 3)$

$$12x - 15 = 5 - 2x - 6$$
$$12x + 2x = 5 + 15 - 6$$
$$14x = 14$$
$$x = 1$$

Exercise 15.2

Solve the following equations:

1. $x - 3 = 5$
2. $2x - 1 = 7$
3. $5x - 8 = 2$
4. $3x + 8 = 5$
5. $2x + 7 = 3$
6. $3x - 4 = x + 6$
7. $3x - 8 = 5x - 20$
8. $23 - x = x + 11$
9. $3(x + 1) = 9$
10. $2(x - 3) - (x - 2) = 5$
11. $5(m + 2) - 3(m - 5) = 29$
12. $4(3p + 2) + 6 = 3(2p - 5) + 5$

Equations Requiring Multiplication and Division

Example 7

Solve $\dfrac{3x}{2} = 12$

Multiplying both sides by 2 gives
$$3x = 24$$

Dividing both sides by 3 gives
$$x = 8$$

When an equation contains fractions, the first step is to get rid of the denominators by multiplying each term in the equation by the LCM of the denominators. The resulting equation is then solved by the methods previously shown.

Example 8

Solve $3p - \dfrac{p}{3} = 8$

Multiplying each term by 3 gives

$$3 \times 3p - \frac{p}{3} \times \frac{3}{1} = 8 \times 3$$

$$9p - p = 24$$

$$8p = 24$$

$$p = \frac{24}{8}$$

$$p = 3$$

Not all equations give a solution which is an integer.

Example 9

Solve $\dfrac{c}{5} = \dfrac{3}{4}$

The LCM of 5 and 4 is 20.

Multiplying each term by 20 gives

$$4c = 15$$

Dividing both sides by 4 gives

$$c = \frac{15}{4}$$

Exercise 15.3

Solve each of the following equations:

1. $\dfrac{x}{5} = 3$

2. $\dfrac{x}{3} = 7$

3. $2x = 14$

4. $\dfrac{2x}{3} = 8$

5. $\dfrac{4x}{5} = 12$

6. $\dfrac{x}{3} = \dfrac{2}{5}$

7. $\dfrac{m}{3} + \dfrac{m}{5} = 2$

8. $\dfrac{x}{5} - \dfrac{x}{3} = 2$

9. $\dfrac{x}{3} + \dfrac{x}{4} + \dfrac{x}{5} = \dfrac{5}{6}$

10. $3x + \dfrac{3}{4} = 2 + \dfrac{2x}{3}$

11. $\dfrac{3x}{2} + 1 = \dfrac{3}{4} + 2x$

12. $\dfrac{x+3}{4} = 2 + \dfrac{(x-3)}{5}$

Making Expressions

Frequently we have to translate information in words into symbols, so making up algebraic expressions.

Example 10

(a) Nisha goes shopping for her mother. She is told to buy 3 pencils at x pence each and 8 pens at y pence each. Make an algebraic expression which will give the total cost of these purchases.

> The cost of 3 pencils at x pence each is $3 \times x = 3x$ pence. The cost of 8 pencils at y pence each is $8 \times y = 8y$ pence. Total cost of these purchases $= 3x + 8y$ pence.

(b) If x apples can be bought for P pence, write down an expression for the cost of y apples.

> x apples cost P pence.
>
> 1 apple costs $\dfrac{P}{x}$ pence.
>
> y apples cost $y \times \dfrac{P}{x} = \dfrac{Py}{x}$.

(c) Two numbers, R and S, are connected by a formula. To find the number S, we start with R, multiply by 7 and then add on 3. Write down a formula for S in terms of R.

$$S = 7 \times R + 3 = 7R + 3$$

Exercise 15.4

1. A box contains x articles each of which weigh 3 kg. They are placed in a box whose weight is 5 kg. Find an expression which gives the total weight of the box and the x articles.

2. r oranges cost P pence. Write down an expression for the cost of s oranges.

3. Ahmed is x years old now. How old was he 5 years ago?

4. Alfred goes shopping and buys 5 pencils at a pence each and 7 pens at b pence each. Find the total cost of Alfred's purchases.

5. Tom Jones works x hours on each weekday except Saturday, when he works y hours. He also works z hours on Sunday.
   (a) Make up an expression for the number of hours worked by Tom Jones in that week.
   (b) If he is paid £q per hour, how much did he earn that week?

6. Philip has £p and Richard has £q. If Philip gives Richard £n, how much money will each have?

7. Halina bought a music centre costing £Y and n records costing X pence each. Find an expression which gives the amount, in pounds, that Halina spent.

8. In one cricket innings, Courtney scored a sixes, b fours and c singles. How many runs did Courtney score?

Solving Word Problems

To construct a simple equation, first represent the quantity to be found by a symbol which is usually x. Then construct an equation which conforms to the data of the problem. Solving this equation gives the value of the quantity to be found.

Example 11

(a) Linda said 'I think of a number, multiply it by five, add on four and divide by 2 and the answer I get is forty-two.' Write this number puzzle as an equation, letting x be the unknown number and then solve the equation.

$$\frac{5x + 4}{2} = 42$$
$$5x + 4 = 84$$
$$5x = 80$$
$$x = 16$$

So the number Linda thought of was 16.

(b) Find three consecutive whole numbers whose sum is 54.

Let the three numbers be x, $x + 1$ and $x + 2$. Then

$$x + x + 1 + x + 2 = 54$$
$$3x + 3 = 54$$
$$x = 17$$

So the three numbers are 17, 18 and 19.

(c) 36 books are bought by a school. Some cost £8 and the remainder cost £10. If the total amount spent is £312, how many of each kind are bought?

Let x be the number of books bought at £8. Then $(36 - x)$ is the number bought at £10.

Cost of x books
at £8 $\qquad = £8x$

Cost of $(36-x)$
books at £10 $\qquad = £10(36-x)$

$\qquad\qquad\qquad = £(360-10x)$

Total cost $= £(8x+360-10x)$

$\qquad\qquad\quad = £(360-2x)$

We are told that the total amount spent is £312, so

$$360-2x = 312$$
$$2x = 48$$
$$x = 24$$

So the number bought at £8 is 24 and the number bought at £10 is 12.

Exercise 15.5

1. Find two consecutive whole numbers whose sum is 37.

2. Roz buys x oranges at 20 pence each. She also buys grapefruit at 30 pence each. She buys 24 fruits altogether which cost her £5.70. Construct an equation from this information and use it to find the number of oranges that Roz bought.

3. John is 3 years older than Emma and the sum of their ages is 33 years. How old is Emma?

4. David said 'I thought of a number, multiplied it by 7 and then added 15. My answer was 78'.

   (a) Write down David's statement in symbols using x as the starting number.

   (b) What is the number that David thought of?

5. (a) Write down the next row of the following number pattern:

1,	2,	4
2,	4,	8
4,	8,	16

   (b) One row of the pattern is

x,	y,	z

   (i) Write down an equation for y in terms of x.

   (ii) If $x+y+z = 224$, what is x?

6. The numbers 20, 21 and 22 are three consecutive whole numbers. Their sum is 63 which is divisible by 3.

   (a) Write down any other three consecutive numbers and show that their sum is divisible by 3.

   (b) Let n stand for the first of three consecutive whole numbers. Write down the other two numbers in terms of n.

   (c) Find the sum of the three numbers in terms of n and factorise the answer.

   (d) The middle term of three consecutive whole numbers is 8. What is the sum of the three numbers?

   (e) If the sum of the three numbers is 39, construct an equation from which n can be found.

   (f) What are the three numbers?

7. The numbers from 1 to 100 are arranged in rows of 5 as shown on page 136. From the array we may choose a 'box' of 4 numbers in a square. Two of these are shown:

   The 8 box which has a total of $8+9+13+14 = 44$.

The 46 box which has a total of
46 + 47 + 51 + 52 = 196.

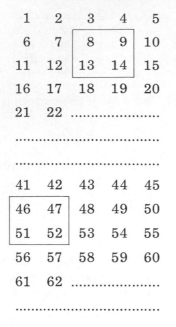

   1   2   3   4   5

   6   7   8   9 10

 11 12 13 14 15

 16 17 18 19 20

 21 22 ....

....

....

 41 42 43 44 45

 46 47 48 49 50

 51 52 53 54 55

 56 57 58 59 60

 61 62 ....

....

....

(a) What is the total of the '76 box'?

(b) (i) Write down in terms of n the
four numbers in the 'n box'.
(ii) Write down the total of the
'n box' in the form $t = \ldots$

(c) Which box has a total of 320?

8. Lena thinks of a number, subtracts 8
from it and multiplies the result by 3.
She says the answer is 21. Taking n as
the starting number, derive an equation
in terms of n and hence find the number Lena thought of.

Formulae

A formula is an equation which shows the
relationship between two or more
quantities.

The statement $E = IR$ is a formula for E
in terms of I and R. The value of E is found
by substituting the given values of I and R.

Example 12

If $v = u + at$, find the value of v when
$u = 20$, $a = 3$ and $t = 5$.

Substituting the given values of
u, a and t we have

$$v = 20 + (3 \times 5)$$
$$= 20 + 15$$
$$= 35$$

Exercise 15.6

1. If $V = Ah$, find the value of V when
$A = 8$ and $h = 4$.

2. The formula $K = Wa + b$ is used in
engineering technology. Find K when
$W = 25$, $a = 3$ and $b = 5$.

3. $S = 90(n - 4)$ is a formula used in
geometry. Find S when $n = 8$.

4. If $P = RT/V$, find the value of P
when $R = 56$, $T = 18$, and $V = 7$.

5. A formula used in physics is
$E = mv^2/2g$. Find the value of E
when $m = 220$, $v = 8$ and $g = 10$.

6. If $A = \frac{1}{2}BH$, find A when $B = 6$
and $H = 7$.

7. If $y = 3t/c$, find y when $t = 12$
and $c = 6$.

Formulae and Equations

Suppose that we are given the formula
$M = P/Q$ and we have to find the value of
Q, given values for M and P. We can do this
by substituting the given values and solving
the resulting equation for Q.

Example 13

Find T from the formula $D = \dfrac{T + 2}{P}$, given

$D = 5$ and $P = 3$.

Substituting the given values we have

$$5 = \frac{T + 2}{3}$$

Multiplying both sides by 3 gives

$$15 = T + 2$$
$$T = 15 - 2$$
$$T = 13$$

Exercise 15.7

1. Find n from the formula $P = 1/n$ when $P = 2$.

2. Find R from the formula $E = IR$ when $E = 20$ and $I = 4$.

3. Find B from the formula $A = BH$ when $A = 12$ and $H = 4$.

4. Find c from the formula $H = abc$ when $H = 40$, $a = 2$ and $b = 5$.

5. Find P from the formula $I = PRT/100$ when $I = 20$, $R = 5$ and $T = 4$.

6. Find D from the formula $C = \pi D$ when $\pi = 3.142$ and $C = 27$.

7. Find r from the formula $A = \pi rl$ when $\pi = 3.142$, $A = 96$ and $l = 12$.

8. Find W from the formula $K = Wa + b$ when $K = 30$, $a = 4$ and $b = 6$.

Transposition of Formulae

Consider again the formula $M = P/Q$. M is called the **subject** of the formula. We may be given several corresponding values of M and Q, and we want to find the corresponding values of P. We could, of course, find these by the method shown in Example 11 but a lot of time and effort would be spent in solving the resulting equations.

Much of this time and effort would be saved if we could express the formula with P as the subject, because we would then only

have to substitute the given values of M and Q in the rearranged formula.

The process of rearranging a formula so that one of the other symbols becomes the subject is called **transposing the formula.** The rules used in transposing formulae are exactly the same as those used in solving linear equations.

Example 14

(a) Transpose the formula $F = ma$ to make a the subject. Dividing both sides of the formula by m gives

$$\frac{F}{m} = \frac{\cancel{m}a}{\cancel{m}}$$

$$\frac{F}{m} = a$$

It is usual to write the subject on the LHS of the formula so we write

$$a = \frac{F}{m}$$

(b) Transpose $x = 3y + 5$ for y.

Subtract 5 from each side, giving

$$x - 5 = 3y$$

Divide both sides by 3 giving

$$\frac{x - 5}{3} = y$$

$$y = \frac{x - 5}{3}$$

Exercise 15.8

Transpose the following formulae:

1. $PV = c$ for V 2. $I = PRT$ for R

3. $E = mgh$ for h 4. $x = a/y$ for a

5. $I = E/R$ for R 6. $P = RT/V$ for T

7. $S = ts/T$ for t 8. $v = 3 + p$ for p

9. $p = 3 - q$ for q 10. $v = u + at$ for a

11. $n = p + 3r$ for r

12. $a = b - cx$ for x

13. $y = mx + c$ for m

14. $T = (D - d)/L$ for D

15. $P = S(C - F)/L$ for L

Inequalities

An **inequality** is a statement that one quantity is greater (or less) than a second quantity.

If we want to say that n is greater than 5 we write

$$n > 5$$

If we want to say that x is less than 7 we write

$$x < 7$$

Thus the symbol $>$ means 'greater than' and the symbol $<$ means 'less than'. Note carefully that the arrow always points to the smaller quantity.

A statement using these signs is called an **inequality**.

Exercise 15.9

Using the symbols $>$, $<$ and $=$, fill in the gap between the following pairs of quantities:

1. 2   5   2. $\frac{1}{2}$ $\frac{1}{3}$

3. 10 mm   1 cm   4. £1   80p

5. 9   1   6. $\frac{2}{3}$ $\frac{4}{9}$

7. -3 -7 8. 7   9

9. x^2 $x^2 - 3$ 10. 2   4

Two more symbols are used in dealing with inequalities. They are

  $\geqslant$ which means 'equal to or greater than'

and

  $\leqslant$ which means 'equal to or less than'

$x \geqslant 2$ means that x is equal to or greater than 2

$n \leqslant 5$ means that n is equal to or less than 5

Solution of Simple Inequalities

The solutions of inequalities may be shown by means of a number line. If $x < 3$, all the possible values of x are shown in Fig. 15.3. The empty circle at the end of the arrowed line shows that $x = 3$ is not included.

Fig. 15.3

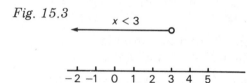

Figure 15.4 shows the solutions for $x \geqslant -4$. Since $x = -4$ is included, the arrowed line ends in a solid circle.

Fig. 15.4

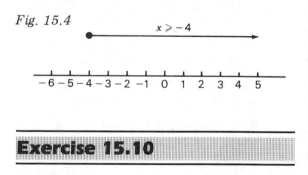

Exercise 15.10

Use number lines to show the solutions for the following inequalities:

1. $x \leqslant 5$ 2. $x > 7$ 3. $x \geqslant 3$

4. $x > -2$ 5. $x < -6$

Example 15

If x has to be a whole number, find the solution for $x \leqslant 3$ and $x > 1$.

  Since x has to be a whole number, the inequalities are represented by distinct points (Fig. 15.5) not continuous lines.

We see that the solution must equal 2 or 3, because the points representing the independent solutions for each inequality overlap. We say that the solution set is {2, 3}.

Fig. 15.5

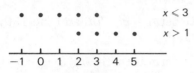

Inequalities may be combined into one statement. For instance, if we have $4 < x$ and $x < 7$, the two statements may be combined to give $4 < x < 7$, which means that x must lie between 4 and 7.

Example 16

If x is an integer, find the solution sets for
(a) $-3 < x < 2$, (b) $-3 \leqslant x \leqslant 2$.

(a) Drawing the number line (Fig. 15.6) we see that the value of x must be the integers from -2 to 1. That is, the solution set is $\{-2, -1, 0, 1\}$.

Fig. 15.6

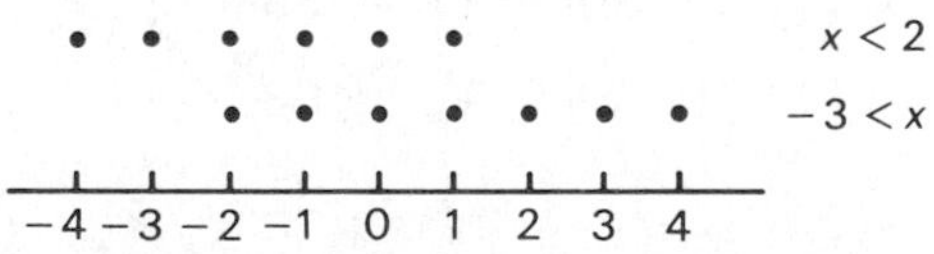

Fig. 15.6 shows a number line from -4 to 4 with $x < 2$ and $-3 < x$.

(b) Drawing the number line (Fig. 15.7) we see that the value of x must be the integers from -3 to 2. Hence the solution set is $\{-3, -2, -1, 0, 1, 2\}$.

Fig. 15.7

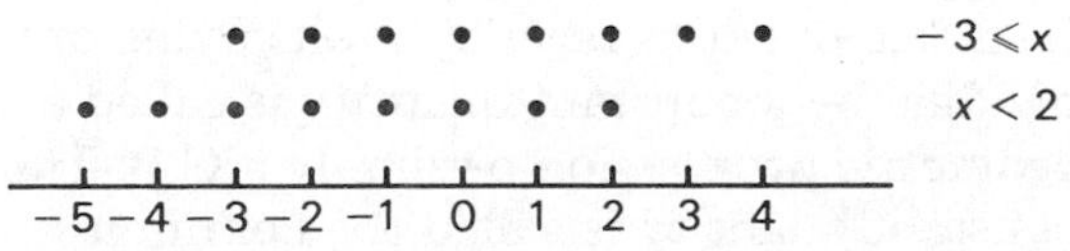

Fig. 15.7 shows a number line from -5 to 4 with $-3 \leqslant x$ and $x < 2$.

Note that when integers are being discussed we use distinct points instead of continuous lines.

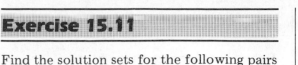

Exercise 15.11

Find the solution sets for the following pairs of inequalities, where x is an integer:

1. $4 < x < 9$ 2. $4 \leqslant x \leqslant 9$

3. $-2 \leqslant x < 3$ 4. $5 \leqslant x \leqslant 8$

5. $-4 < x < 5$

Inequations

$$5x - 3 = 12 \text{ is an equation}$$
$$5x - 3 > 12 \text{ is an inequation}$$

The rules for solving equations also apply to inequations, namely:

(1) The same term may be added to or subtracted from both sides of an inequation or an equation.

(2) Multiplying or dividing both sides of an equation or inequation by the same **positive** number leaves the equation or inequation unaltered.

Example 17

Solve the inequation $5x + 17 > 2x + 29$.

Bringing the terms in x to the LHS and the other terms to the RHS gives

$$5x - 2x > 29 - 17$$
$$3x > 12$$
$$x > 4$$

If both sides of an inequation are multiplied by a **negative number** the inequality sign must be reversed. For instance if

$$-x > 4$$

the inequality is easier to understand if we make x positive by multiplying both sides of the equation by -1 giving

$$x < -4$$

Exercise 15.12

Solve the following inequations:

1. $3x \geqslant 12$ 2. $x - 4 < 1$

3. $5 - x > 7$ 4. $2x - 7 \leqslant 9$

5. $5x - 3 < 2x + 15$

6. $5(x - 2) \geqslant 15$

7. $5(x - 2) - 3(x - 5) \leqslant 29$

8. $4(x - 5) > 7 - 5(3 - 2x)$

Number Sequences

A set of numbers which are connected by some definite law are called a sequence or a progression. The following sets of numbers are examples of sequences:

$$1, 3, 5, 7, \ldots$$

(Each term is obtained by adding 2 to the previous term.)

$$3, 9, 27, \ldots$$

(Each term is obtained by multiplying the previous term by 3.)

Sequence in Arithmetic Progression

A sequence in which each term is obtained by adding or subtracting a constant amount is called an **arithmetic progression** (often abbreviated to A.P.). Thus in the sequence 1, 4, 7, 10, ... the difference between each term and the preceding one is 3. We call this the **common difference.** Thus for the sequence 5, 10, 15, 20, ... the common difference is 5 and hence this is a sequence in A.P.

Suppose the first term of a sequence in A.P. is a and the common difference is d then the sequence can be written as

$$a, \ a + d, \ a + 2d, \ a + 3d, \ldots$$

The first term is a.
The second term is $a + d$.
The third term is $a + 2d$.
The fourth term is $a + 3d$.

We notice that the coefficient of d is always one less than the number of the term. Thus:

The 8th term $= a + 7d$ and the 19th term $= a + 18d$.

The nth term is therefore $a + (n - 1)d$.

Example 18

(a) Find the 7th and 16th terms of the sequence 3, 8, 13, ...

Here we have $a = 3$ and the common difference is $d = 5$.

7th term $= a + 6d = 3 + 6 \times 5 = 33$.

16th term $= a + 15d = 3 + 15 \times 5 = 78$.

(b) Find the number of the term which is 65 in the series 2, 5, 8, ...

Here $a = 2$ and $d = 3$.

Let the nth term be 65. Then

$$2 + 3(n - 1) = 65$$
$$3(n - 1) = 63$$
$$n - 1 = 21$$
$$n = 22$$

So the 22nd term in the sequence equals 65.

Sequence in Geometric Progression

A sequence in which each term is obtained from the previous term by multiplying or dividing by a constant quantity is called a **geometric progression** or simply a G.P. The constant quantity is called the **common ratio.** In the sequence 2, 4, 8, 16, ... each successive term is formed by multiplying the previous term by 2. The sequence is therefore in G.P. with a common ratio of 2.

Suppose that the first term of a sequence in G.P. is a and the common ratio is r. The sequence can be represented by a, ar, ar^2, ar^3, ...

The first term is a.
The second term is ar.
The third term is ar^2.
The fourth term is ar^3.
The nth term is ar^{n-1}.

We notice that the index of r is always one less than the number of the term, so that the ninth term is ar^8.

Example 19

Find the 7th term of the sequence 2, 6, 18, ...

> The first term is 2 and the common ratio is 3, i.e. $a = 2$ and $r = 3$.
>
> 7th term $= ar^6 = 2 \times 3^6$
>
> $\qquad = 1458$

Patterns

Number sequences are often used to form patterns of various kinds. The most frequently used number sequences used are:

(1) **Square numbers** whose sequence is 1, 4, 9, 16, ..., i.e. 1^2, 2^2, 3^2, 4^2. If the first number in the sequence is 1 then the nth number in the sequence is n^2. So the 20th term in the sequence is $20^2 = 400$.

(2) **Rectangular numbers** which can be represented as a pattern of dots in the form of a rectangle:

$6 = 3 \times 2$ $24 = 6 \times 4$ $24 = 8 \times 3$

Note that 1 is not regarded as being a rectangular number. The sequence of rectangular numbers is 4, 6, 8, 9, 10, ...

(3) **Triangular numbers** which can be represented as a pattern of dots in the form of a triangle:

3   6   10

The sequence of triangular numbers is 1, 3, 6, 10, 15, ... Note that $3 = 2 + 1$, $6 = 3 + 2 + 1$, $10 = 4 + 3 + 2 + 1$, etc.

Example 20

Fig. 15.8 shows patterns of models made by using matches.

(a) Write down the rule that could be used to find a model containing any number of patterns.

(b) Find the number of matches needed to make a 9 pattern model.

(c) Use your rule to find the number model which uses 116 matches.

> The number of matches used in the first model is 6. The number used in the second model is 11 and the number used in the third model is 16. Thus the number of matches used forms the sequence 6, 11, 16, ... which is an A.P. with $a = 6$ and $d = 5$.

Fig. 15.8

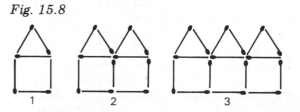

(a) The rule is $a + (n - 1)d$ where n is the number of the model.

(b) For a 9 model $n = 9$ and

> Number of matches used $= 6 + 5(9 - 1)$
>
> $\qquad\qquad\qquad = 6 + 40 = 46$

(c) We have to solve the following equation to find n:

$$5 + 4(n-1) = 113$$
$$4(n-1) = 108$$
$$n - 1 = 27$$
$$n = 28$$

So the 28th model will use 113 matches.

Exercise 15.13

1. Find the 15th term of the sequence 3, 6, 9, ...

2. Find the 8th term of the sequence 8, 13, 18, ...

3. Find the 8th term of the sequence 2, 4, 8, ...

4. Find the 9th term of the sequence 1, 3, 9, ...

5. Below is a number pattern:

	42	44	46	48	50	52	
	40	14	16	18	20	27	
	38	12	2	4	22	56	
78	36	10	8	6	24	58	
76	34	32	30	28	26	60	
74	72	70	68	66	64	62	

(a) Write down, in numerical order, the numbers with squares around them and then find the next two numbers in the sequence.

(b) Copy the above table and write these two numbers in their correct positions on the grid.

6. Jackie forms the patterns shown in Fig. 15.9 by using sugar lumps.

(a) Write down a rule whereby Jackie could find the number of sugar lumps in a pattern of sugar lumps.

(b) Use this rule to find the number of sugar lumps in the 9th pattern.

Fig. 15.9

7. Philip made the equilateral triangle arrangements shown in Fig. 15.10 by using matchsticks.

(a) Draw the fourth and fifth arrangements.

(b) Find the number of matchsticks needed to make the 6th equilateral triangle arrangement.

Fig. 15.10

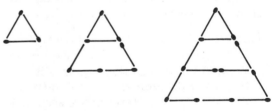

8. Matthew uses matchsticks to make the grids shown in Fig. 15.11.

(a) Copy and complete the following table:

Grid number	Total number of matchsticks
1	2 × 2 = 4
2	4 × 3 = 12
3	6 × 4 = 24
4	
5	
6	
7	14 × 8 = 112

(b) Find a formula in terms of n, the number of matchsticks needed to form grid n (Gn).

(c) Use your formula to find the number of matchsticks needed to form G19.

(d) Matthew has 364 matches. What number grid could he form with them?

Fig. 15.11

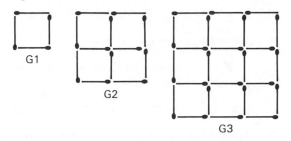

G1

G2

G3

9. Fig. 15.12 shows three arrangements for a display of tiles.

(a) If the sequence is to be maintained obtain a formula for the number of white tiles in the nth arrangement.

(b) Find the number of white tiles in the tenth arrangement.

Fig. 15.12

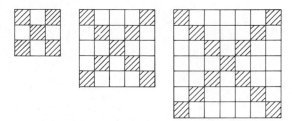

10. Stan is making a pattern of tiles by surrounding black tiles by white tiles. The first three patterns are shown in Fig. 15.13.

(a) Write down a formula that Stan could use to find the number of white tiles needed to surround n black tiles.

(b) Copy and complete the table below to show the number of tiles needed for up to 9 black tiles.

Number of black tiles	1	2	3	4	5	6	7	8	9
Number of white tiles	8	10							

(c) If Stan has 68 white tiles, what is the greatest number of black tiles that can be surrounded?

Fig. 15.13

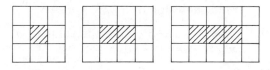

Miscellaneous Exercise 15

1. If $v = u + at$, find the value of v when $u = 40$, $t = 10$ and $a = 3$.

2. The area of a trapezium is found by using the formula $A = \frac{1}{2}(a + b)h$. Find the value of A when $a = 3$, $b = 5$ and $h = 7$.

3. Find the value of C in the formula $C = 2\pi r$ when $\pi = 3.14$ and $r = 10$.

4. If $A = \pi r^2$, find the value of A when $\pi = 3.14$ and $r = 5$.

5. If $V = \frac{1}{3}\pi r^2 h$ find V when $\pi = 3.14$, $r = 10$ and $h = 6$.

6. If x is a positive integer such that $10 < x < 100$, state the least positive value of x.

7. Use one of the symbols $> = <$ in each of the following to make true statements:

    (a) $5^2 - 3^2 \quad 4^2$ **(b)** $9^2 - 3^2 \quad 3^2$

    (c) $\frac{5}{9} - \frac{1}{3} \quad \frac{2}{3}$ **(d)** $\sqrt{14.4} \quad 1.2$

    (e) $0.5^2 \quad 0.5^3$

8. Solve for x, $2(x - 1) - 4(x + 1) = 0$.

9. Solve the equations

    (a) $3x - 8 = 7$

    (b) $-(x - 5) = 2$

10. Solve the equation $2(4x - 3) = 18$.

11. Solve the equations
    (a) $2x + 4 = 5$
    (b) $(8/x) + 1 = 3$

12. Factorise
    (a) $p^2 - 5p$
    (b) $2x^2x + yx^2$

13. Consider the formula $S = 180(n - 2)$. Find the value of S when $n = 8$.

14. If $P = 2(a + b)$, find the value of a when $P = 10$ and $b = 3$.

15. Transpose the formula $v = u + at$ to make t the subject.

16. Use the formula $P = q + (n - 1)p$ to find the value of P when $q = 6$, $n = 5$ and $p = 4$.

17. Transpose the formula $T = 20 + \frac{1}{3}N$ to make N the subject.

Mental Test 15

Try to answer the following questions without writing anything down except the answer.

1. If $3x = 9$, what is the value of x?

2. If $\dfrac{x}{2} = 4$, what is the value of x?

3. If $y + 5 = 8$, what is the value of y?

4. Solve the equation $x - 4 = 2$.

5. If $F = rs$, work out the value of F when $r = 2$ and $s = 5$.

6. Transpose the formula $MN = p$ for N.

7. Transpose the formula $x = a/y$ to make a the subject.

8. Solve the equation $8x - 5x = 12$.

Simultaneous and Quadratic Equations

Look at the equation

$$3x + 6 = 12$$

This equation is easy to solve and the solution is

$$x = 2$$

Now look at the equation

$$3x + 2y = 12$$

We cannot solve this equation because there are two unknown quantities, x and y. We need another equation with x and y in which the values of x and y are the same as in the first equation. Thus we may have

$$3x + 2y = 12 \qquad (1)$$
$$5x - 2y = 4 \qquad (2)$$

Because x and y have the same value in both equations, they are called **simultaneous equations**.

Elimination Method in Solving Simultaneous Equations

Since we are familiar with solving simple equations in one variable, simultaneous equations may be solved by getting rid of one of the unknown quantities, that is, by eliminating one of the unknown quantities, thereby forming an equation in one variable.

Example 1

Solve the equations

$$3x + 2y = 12 \qquad (1)$$
$$5x - 2y = 4 \qquad (2)$$

Since we have the same coefficient of y in both equations, we can get rid of the y term by adding the two equations:

$$(3x + 2y) + (5x - 2y) = 12 + 4$$
$$3x + 2y + 5x - 2y = 16$$
$$8x = 16$$
$$x = 2$$

To find the value of y we substitute for x in either of the original equations. Thus, substituting for x in equation (1) gives

$$(3 \times 2) + 2y = 12$$
$$6 + 2y = 12$$
$$2y = 6$$
$$y = 3$$

To check the values of x and y, substitute them in equation (2). There is no point in using equation (1) because this was used to find the value of y.

$$\text{LHS} = (5 \times 2) - (2 \times 3)$$
$$= 10 - 6$$
$$= 4$$
$$= \text{RHS}$$

Since the LHS and the RHS are the same, the solutions are correct.

Example 2

Solve the equations

$$4x + 5y = 30 \qquad (1)$$

$$4x + 3y = 26 \qquad (2)$$

Because we have the same coefficient of x in both equations, we can get rid of x by subtracting equation (2) from equation (1):

$$(4x + 5y) - (4x + 3y) = 30 - 26$$

$$4x + 5y - 4x - 3y = 4$$

$$2y = 4$$

$$y = 2$$

To find the value of x, substitute for y in equation (1):

$$4x + (5 \times 2) = 30$$

$$4x + 10 = 30$$

$$4x = 20$$

$$x = 5$$

Thus the solutions are $x = 5$ and $y = 2$.

Checking the solutions in equation (2):

$$LHS = (4 \times 5) + (3 \times 2)$$

$$= 20 + 6$$

$$= 26 = RHS$$

Because the LHS equals the RHS, the solutions are correct.

Exercise 16.1

Solve the following pairs of equations for x and y and check their solutions:

1. $x + y = 8$
   $x - y = 2$

2. $x + 3y = 10$
   $2x - 3y = 2$

3. $3x + 4y = 25$
   $3x + 2y = 17$

4. $5x - 2y = 1$
   $4x - 2y = 0$

5. $6x + 3y = 15$
   $6x - 5y = 7$

6. $4x - 3y = 14$
   $2x - 3y = 4$

7. $x + 2y = 10$
   $5x + 2y = 26$

8. $3x + 2y = 20$
   $3x + 4y = 34$

Example 3

Solve the equations

$$3x + 5y = 26 \qquad (1)$$

$$2x + 3y = 16 \qquad (2)$$

If we multiply equation (1) by 2 and equation (2) by 3, we shall have the same coefficient of x in each equation. We can then eliminate x by subtracting equation (2) from equation (1).

Alternatively, if we multiply equation (1) by 3 and equation (2) by 5, we shall have the same coefficient of y in each equation. We can then eliminate y by subtracting equation (2) from equation (1).

You have to decide which of the variables x or y you are going to eliminate. We will eliminate x.

Multiplying equation (1) by 2 and equation (2) by 3:

$$6x + 10y = 52 \qquad (3)$$

$$6x + 9y = 48 \qquad (4)$$

Subtracting equation (4) from equation (3):

$$(6x + 10y) - (6x + 9y) = 52 - 48$$

$$6x + 10y - 6x - 9y = 4$$

$$y = 4$$

Substituting for y in equation (1):

$$3x + (5 \times 4) = 26$$

$$3x + 20 = 26$$

$$3x = 6$$

$$x = 2$$

The solutions are $x = 2$ and $y = 4$.

Checking in equation (2):

$$LHS = (2 \times 2) + (3 \times 4)$$

$$= 4 + 12$$

$$= 16$$

$$= RHS$$

Since the LHS equals the RHS, the solutions are correct.

Example 4

Solve the equations

$$5x - 4y = 13 \qquad (1)$$
$$7x - 2y = 29 \qquad (2)$$

This time we can eliminate y by multiplying equation (2) by 2:

$$14x - 4y = 58 \qquad (3)$$

Subtracting equation (1) from equation (3):

$$(14x - 4y) - (5x - 4y) = 58 - 13$$
$$14x - 4y - 5x + 4y = 45$$
$$9x = 45$$
$$x = 5$$

Substituting for x in equation (1):

$$(5 \times 5) - 4y = 13$$
$$25 - 4y = 13$$
$$-4y = -12$$
$$y = 3$$

The solutions are $x = 5$ and $y = 3$.

Checking in equation (2):

$$\text{LHS} = (7 \times 5) - (2 \times 3)$$
$$= 35 - 6$$
$$= 29$$
$$= \text{RHS}$$

Hence the solutions are correct.

Exercise 16.2

Solve the following pairs of equations for x and y and check their solutions:

1. $x + 3y = 11$
   $2x + 5y = 19$

2. $3x + 2y = 21$
   $2x - y = 7$

3. $5x - 4y = 7$
   $7x - 2y = 17$

4. $3x + 5y = 31$
   $2x + 3y = 19$

5. $5x + 2y = 9$
   $3x + 5y = 13$

6. $2x + 5y = 27$
   $3x + 2y = 13$

7. $3x + 2y = 13$
   $2x + 3y = 7$

8. $7x - 4y = 8$
   $3x - 5y = -13$

The Product of Two Binomial Expressions

A binomial expression consists of *two terms*. Thus $3x + 5$, $a + b$, $2x + 3y$ and $4p - q$ are all binomial expressions.

To find the product of $(a + b)(c + d)$ consider the diagram (Fig. 16.1).

Fig. 16.1

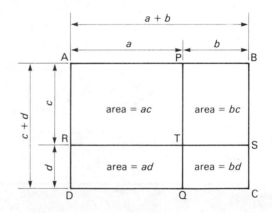

In Fig. 16.1 the rectangular area ABCD is made up as follows:

$$\text{ABCD} = \text{APTR} + \text{RTQD} + \text{PBST} + \text{TSQC}$$

i.e. $(a + b)(c + d) = ac + ad + bc + bd$

It will be noticed that the expression on the right-hand side is obtained by multiplying each term in the one bracket by each term in the *other* bracket. The process is illustrated below where each pair of terms connected by a line are multiplied together.

$$(a + b)(c + d) = ac + ad + bc + bd$$

Example 5

(a) $(x + 3)(x + 2) = x \times x + 3 \times x + x \times 2$
$$+ 3 \times 2$$
$$= x^2 + 5x + 6$$

(b) $(x - 4)(x + 3) = x \times x - 4 \times x + x \times 3$
$$- 4 \times 3$$
$$= x^2 - x - 12$$

(c) $(x - 2)(x - 5) = x \times x - 2 \times x - x \times 5$
$$- 2 \times (-5)$$
$$= x^2 - 7x + 10$$

(d) $(3x + 2y)^2 = (3x + 2y)(3x + 2y)$
$$= 3x \times 3x + 2y \times 3x$$
$$+ 3x \times 2y + 2y \times 2y$$
$$= 9x^2 + 6xy + 6xy + 4y^2$$
$$= 9x^2 + 12xy + 4y^2$$

(e) $(5x - 2y)(5x + 2y) = 5x \times 5x$
$$- 2y \times 5x$$
$$+ 5x \times 2y$$
$$+ (-2y) \times 2y$$
$$= 25x^2 - 10xy$$
$$+ 10xy - 4y^2$$
$$= 25x^2 - 4y^2$$

Exercise 16.3

Remove the brackets from each of the following:

1. $(x + 1)(x + 3)$ 2. $(x + 2)(x + 4)$

3. $(x + 5)(x + 1)$ 4. $(x + 3)(x + 6)$

5. $(x - 1)(x - 2)$ 6. $(x - 3)(x - 4)$

7. $(3x - 2)(x - 5)$ 8. $(5x - 4)(2x - 3)$

9. $(3x + 1)(2x - 3)$ 10. $(x - 3)^2$

11. $(5x + 3y)^2$ 12. $(2x - 5)(2x + 5)$

Solving Polynomial Equations

Many equations contain powers of the unknown quantity. For instance

$$5x^2 = 12, \quad x^3 = 47 \quad \text{and} \quad 3x^2 - x = 5$$

Such equations are called **polynomials**. (When the highest power of the unknown quantity is 2, these are called **quadratic** equations.) Polynomials can be solved by using a method called **trial and improvement**.

We first make a guess at the solution. Then by using this initial guess we make a second approximation which is an improvement on the initial guess. Continuing in this way we eventually arrive at a solution which is as accurate as we desire.

Example 6

Margaret's calculator has no square root key. So to find $\sqrt{11}$ correct to 1 decimal place she solves the equation $x^2 = 11$ by trial and improvement. Her two initial guesses are

Try $x = 3.1$: $x^2 = (3.1)^2$
$$= 9.61 \text{ which is too small.}$$
Try $x = 3.4$: $x^2 = (3.4)^2$
$$= 11.56 \text{ which is too large.}$$

Continue in this way until you find the value of $\sqrt{11}$ correct to 1 decimal place.

So the value of x must lie between 3.1 and 3.4.

Try $x = 3.3$: $x^2 = (3.3)^2 = 10.89$ which is too small.

The value of x must lie between 3.3 and 3.4.

Try $x = 3.32$: $x^2 = (3.32)^2 = 11.02$ which is too large.

So the value of x lies between 3.3 and 3.32, and so

$$\sqrt{11} = 3.3 \text{ correct to 1 decimal place.}$$

Example 7

Peter wishes to solve the equation
$x^2 - 2x = 10$ using trial and improvement.

Peter starts by trying $x = 4.1$ and $x = 4.4$. Continue Peter's working until the value of x correct to 2 decimal places is found.

Try $x = 4.1$: $(4.1)^2 - 2 \times 4.1 = 8.61$ which is too small.

Try $x = 4.4$: $(4.4)^2 - 2 \times 4.4 = 10.56$ which is too large.

The solution must lie between 4.1 and 4.4, so

Try $x = 4.3$: $(4.3)^2 - 2 \times 4.3 = 9.89$ which is too small.

The solution must lie between $x = 4.3$ and $x = 4.4$, so

Try $x = 4.33$: $(4.33)^2 - 2 \times 4.33$ $= 10.09$ which is too large.

The solution must lie between 4.3 and 4.33, so

Try $x = 4.31$: $(4.31)^2 - 2 \times 4.31$ $= 9.96$ which is too small.

The solution must lie between $x = 4.31$ and 4.33, so

Try $x = 4.32$: $(4.32)^2 - 2 \times 4.32$ $= 10.02$ which is too large.

The solution must lie between $x = 4.31$ and $x = 4.32$, so

Try $x = 4.316$: $(4.316)^2 - 2 \times 4.316$ $= 9.996$ which is too small.

Therefore, correct to 2 decimal places the solution is $x = 4.32$.

Exercise 16.4

1. A pupil has a calculator which has no square root key. He wishes to find $\sqrt{15}$ by using trial and improvement. To do this he uses the equation $x^2 = 15$. Her two initial guesses are:

   $x = 3.8$: $(3.8)^2 = 14.44$ which is too small.

   $x = 4.1$: $(4.1)^2 = 16.81$ which is too large.

Continue the calculation until you find $\sqrt{15}$ correct to 1 decimal place.

2. A second pupil wishes to find $\sqrt{33}$ without using the square root key on his calculator. To do this he uses the equation $x^2 = 33$ and his two initial estimates are:

   $x = 5.8$: $(5.8)^2 = 33.54$ which is too large.

   $x = 5.6$: $(5.6)^2 = 31.36$ which is too small.

Using a calculator continue the calculation until you find the value of $\sqrt{33}$ correct to 2 decimal places.

3. Mandy wishes to solve the equation $a \times a \times a = 29\,791$. To try to find the value of a she makes two estimates:

   $$29 \times 29 \times 29 = 24\,389$$
   $$33 \times 33 \times 33 = 35\,937$$

Continue Mandy's calculation to find the exact value of a.

4. Mark wishes to solve the equation $x^2 + x = 15$ using trial and improvement. To try to find the solution he makes two estimates which are:

   Try $x = 3.2$: $x^2 + x = (3.2)^2 + 3.2$ $= 13.4$ (too small).

   Try $x = 3.6$: $x^2 + x = (3.6)^2 + 3.6$ $= 16.6$ (too large).

Continue in this way until you find the value of x correct to 1 decimal place.

5. Percy wants to solve the equation $x^2 - 5x - 8 = 0$ by using trial and improvement. He makes two initial estimates which are:

   $x = 6.4$: $(6.4)^2 - 5 \times 6.4 - 8$ $= 0.96$ which is too large.

   $x = 6.2$: $(6.2)^2 - 5 \times 6.2 - 8$ $= -0.56$ which is too small.

Continue Percy's calculation until the solution correct to 1 decimal place is found.

Flow Charts

Frequently a series of operations or instructions has to be carried out. A simple and clear way of doing this is by using a **flow chart**.

Example 8

(a) Fig. 16.2 shows a flow chart. Work out the answer to the instructions given.

Fig. 16.2

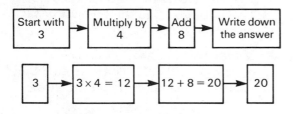

So the required answer is 20.

Note that when S is the starting number and F the finishing number these instructions can be written as $F = S \times 4 + 8$.

(b) Draw a flow chart to represent the equation $y = 3(x + 5)$ and use it to find the value of y when $x = 2$.

The flow chart is shown in Fig. 16.3.

Fig. 16.3

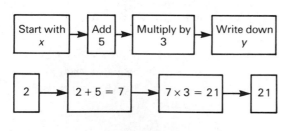

So when $x = 2$, $y = 21$.

Reverse flow charts are useful in solving certain numerical problems. In drawing the reverse flow diagram remember that + signs become −, − signs become +, × signs become ÷ and ÷ signs become ×.

Example 9

(a) Draw the flow chart for $y = \dfrac{x + 4}{5}$ and hence draw a reverse flow chart and use it to find the value of x when $y = 3$.

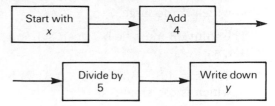

The reverse flow chart is then

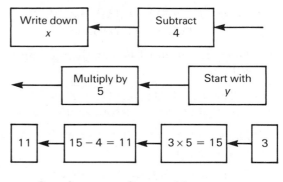

So when $y = 3$, $x = 11$.

(b) Mandy thinks of a number. She multiplies it by 3 and then adds 7. She ends up with the number 43. What number did she think of?

Let x be the number she thought of then the flow chart is as follows:

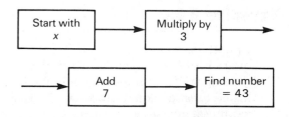

The reverse flow chart is then

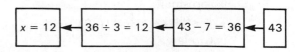

Flow Charts with Loops

Example 10

Fig. 16.4 shows a flow chart for a calculation to be done with a calculator.

Fig. 16.4

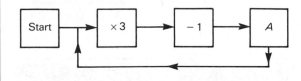

The output number A is always fed back to provide the next input (start number). The process repeats for ever. Describe what happens when the starting value is 4.

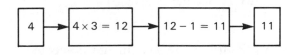

So the next starting value is 11 which gives

$$A = 11 \times 3 - 1 = 32$$

With a starting value of 32

$$A = 32 \times 3 - 1 = 95$$

and so on. So the sequence generated by this looped flow chart is 11, 32, 95, ...

Exercise 16.5

1. Work out the answer to these instructions:

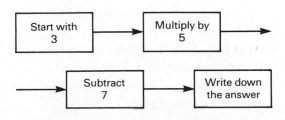

2. Work out the answer to these instructions:

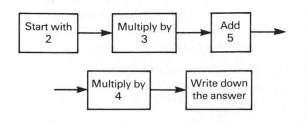

3. Draw a flow chart to represent the equation $y = 3x + 5$ and use it to find the value of y when $x = 6$.

4. Draw a flow chart to represent the equation $y = \dfrac{x + 6}{7}$, and use it to find the value of y when $x = 8$.

5. Draw a flow chart for the equation $y = 4(x + 7)$.

6. Draw a flow chart for the equation $y = \dfrac{7x + 3}{5}$ and hence draw a reverse flow chart and use it to find the value of x when $y = 9$.

7. Martin thinks of a number. He multiplies it by 3 and adds 6. He says that the answer is 30. What is the number Martin thought of?

8. I think of a number, double it and subtract 5. My answer is 13. Write down the number I am thinking of.

9. Here is a flow chart for changing dollars into pounds.

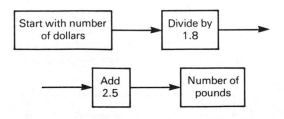

Use the flow chart to find the number of pounds that Dennis gets for $450.

10. One way of changing pints into litres is to multiply the number of pints by 6 and divide by 10. Draw a reverse flow chart for this information and use it to work out the number of pints equivalent to 12 litres.

11. Here are some instructions.

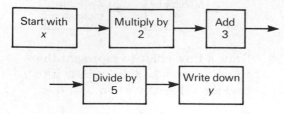

Form an algebraic expression for these instructions.

12. Form an algebraic expression for the instructions given in the flow chart shown below.

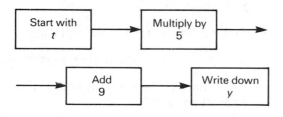

Find the value of the expression when $t = 4$.

13. The flow chart for a calculation to be done with a calculator is shown below.

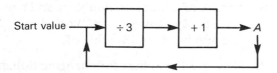

Obtain the first five values of A, starting with 6.

14. From the looped flow chart shown below, work out the first three values of A taking $x = 1$ as the first start value.

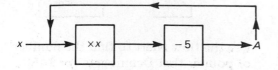

15. Using the looped flow diagram shown below generate the first 3 terms of the sequence of numbers given that the initial start value is 2.

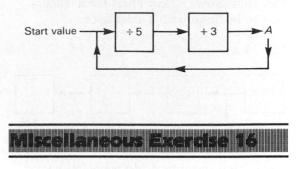

Miscellaneous Exercise 16

1. Write without brackets:

(a) $(x + 5)(x + 2)$

(b) $(x - 6)(x - 3)$

(c) $(x + 3)(x - 7)$

Solve the following pairs of simultaneous equations:

2. $4x - y = 5$
$2x + y = -2$

3. $3x - 2y = 5$
$x - 2y = -1$

4. $4x - 3y = 15$
$2x + 3y = 3$

5. If $p - q = 8$ and $2p - q = 20$, calculate the value of p.

6. Given the two equations

$$3x + 2y = 9$$
$$2x + 3y = 16$$

(a) by adding the two equations find the value of $x + y$,

(b) by subtracting the two equations find the value of $x - y$,

(c) hence, or otherwise, solve the equations for x and y.

7. Solve the following equations for x and y:

$$2x - 3y = 17$$

$$2x + 3y = 11$$

8. Dave thinks of a number. He divides it by 5 and adds 3. He says the result is 7.

(a) Draw a flow chart of this information.

(b) Hence draw a reverse flow chart.

(c) Use the reverse flow chart to find the number Dave thought of.

9. Iris wishes to solve the equation $x^2 - 7x = 19$ by using trial and improvement. Iris starts her calculation as

Try $x = 8.8$: LHS $= (8.8)^2 - 7 \times 8.8$
$= 15.84$ which is too small.

Try $x = 9.2$: LHS $= (9.2)^2 - 7 \times 9.2$
$= 20.24$ which is too big.

Continue Iris's calculation until the solution is found correct to 2 significant figures.

10. The time, t, an experiment takes depends upon the temperature, $T\,^\circ$C, and is given by the formula

$$t = T^2 - T + 1$$

(a) Calculate the time the experiment takes when the temperature is (i) $-2\,^\circ$C, (ii) $5\,^\circ$C.

(b) If the time taken is 2 seconds, use trial and improvement to find the temperature given that an initial guess of $T = 1.6$ is appropriate. State T correct to 2 decimal places.

Angles and Straight Lines

Some Definitions

(1) A **point** has position but no size, although to show it on paper it obviously must be given some size.

(2) If two points A and B are chosen (Fig. 17.1) then only one straight line can contain them.

Fig. 17.1

A _____ B

(3) If the two points chosen are the end points of the line then AB is called a line segment (Fig. 17.2).

Fig. 17.2

A _____ B

Angles

When two lines meet at a point they form an angle. The size of the angle depends upon the amount of opening between the two lines. It does not depend upon the lengths of the lines forming the angle. In Fig. 17.3 the angle *A* is larger than the angle *B* despite the fact that the arms are shorter.

Fig. 17.3

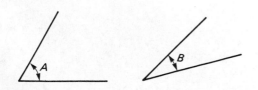

Measurement of Angles

An angle may be looked upon as the amount of rotation or turning. In Fig. 17.4 the line OA has been turned about O until it takes up the position OB. The angle through which the line OA has turned is the amount of opening between the lines OA and OB.

Fig. 17.4

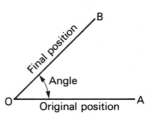

If the line OA is rotated further until it returns to its starting position it will have completed one revolution. Therefore we can measure an angle as a fraction of a revolution.

Fig. 17.5 shows a circle divided into 36 equal parts. The first division is split into 10 equal divisions so that each small division is $\frac{1}{360}$ of a complete revolution. This small division is called a **degree**. Therefore

$$360 \text{ degrees } = 1 \text{ revolution}$$

which is written for brevity as

$$360° = 1 \text{ rev}$$

Fig. 17.5

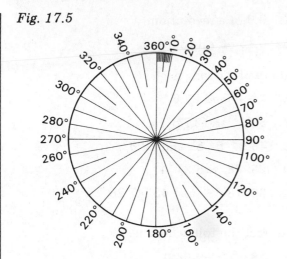

For many purposes the degree is too large a unit and it is divided into minutes and seconds such that

$$60 \text{ seconds} = 1 \text{ minute}$$

$$60 \text{ minutes} = 1 \text{ degree}$$

$$1 \text{ revolution} = 360 \text{ degrees}$$

An angle of 25 degrees 7 minutes 30 seconds would be written $25°7'30''$. Frequently angles are stated in degrees and decimals of a degree. A typical angle might be written $36.7°$.

It is possible on some calculators (for instance the Casio fx-31) to change an angle in degrees minutes and seconds to decimals by using a special key usually marked $\circ$ ' '' .

Example 1

(a) Find the angle, in degrees, corresponding to one-third of a revolution.

$$1 \text{ rev} = 360°$$

$$\tfrac{1}{3} \text{ rev} = \tfrac{1}{3} \times 360°$$

$$= 120°$$

(b) Find the angle, in degrees, corresponding to 0.7 of a revolution.

$$0.7 \text{ rev} = 0.7 \times 360°$$

$$= 252°$$

Example 2

(a) Add together $46.3°$ and $36.9°$.

$$\begin{array}{r} 46.3° \\ 36.9° + \\ \hline 83.2° \end{array}$$

(b) Subtract $38.75°$ from $65.62°$.

$$\begin{array}{r} 65.62° \\ 38.75° - \\ \hline 26.87° \end{array}$$

(c) Add $22°35'39''$ and $49°42'12''$.

$$\begin{array}{r} 22°35'39'' \\ 49°42'12'' + \\ \hline 72°17'51'' \end{array}$$

(d) Subtract $17°49'$ from $39°27'$.

$$\begin{array}{r} 39°27' \\ 17°49' - \\ \hline 21°38' \end{array}$$

(e) Convert $42°15'18''$ into degrees and decimals of a degree.

Method 1 using a calculator:

Input	Display
42	42.
$\circ$ ' ''	42.
15	15.
$\circ$ ' ''	42.25
18	18.
$\circ$ ' ''	42.255

So $42°15'18'' = 42.255°$

Method 2 by ordinary calculation:

If your calculator does not have a decimalisation key the following method may be used.

$$18'' = \frac{18}{60}$$

$$= 0.3'$$

$$15'18'' = 15.3' = \frac{15.3}{60}$$

$$= 0.255°$$

$$42°15'18'' = 42.255°$$

(f) Convert 43.8723° into degrees minutes and seconds.

Method 1 using a calculator:

Input	Display
43.8723	43.8723
INV	43.8723
° ' ''	43°52'20''

Method 2 by ordinary calculation:

If your calculator does not have a decimalisation key the following method may be used.

$$43.8723° = 43° + (0.8723 \times 60)'$$

$$= 43°52.338'$$

$$= 43°52' + (0.338 \times 60)''$$

$$= 43°52'20.28''$$

$$= 43°52'20'' \text{ correct to the nearest second.}$$

Exercise 17.1

Find the angle in degrees corresponding to each of the following:

1. $\frac{2}{3}$ of a revolution.

2. $\frac{3}{8}$ of a revolution.

3. 0.2 of a revolution.

4. 0.9 of a revolution.

5. 0.15 of a revolution.

Add the following angles:

6. 18.9° and 27.6°.

7. 39.2° and 17.8°.

8. 43.2°, 54.5° and 72.6°.

9. 25°38' and 27°4'

10. 36°18'39'' and 43°47'26''

Subtract the following angles:

11. 18.3° from 54.6°.

12. 37.9° from 46.1°.

13. 48°19' from 72°7'

14. 0°7'15'' from 18°4'9''

15. Convert the following to degrees and decimals of a degree correct to 4 d.p. :
    (a) 12°23'   (b) 35°39'43''
    (c) 71°54'6''

16. Convert the following to degrees, minutes and seconds correct to the nearest second:
    (a) 37.1827°   (b) 71.5566°
    (c) 12.4678°

Types of Angles

An **acute angle** is an angle of less than 90° (Fig. 17.6).

Fig. 17.6

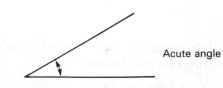

Acute angle

A **right angle** is an angle equal to 90° or $\frac{1}{4}$ of a revolution (Fig. 17.7). Note carefully how a right angle is marked.

Fig. 17.7

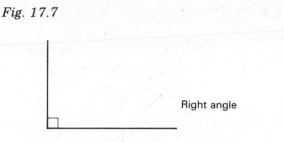

Right angle

An **obtuse angle** is an angle between $90°$ and $180°$ (Fig. 17.8).

Fig. 17.8

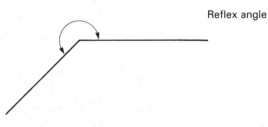

Obtuse angle

A **reflex angle** is an angle greater than $180°$ (Fig. 17.9).

Fig. 17.9

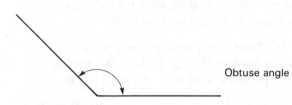

Reflex angle

Complementary angles are angles whose sum is $90°$. Thus $18°$ and $72°$ are complementary angles because $18° + 72° = 90°$.

Supplementary angles are angles whose sum is $180°$. Thus $103°$ and $77°$ are supplementary angles because $103° + 77° = 180°$.

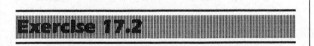
Exercise 17.2

1. Look at each of the angles in Fig. 17.10. State which are acute, which are obtuse and which are reflex.

Fig. 17.10

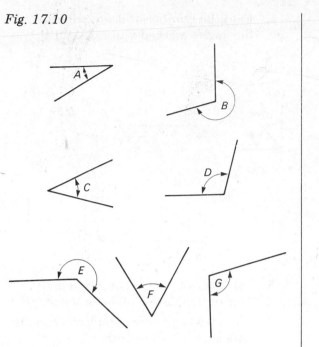

2. State the angle, in degrees, corresponding to each of the following:
   (a) $\frac{2}{3}$ of a right angle
   (b) $\frac{1}{4}$ of a right angle
   (c) 2 right angles
   (d) $1\frac{1}{2}$ right angles
   (e) 0.6 of a right angle
   (f) 0.35 of a right angle.

3. In each of the diagrams of Fig. 17.11, the angles at the centre of each one are of equal size. Work out the number of degrees in each of the angles.

Fig. 17.11

(a)   (b)

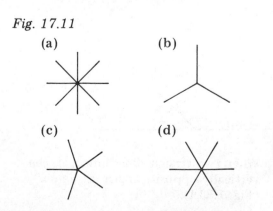

(c)   (d)

4. Find the number of degrees in each of the angles marked x in Fig. 17.12.

Fig. 17.12

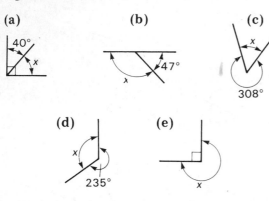

(a)   **(b)**   **(c)**

(d)   **(e)**

5. (a) Two angles are complementary. One of them is $57°$. What is the other?

(b) Two angles are complementary. If one is $34°$, find the other.

(c) Two angles are supplementary. One is $27°$, what is the other?

(d) Angles A and B are supplementary. If $A = 98°$, what is the size of B?

Properties of Angles and Straight Lines

(1) The total angle on a straight line is $180°$, i.e. two right angles. The angles A and B in Fig. 17.13 are called **adjacent angles** on a straight line and their sum is $180°$.

Fig. 17.13

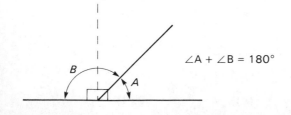

$$\angle A + \angle B = 180°$$

(2) When two straight lines intersect, the **vertically opposite angles** are equal (Fig. 17.14).

Fig. 17.14

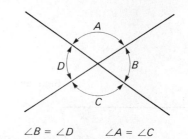

$$\angle B = \angle D \qquad \angle A = \angle C$$

Parallel Lines

Two lines in a plane that have no points in common, no matter how far they are produced, are called **parallel lines.**

(1) When two parallel lines are cut by a transversal (Fig. 17.15), the **corresponding angles** are equal. That is

$$a = l \quad b = m \quad c = p \quad d = q$$

Fig. 17.15

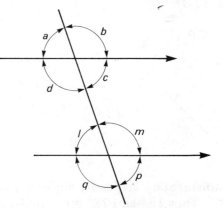

(2) The **alternate angles** are also equal. That is

$$d = m \quad \text{and} \quad c = l$$

(3) The **interior angles** are supplementary. That is

$$d + l = 180° \quad \text{and} \quad c + m = 180°$$

Example 3

In Fig. 17.16, AB and CD are two parallel straight lines. Find the size of the angles marked a, b, c, d and e.

Fig. 17.16

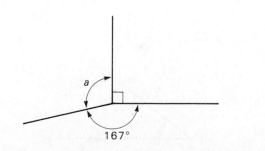

$a = 64°$ (vertically opposite angles)

$b = 180° - 64°$

    $= 116°$ (sum of the angles on a straight line equals 180°)

$c = 180° - 55° - 81°$

    $= 44°$ (sum of the angles on a straight line = 180°)

$d = 81°$ (AB $\parallel$ CD, corresponding angles)

$e = 55°$ (AB $\parallel$ CD, alternate angles)

(Note the symbol $\parallel$ which means 'is parallel to'.)

Exercise 17.3

1. Find the size of the angle a in Fig. 17.17.

Fig. 17.17

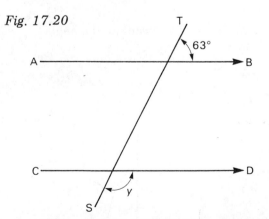

2. In Fig. 17.18, find the size of the angles marked x and y.

Fig. 17.18

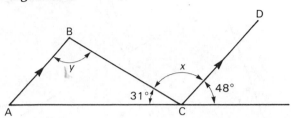

3. In Fig. 17.19, AB and CD are parallel. Calculate the size of the angles marked m, n, p and q.

Fig. 17.19

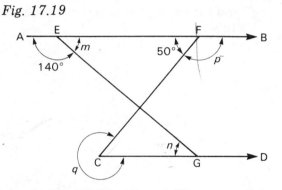

4. In Fig. 17.20, AB and CD are parallel lines crossed by the line ST. Find the size of the angle marked y.

Fig. 17.20

5. ABC and DEF (Fig. 17.21) are two parallel straight lines with BD, EC and BF transversals. CE and BF intersect at 90°. Find the size of the angles marked p, q, r and s.

Fig. 17.21

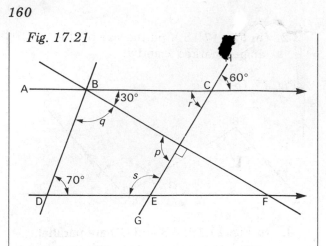

6. In Fig. 17.22, AB is parallel to CD. Find the size of the angles marked x and y.

Fig. 17.22

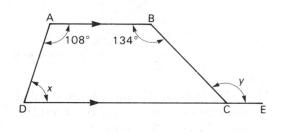

7. In Fig. 17.23, calculate the angle marked a.

Fig. 17.23

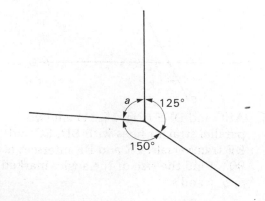

8. In Fig. 17.24, ABCDE and FGH are two parallel straight lines with BKG, FKCJ and GDJ transversals. Find the size of the angles marked u, v, w, x, y and z.

Fig. 17.24

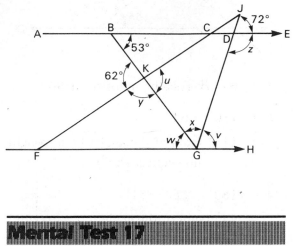

Mental Test 17

Try to answer the following questions without writing anything down except the answer.

1. Express a degree as a fraction of a complete revolution.

2. Find the angle in degrees corresponding to one-sixth of a revolution.

3. Find the angle in degrees corresponding to 0.3 of a revolution.

4. What is the angle in Fig. 17.25 called?

Fig. 17.25

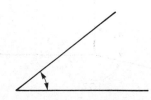

5. Two angles are complementary. One is 60°. What is the size of the other?

6. Two angles are supplementary. One is 130°. What is the size of the other?

7. Find the number of degrees in each of the angles marked *x* in Fig. 17.26.

Fig. 17.26

(a)   **(b)**   **(c)**

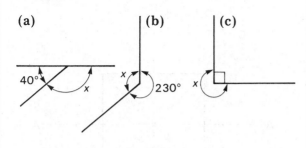

8. In Fig. 17.27, write down the size of the angles *A* and *B*.

Fig. 17.27

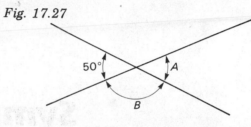

9. In Fig. 17.28, write down the size of the angles marked *b*, *c*, *d*, *e*, *f*, *g* and *h*.

Fig. 17.28

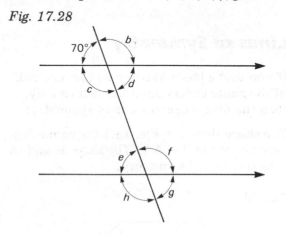

10. Add 35.3° and 44.7°.

Symmetry

Lines of Symmetry

If you fold a piece of paper so that one half of the paper covers the other half exactly, then the fold is called a **line of symmetry**.

The shape shown in Fig. 18.1 is symmetrical only about the line AA'. The shape is said to have one line of symmetry.

Fig. 18.1

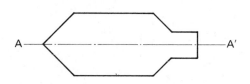

The shape shown in Fig. 18.2 is symmetrical about the lines XX' and YY'. The shape therefore has two lines of symmetry.

Fig. 18.2

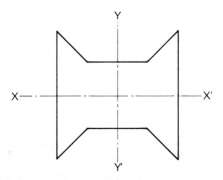

It is possible for a shape to have several lines of symmetry. For instance, the square shown in Fig. 18.3 has four lines of symmetry AA', BB', CC' and DD'.

Fig. 18.3

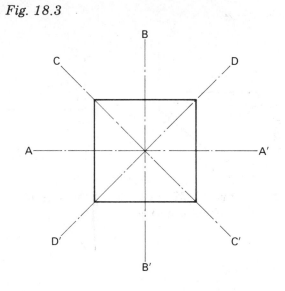

Some shapes have no lines of symmetry. The parallelogram shown in Fig. 18.4 is an example.

Fig. 18.4

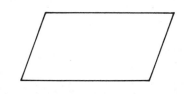

From the examples given above it can be seen that a line of symmetry may be horizontal, vertical or oblique. Symmetry is also discussed on page 339.

Exercise 18.1

1. Each of the shapes shown in Fig. 18.5 has one line of symmetry. Copy the shapes on squared paper and then draw the line of symmetry.

Fig. 18.5

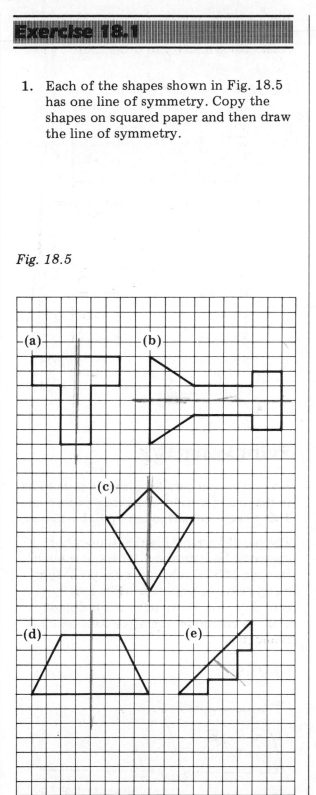

2. Fig. 18.6 shows a number of half-shapes with the line of symmetry indicated in chain dot. Draw the complete symmetrical shape using squared paper.

Fig. 18.6

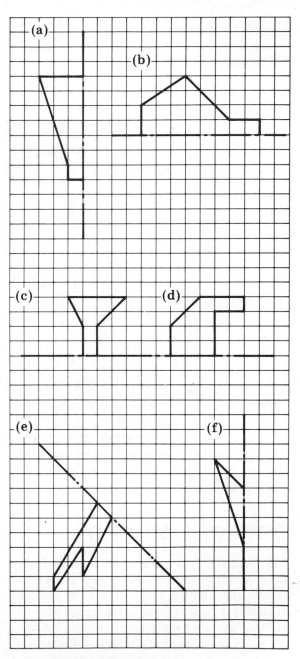

3. Each of the shapes shown in Fig. 18.7 has two lines of symmetry. Using squared paper copy each of these shapes and then draw the two lines of symmetry.

Fig. 18.7

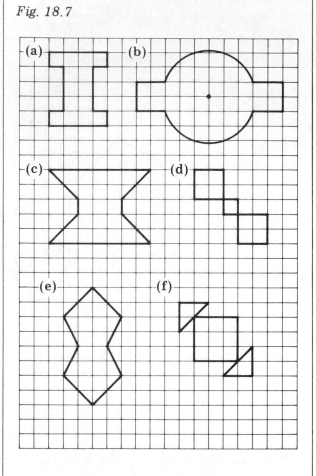

4. Each of the capital letters M, K, T, C, X and H has one or more lines of symmetry. Sketch the letters and draw all the lines of symmetry for each of the letters.

Rotational Symmetry

Fig. 18.8 shows a square ABCD whose diagonals intersect at O. If we rotate the square through 90°, 180° and 270° the square does not appear to have moved (unless we label the corners A, B, C and D, when the change is apparent). Because

there are four positions where it appears not to have moved, we say that the square has **rotational symmetry of order four.**

Fig. 18.8

Original position and rotated through 360°

Rotated through 90°

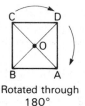

Rotated through 180°

Rotated through 270°

Point Symmetry

Rectangles and parallelograms appear to be in the same position when rotated through 180°. They are said to have **point symmetry.**

The parallelogram (Fig. 18.9) does not appear to have moved after being rotated through 180° and hence it has point symmetry, but the trapezium (Fig. 18.10) appears upside down after a rotation of 180°. A trapezium therefore has no point symmetry.

Fig. 18.9

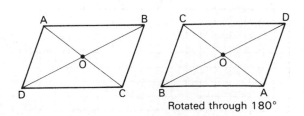

Rotated through 180°

Fig. 18.10

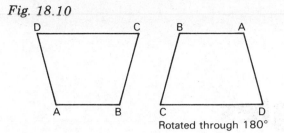

Rotated through 180°

Planes of Symmetry

If a solid figure, such as a sphere, is cut into two equal parts as shown in Fig. 18.11, the plane of the cut is called a **plane of symmetry.** The cuboid shown in Fig. 18.12 has been cut into equal parts by the plane ABCD. Hence the plane ABCD is a plane of symmetry for the cuboid. A cuboid has two other planes of symmetry and we say that a cuboid has three planes of symmetry.

Fig. 18.11

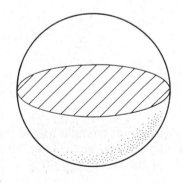

Fig. 18.12

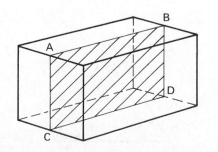

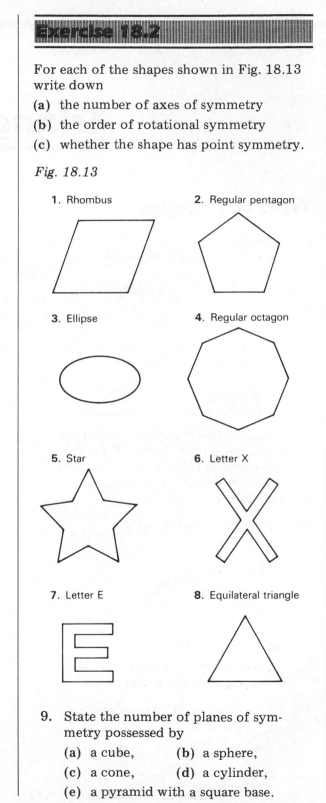

Exercise 18.2

For each of the shapes shown in Fig. 18.13 write down

(a) the number of axes of symmetry

(b) the order of rotational symmetry

(c) whether the shape has point symmetry.

Fig. 18.13

1. Rhombus

2. Regular pentagon

3. Ellipse

4. Regular octagon

5. Star

6. Letter X

7. Letter E

8. Equilateral triangle

9. State the number of planes of symmetry possessed by

(a) a cube,   (b) a sphere,

(c) a cone,   (d) a cylinder,

(e) a pyramid with a square base.

19 Triangles

Types of Triangles

(1) An **acute-angled triangle** (Fig. 19.1) has each of its angles less than 90°.

Fig. 19.1   Acute-angled triangle

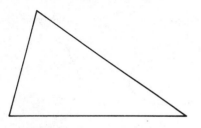

(2) A **right-angled triangle** has one of its angles equal to 90° (Fig. 19.2). The side opposite to the right angle is the longest side and it is called the **hypotenuse**.

Fig. 19.2   Right-angled triangle

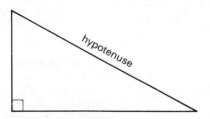

(3) An **obtuse-angled triangle** (Fig. 19.3) has one angle greater than 90°.

Fig. 19.3   Obtuse-angled triangle

∠A greater than 90°

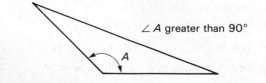

(4) A **scalene triangle** has all three sides of different length and all three angles of different size.

(5) An **isosceles triangle** has two sides and two angles equal (Fig. 19.4). The equal sides lie opposite to the equal angles.

Fig. 19.4   Isosceles triangle

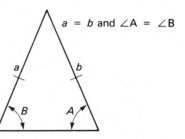

$a = b$ and ∠A = ∠B

(6) An **equilateral triangle** has all its sides and all its angles equal (Fig. 19.5). Each angle of the triangle is 60°.

Fig. 19.5   Equilateral triangle

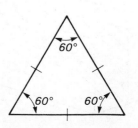

60°
60°   60°

Look at the triangles shown in Fig. 19.6 and decide:

1. Which are equilateral triangles.

2. Which are obtuse-angled triangles.

3. Which are scalene triangles.

4. Which are isosceles triangles.

5. Which are right-angled triangles.

Fig. 19.6

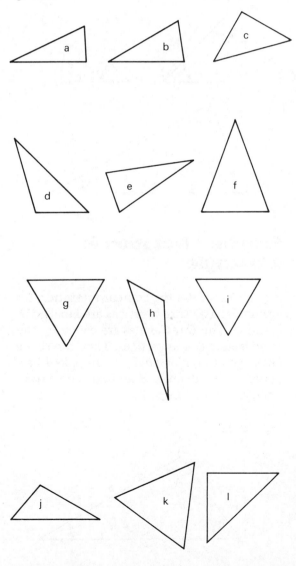

Angle Properties of Triangles

(1) The sum of the angles of a triangle is 180° (Fig. 19.7).

Fig. 19.7

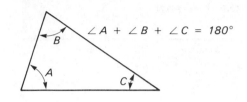

$\angle A + \angle B + \angle C = 180°$

Example 1

Two of the angles of a triangle are $35°$ and $72°$. What is the size of the third angle?

$$\text{Third angle} = 180° - (35° + 72°)$$
$$= 180° - 107°$$
$$= 73°$$

(2) When the side of a triangle is produced, the exterior angle so formed is equal to the sum of the opposite interior angles (Fig. 19.8).

Fig. 19.8

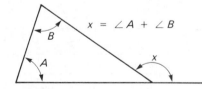

$x = \angle A + \angle B$

Example 2

In Fig. 19.9, find the size of the angle marked y.

Fig. 19.9

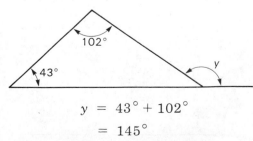

$$y = 43° + 102°$$
$$= 145°$$

Exercise 19.2

1. In each of the triangles listed in the table below, the size of two angles is given. Find by calculation the size of the third angle.

Triangle	Angles		
A	28°	67°	
B	37°		82°
C	80°	80°	
D		90°	28°
E	104°		63°
F		60°	30°

2. In Fig. 19.10, x is the exterior angle and A and B are the opposite interior angles. For each of the triangles listed below, find the size of the angle x.

Fig. 19.10

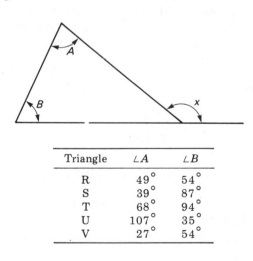

Triangle	∠A	∠B
R	49°	54°
S	39°	87°
T	68°	94°
U	107°	35°
V	27°	54°

3. Using Fig. 19.10, find the missing angle for each of the triangles listed below.

Triangle	∠A	∠B	x
G	39°	73°	
H	41°		108°
I	65°		89°
J	54°		165°
K	78°	90°	
L		86°	97°

4. Find each of the angles marked x and y in Fig. 19.11.

Fig. 19.11

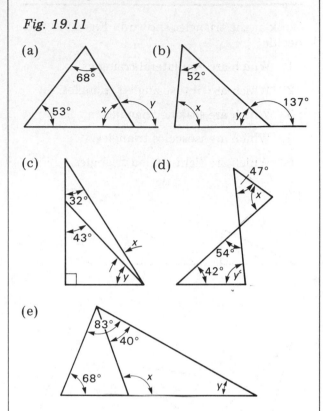

(a)

(b)

(c)

(d)

(e)

Standard Notation for a Triangle

Fig. 19.12 shows the standard notation for a triangle. The three vertices are labelled A, B and C. The three angles are called by the same letters as the vertices. Then the side a lies opposite to the angle A, the side b lies opposite to the angle B and the side c lies opposite to the angle C.

Fig. 19.12

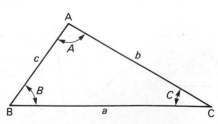

Constructing Triangles

(1) *To construct a triangle given the lengths of the three sides.*

Example 3

Construct the triangle ABC given that $a = 6$ cm, $b = 3$ cm and $c = 4$ cm.

Draw BC = 6 cm (Fig. 19.13), then with centre B and radius 4 cm using compasses draw an arc. With centre C and radius 3 cm, draw a circular arc to cut the first arc at A. Join A and B and also A and C, then ABC is the required triangle.

Fig. 19.13

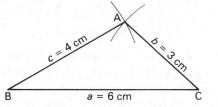

(2) *To construct a triangle given two sides and the angle included between the two sides.*

Example 4

Construct the triangle ABC given that $b = 5$ cm, $c = 6$ cm and $A = 65°$.

Draw AB = 6 cm (Fig. 19.14). Using a protractor draw AX such that ∠BAX = 65°. Along AX̀ mark off AC = 5 cm, and join B and C. Then ABC is the required triangle.

Fig. 19.14

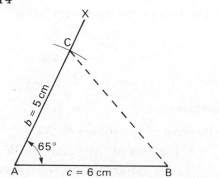

(3) *To construct a triangle given the lengths of two of the sides and an angle which is not the angle included between the two sides.*

Example 5

Construct the triangle ABC given that $a = 5$ cm, $b = 6$ cm and $B = 70°$.

Draw BC = 5 cm (Fig. 19.15). Using a protractor draw BX such that ∠CBX = 70°. With centre C and a radius of 6 cm, use compasses to draw an arc to cut BX at A. Join CA, then ABC is the required triangle.

Fig. 19.15

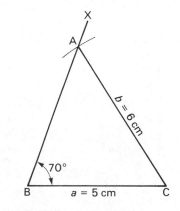

(4) *To construct a triangle given two angles and the length of the side opposite to the third angle of the triangle.*

Example 6

Construct the triangle ABC given that $a = 4.7$ cm, $B = 54°$ and $C = 46°$.

Draw BC = 4.7 cm (Fig. 19.16). At B draw a line BX making an angle of 54° with BC. At C draw a line CY making an angle of 46° with BC. The point of intersection of BX and CY is A, and ABC is the required triangle.

Fig. 19.16

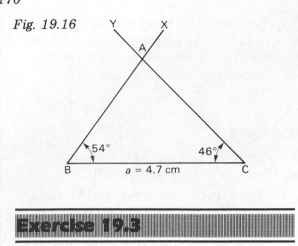

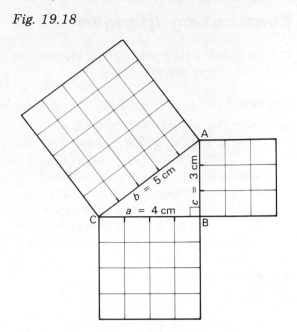

Fig. 19.18

Exercise 19.3

Construct each of the triangles ABC listed in the table below. Then measure the lengths of the sides and the sizes of the angles not given. In the table the lengths of *a*, *b* and *c* are stated in centimetres.

Triangle	$\angle A$	$\angle B$	$\angle C$	a	b	c
1	75°	34°		10		
2	19°		105°			11
3	116°		18°	8.5		
4	50°			7		9
5				5	8	7
6	43°			7	9	
7	62°				9	7
8		84°		5	6	
9				7	6	10
10			85°		7	9
11		60°	70°	8		
12	110°	40°				7

The Right-Angled Triangle

In any right-angled triangle the hypotenuse is the longest side and it always lies opposite to the right angle. Thus in Fig. 19.17 the side AC is the hypotenuse because it lies opposite to the right angle at B.

Fig. 19.17

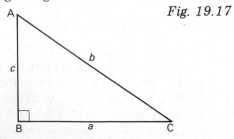

Fig. 19.18 shows a triangle ABC which is right angled at B. The side AB is 3 cm long and the side BC is 4 cm long. By accurately constructing the triangle the side AC is found to be 5 cm long. By constructing squares on each of the three sides we find that:

the area of the square on AC is
$$5 \times 5 = 5^2$$
$$= AC^2$$
$$= 25 \, cm^2$$

the area of the square on BC is
$$4 \times 4 = 4^2$$
$$= BC^2$$
$$= 16 \, cm^2$$

the area of the square on AB is
$$3 \times 3 = 3^2$$
$$= AB^2$$
$$= 9 \, cm^2$$

Therefore

the area of the square on AC
= the area of the square on BC
+ the area of the square on AB

That is

the area of the square on the hypotenuse

= the sum of the squares on the other

two sides

No matter how many right-angled triangles we draw we will always find this statement to be true. This statement is known as Pythagoras' theorem. For Fig. 19.18,

$$AC^2 = BC^2 + AB^2 \quad \text{or} \quad b^2 = a^2 + c^2$$

Example 7

In $\triangle ABC$, $A = 90°$, $b = 3$ cm and $c = 4$ cm. Find a.

The triangle is sketched in Fig. 19.19 and we see that a is the hypotenuse because it lies opposite to the right angle at A. Hence

$$a^2 = b^2 + c^2$$
$$a^2 = 3^2 + 4^2$$
$$= 9 + 16$$
$$a = \sqrt{25}$$
$$= 5 \text{ cm}$$

Fig. 19.19

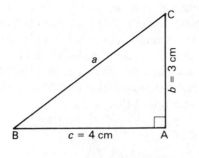

It is worth remembering that triangles with sides of $3:4:5$, $5:12:13$ and $7:24:25$ are right-angled triangles. Multiples of these lengths also give right-angled triangles. Hence triangles with sides 10, 24, 26 and 14, 48, 50 are right-angled.

Example 8

In $\triangle ABC$, $A = 90°$, $a = 6.4$ cm and $b = 5.2$ cm. Find the length of the side c.

The triangle is sketched in Fig. 19.20 and we see that a is the hypotenuse. Hence

$$a^2 = b^2 + c^2$$
$$c^2 = a^2 - b^2$$
$$= 6.4^2 - 5.2^2$$
$$= 40.96 - 27.04$$
$$= 13.92$$
$$c = \sqrt{13.92}$$
$$= 3.731 \text{ cm}$$

Fig. 19.20

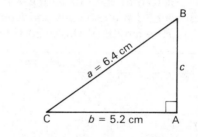

Exercise 19.4

1. Fig. 19.21 shows several right-angled triangles. For each triangle, name the hypotenuse.

Fig. 19.21

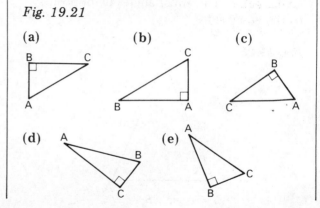

2. Fig. 19.22 shows a number of right-angled triangles. For each triangle, calculate the side marked x.

Fig. 19.22

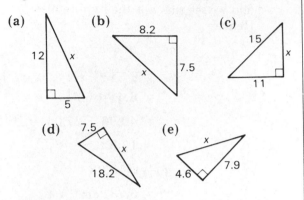

3. For each of the triangles ABC listed in the table below, the lengths, in centimetres, of two of the sides are given. Sketch each of the triangles and then work out the length of the third side.

Triangle	$\angle A$	$\angle B$	$\angle C$	a	b	c
(a)	$90°$				4	5
(b)			$90°$	8	11	
(c)		$90°$			15	8
(d)			$90°$	9		17
(e)	$90°$				6	3

Properties of the Isosceles Triangle

It will be remembered that an isosceles triangle (Fig. 19.23) has two sides and two angles equal. The equal angles lie opposite to the equal sides.

Fig. 19.23

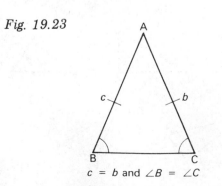

$c = b$ and $\angle B = \angle C$

Example 9

In Fig. 19.24, find the size of the angles A and C.

Fig. 19.24

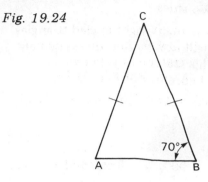

Since $AC = BC$, ABC is isosceles. Hence

$$A = B$$
$$= 70°$$

Since the sum of the angles of a triangle equals $180°$,

$$C = 180° - (70° + 70°)$$
$$= 180° - 140°$$
$$= 40°$$

Looking at Fig. 19.25, it is clear that the isosceles triangle ABC is symmetrical about the line AD. Hence the line AD bisects the apex angle A (i.e. divides the angle A into two equal angles). AD also bisects the base BC of the triangle (i.e. $BD = DC$) and finally AD is perpendicular to BC (i.e. it makes an angle of $90°$ with BC). Note that AD is the perpendicular height or altitude of the triangle ABC.

Fig. 19.25

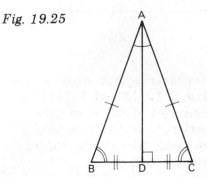

Example 10

△ABC (Fig. 19.26) is isosceles with altitude AD = 10.2 cm and base BC = 8.4 cm. Calculate the length of the equal sides.

Fig. 19.26

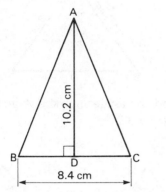

Since AD is the perpendicular dropped from the apex A,

$$BD = CD$$
$$= \tfrac{1}{2}BC$$
$$= \tfrac{1}{2} \times 8.4 \text{ cm}$$
$$= 4.2 \text{ cm}$$

In △ABD, AB is the hypotenuse and by Pythagoras' theorem,

$$AB^2 = BD^2 + AD^2$$
$$= 4.2^2 + 10.2^2$$
$$= 17.64 + 104.04$$
$$= 121.68$$
$$AB = \sqrt{121.68}$$
$$= 11.03 \text{ cm}$$

Hence the two equal sides AB and AC are 11.03 cm long.

Properties of the Equilateral Triangle

It will be recalled that an equilateral triangle has all its sides equal in length and all its angles equal in size. As can be seen from

Fig. 19.27, the equilateral triangle ABC has three axes of symmetry AD, BE and CF. Hence AD is perpendicular to BC, BE is perpendicular to AC and CF is perpendicular to AB. Since ∠BAC = ∠ACB = ∠ABC = 60°, each of the angles BAD, CAD, ACF, BCF, ABE and EBC equals 30°.

Fig. 19.27

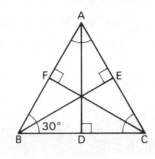

Example 11

△ABC (Fig. 19.28) is equilateral with sides 8 cm long. Calculate the altitude BD of this triangle. Since BD is the perpendicular dropped from the vertex B,

$$AD = CD$$
$$= \tfrac{1}{2}AC$$
$$= \tfrac{1}{2} \times 8 \text{ cm}$$
$$= 4 \text{ cm}$$

Fig. 19.28

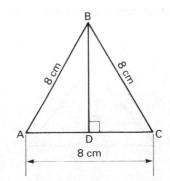

In △ADB, AB is the hypotenuse and by Pythagoras' theorem,

$$AB^2 = AD^2 + BD^2$$
$$BD^2 = AB^2 - AD^2$$
$$= 8^2 - 4^2$$
$$= 64 - 16$$
$$= 48$$
$$BD = \sqrt{48}$$
$$= 6.928 \text{ cm to 4 sig. fig.}$$

Hence the altitude of ABC is approx. 6.9 cm.

Exercise 19.5

1. Find the altitude AD of △ABC (Fig. 19.29).

Fig. 19.29

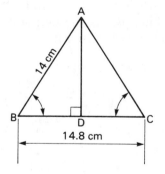

2. Find the side marked a in Fig. 19.30.

Fig. 19.30

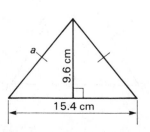

3. Find the length of the base, BC, of △ABC in Fig. 19.31.

Fig. 19.31

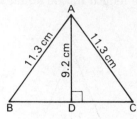

4. Find the angles marked A and B in Fig. 19.32.

Fig. 19.32

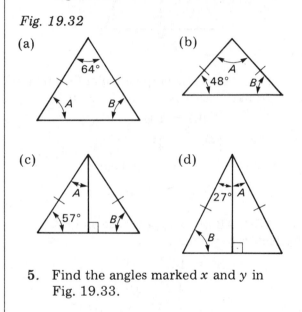

5. Find the angles marked x and y in Fig. 19.33.

Fig. 19.33

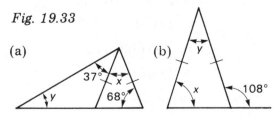

6. An equilateral triangle has sides 12 cm long. Calculate its altitude.

7. In △ABC, $a = 12$ cm, $b = 12$ cm and $c = 10$ cm. Calculate its altitude.

8. What is the vertical height of an equilateral triangle whose sides are each 9 cm long?

Miscellaneous Exercise 19

1. In Fig. 19.34 calculate the length of PQ.

Fig. 19.34

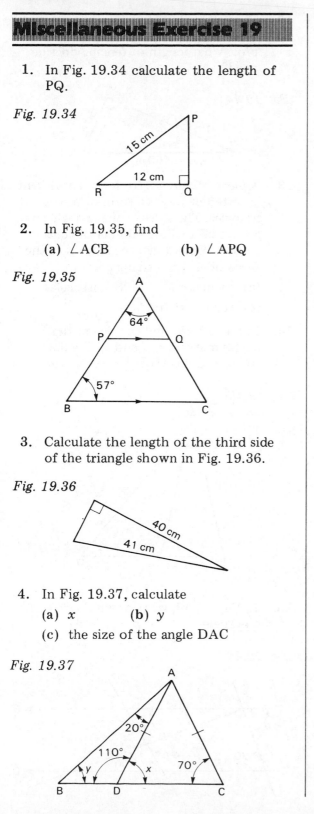

2. In Fig. 19.35, find
   (a) ∠ACB   (b) ∠APQ

Fig. 19.35

3. Calculate the length of the third side of the triangle shown in Fig. 19.36.

Fig. 19.36

4. In Fig. 19.37, calculate
   (a) x (b) y
   (c) the size of the angle DAC

Fig. 19.37

5. In Fig. 19.38 find
   (a) x (b) y (c) z
   (d) w (e) t

Fig. 19.38

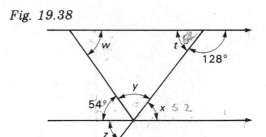

6. A rectangle is 9 cm long and 6 cm wide. Find the length of its diagonal.

7. Find the height of the isosceles triangle shown in Fig. 19.39.

Fig. 19.39

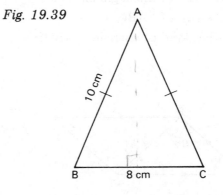

8. In Fig. 19.40, calculate the length of PQ.

Fig. 19.40

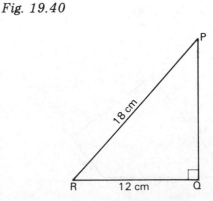

9. In Fig. 19.41, calculate the length of AB.

Fig. 19.41

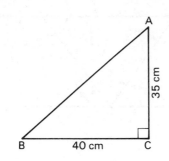

10. In Fig. 19.42, find

   (a) the size of the angle marked x

   (b) the size of the angle marked y

   (c) the size of $\angle ABC$

   (d) What kind of triangle is ABD?

Fig. 19.42

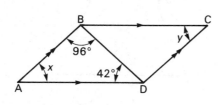

11. Fig. 19.43 shows a template. Calculate the length marked x which is needed to check the template.

Fig. 19.43

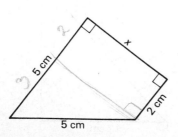

12. Fig. 19.44 shows a triangular piece of land. The owner wishes to fence the plot. What length of fencing does he need?

Fig. 19.44

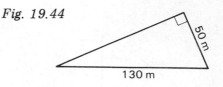

13. A piece of wire 12 cm long can be bent in different ways to form different triangles. The sides of the triangle must always be a whole number of centimetres. Write down the lengths of the three sides if the triangle is

   (a) equilateral   **(b)** isosceles

   (c) right-angled

14. In Fig. 19.45, find the size of the angles marked x, y and z. The line AC bisects $\angle BAD$.

Fig. 19.45

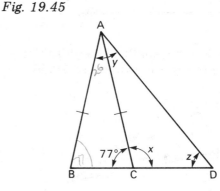

15. In Fig. 19.46, write down the size of the angles marked p, q, r and s.

Fig. 19.46

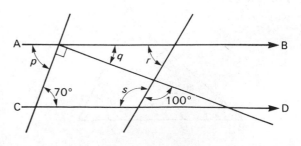

16. In Fig. 19.47, write down the size of the angles marked *m*, *n*, *p* and *q*.

Fig. 19.47

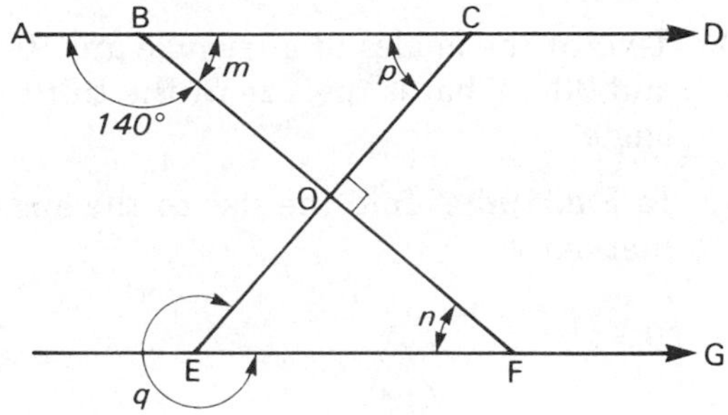

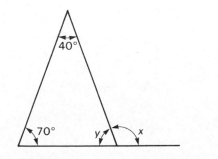

17. Triangle ABC has *a* = *b* = 12 cm and *c* = 8 cm. Sketch the triangle and then calculate its altitude.

Mental Test 19

Try to answer the following questions without writing anything down except the answer.

1. Two angles of a triangle are 40° and 60°. What is the size of the third angle?

2. An isosceles triangle has an apex angle of 20°. What is the size of each of the other two angles?

3. In Fig. 19.48, what is the size of the angle marked *x*?

Fig. 19.48

4. In Fig. 19.48, what is the size of the angle marked *y*?

5. Using Fig. 19.49, find the missing angle for each of the triangles listed below.

Triangle	∠A	∠B	*x*
(a)	20°	40°	?
(b)	100°	20°	?
(c)	40°	?	100°
(d)	?	80°	150°

Fig. 19.49

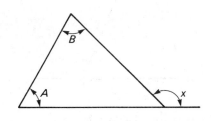

6. Using Fig. 19.50, find the length of the missing side for each of the triangles listed below.

Triangle	*a*	*b*	*c*
(a)	6	8	?
(b)	?	12	20
(c)	?	5	13
(d)	15	36	?
(e)	7	?	25
(f)	21	72	?

Fig. 19.50

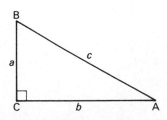

7. In Fig. 19.51, △ABC is isosceles. Find the missing angles for each of the triangles listed below.

Triangle	A	B	C	x
(a)	70°	?	?	?
(b)	?	?	60°	?
(c)	?	30°	?	?
(d)	?	?	?	100°
(e)	?	?	80°	?
(f)	?	?	?	120°
(g)	45°	?	?	?

Fig. 19.51

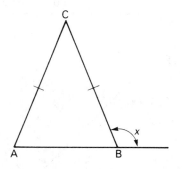

8. What is the triangle shown in Fig. 19.52 called?

Fig. 19.52

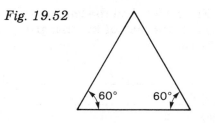

9. Fig. 19.53 shows an isosceles triangle. Find the size of the angles x and y.

Fig. 19.53

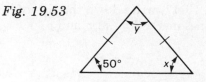

10. Two of the angles of a triangle are 70° and 80°. What is the size of the third angle?

11. In Fig. 19.54, find the size of the angle marked x.

Fig. 19.54

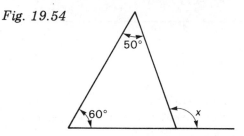

12. In Fig. 19.55, find the size of the angle marked a.

Fig. 19.55

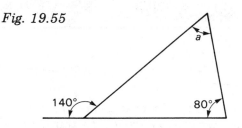

13. What is the longest side of a right-angled triangle called?

14. Two sides of a right-angled triangle are 5 and 12 cm long. What is the length of the third side which is the hypotenuse?

21 Quadrilaterals

Introduction

A **quadrilateral** is any plane figure bounded by four straight lines.

If we draw a straight line from one corner to the opposite corner of a quadrilateral (Fig. 21.1) the quadrilateral is divided into two triangles. We have seen in Chapter 19 that the sum of the angles of a triangle is $180°$. Hence *the sum of the angles of a quadrilateral is $360°$.*

Fig. 21.1

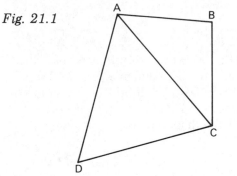

Example 1

Fig. 21.2 shows a quadrilateral. Find the size of the angle marked x.

Fig. 21.2

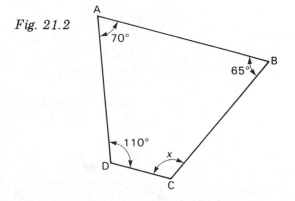

$$x + 70° + 65° + 110° = 360°$$
$$x + 245° = 360°$$
$$x = 115°$$

The Parallelogram

The **parallelogram** has both pairs of opposite sides parallel (Fig. 21.3a).

As shown in Chapter 18, a parallelogram has point symmetry. That is, if we rotate the parallelogram about O (Fig. 21.3b) through $180°$ it does not appear to have moved. However, we see that the vertices A, B, C and D become C', D', A' and B' respectively.

Fig. 21.3

(a)　　　　　　　　　　(b)

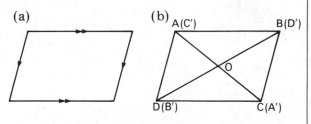

From Fig. 21.3b we see that:

(1) AB = CD and AD = BC. That is, the sides which are opposite to each other are equal in length.

(2) ∠ABC = ∠ADC and ∠DAB = ∠BCD. That is, the angles which are opposite to each other are equal in size.

(3) AO = OC and BO = OD. That is, the diagonals bisect each other.

(4) Triangles ACD and ABC are congruent, as are triangles BCD and ABD.

Example 2

Fig. 21.4 shows a parallelogram. Find the size of angles marked w, x, y and z.

Fig. 21.4

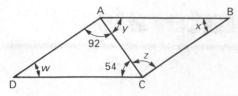

In $\triangle ACD$,

$$w + 92° + 54° = 180° \text{ (sum of the angles of a triangle)}$$

$$w + 146° = 180°$$

$$w = 180° - 146°$$

$$= 34°$$

In the parallelogram ABCD,

$$x = w$$

$$= 34° \quad \text{(opposite angles of a parallelogram are equal)}$$

$$y = 54° \quad \text{(AB}\|\text{CD, alternate angles)}$$

$$z + 54° = 92° + y \quad \text{(opposite angles of parallelogram are equal)}$$

$$z = 92° + 54° - 54°$$

$$= 92°$$

Alternatively

$$z = 92° \text{ (BC} \| \text{AD, alternate angles)}$$

The Rhombus

The **rhombus** is a parallelogram with all its sides equal in length. Hence it possesses all the properties of a parallelogram. Whereas the parallelogram has no axes of symmetry, the rhombus is symmetrical about each of its diagonals. This leads to two further properties (Fig. 21.5):

(1) Triangles ABD and BCD are congruent. Hence $\angle ABD = \angle DBC$ and $\angle ADB = \angle BDC$. That is, the diagonal bisects the angles through which it passes. Likewise the diagonal AC also bisects the angles through which it passes.

(2) From Fig. 21.5 it can be seen that the triangles AOB, BOC, COD and AOD are all congruent. Hence the four angles at O are all equal to 90°. This means that the diagonals of a rhombus bisect at right angles.

Fig. 21.5

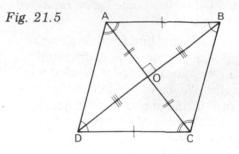

Example 3

ABCD (Fig. 21.6) is a rhombus whose sides are 12 cm long. The diagonal BD is 11.3 cm long. If $\angle ABD = 70°$, find the size of angles BDC, ADC and BAD and calculate the length of the diagonal AC.

Fig. 21.6

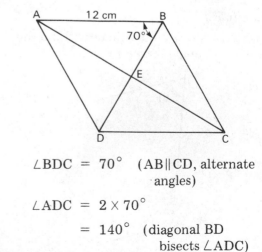

$$\angle BDC = 70° \quad \text{(AB}\|\text{CD, alternate angles)}$$

$$\angle ADC = 2 \times 70°$$

$$= 140° \quad \text{(diagonal BD bisects } \angle ADC)$$

$$\angle BAD + \angle ABC = 180°$$
$$\angle ABC = 2 \times \angle ABD$$
$$= 2 \times 70°$$
$$= 140°$$
$$\angle BAD = 180° - 140°$$
$$= 40°$$

In $\triangle ABE$,

$$\angle AEB = 90° \text{ and } AE = 5.65 \text{ cm}$$

(diagonals of a rhombus bisect at right angles)

Also $AB = 12$ cm (given). Using Pythagoras' theorem,

$$AE^2 = AB^2 - BE^2$$
$$= 12^2 - 5.65^2$$
$$= 144 - 31.9$$
$$= 112.1$$
$$AE = \sqrt{112.1}$$
$$\cong 10.59 \text{ cm}$$
$$AC = 2 \times AE$$
$$\cong 2 \times 10.59$$
$$= 21.18 \text{ cm}$$

The Rectangle

A **rectangle** is a parallelogram with each of its vertex angles equal to 90°. Hence it possesses all the properties of a parallelogram.

Looking at Fig. 21.7 we see that triangles ACD and BCD are congruent. Hence the diagonals AC and BD are equal in length.

Fig. 21.7

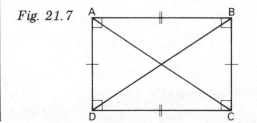

Example 4

A rectangle has sides 8 cm and 12 cm in length.

Calculate the length of its diagonals.

Fig. 21.8

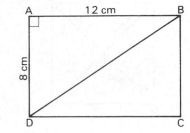

In $\triangle ABD$ (Fig. 21.8), by Pythagoras' theorem,

$$BD^2 = AB^2 + AD^2$$
$$= 12^2 + 8^2$$
$$= 144 + 64$$
$$= 208$$
$$BD = \sqrt{208}$$
$$\cong 14.4$$

Hence the length of the diagonals is 14.4 cm.

The Square

A **square** is a rectangle with all its sides equal in length. It has all the properties of a parallelogram, rhombus and rectangle. A square is symmetrical about the four axes shown in Fig. 21.9.

Fig. 21.9

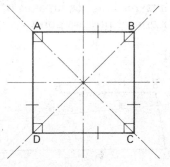

Example 5

A square has diagonals 8 cm long. Find the length of the sides of the square.

Fig. 21.10

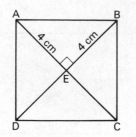

Since the diagonals bisect at right angles (Fig. 21.10), in $\triangle AEB$ by Pythagoras' theorem,

$$AB^2 = AE^2 + BE^2$$
$$= 4^2 + 4^2$$
$$= 16 + 16$$
$$= 32$$
$$AB = \sqrt{32}$$
$$\simeq 5.7$$

Hence the sides of the square are 5.7 cm long.

The Trapezium

A **trapezium** is a quadrilateral with one pair of sides parallel (Fig. 21.11). Generally speaking a trapezium has no axes of symmetry nor does it possess point symmetry.

Fig. 21.11

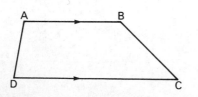

An **isosceles trapezium** has its two non-parallel sides equal in length (Fig. 21.12) and it has the one axis of symmetry shown in the diagram.

Fig. 21.12

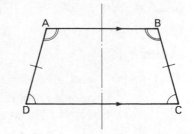

In the isosceles trapezium shown in Fig. 21.12,
$\angle BCD = \angle ADC$ and $\angle DAB = \angle CBA$.
Also $\angle CDA + \angle DAB = 180°$
and $\angle DCB + \angle ABC = 180°$.

Example 6

(a) ABCD (Fig. 21.13) is a trapezium with AB parallel to CD. Given the angles shown in the diagram, calculate the angle marked q.

Fig. 21.13

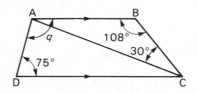

$q + 108° + 30° + \angle ACD + 75°$
$= 360°$ (sum of the angles of a quadrilateral)

Hence $q = 147° - \angle ACD$

Since $AB \parallel CD$,

$\angle ACD = \angle BAC$ (alternate angles)
$= 180° - 108° - 30°$
$= 42°$

So $q = 147° - 42°$
$= 105°$

(b) In Fig. 21.14, ABCD is an isosceles trapezium with AD = BC.
Find the size of the angles marked x and y.

Fig. 21.14

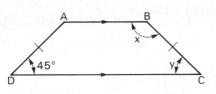

Since ABCD is an isosceles trapezium,

$$y = 45°$$

Also,

$$x + y = 180°$$
$$x + 45° = 180°$$
$$x = 135°$$

The Kite

The **kite** is a quadrilateral having two pairs of adjacent sides equal in length (Fig. 21.15). It is symmetrical about the line BD, and hence this line bisects the angles ADC and ABC. The angles DAB and DCB are also equal and the diagonals AC and BD intersect at right angles.

Fig. 21.15

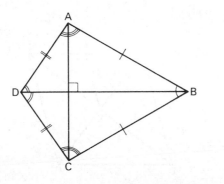

Example 7

Fig. 21.16 shows a kite ABCD. Given that ∠DAC = 15° and ∠CBD = 24°, find the size of the angles ACD, ADC and BAD.

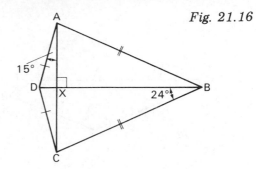

Fig. 21.16

Since AD = DC, △ADC is isosceles and therefore

$$∠DAC = ∠ACD$$
$$= 15°$$
$$∠ADC = 180° - (15° + 15°)$$
$$= 180° - 30°$$
$$= 150°$$

Since ∠CBD = 24°, ∠ABD = 24°

$$∠BAX = 180° - (90° + 24°)$$
$$= 180° - 114°$$
$$= 66°$$
$$∠BAD = 66° + 15°$$
$$= 81°$$

Exercise 21.1

1. Fig. 21.17 shows a kite. Find the angles marked x, y and z.

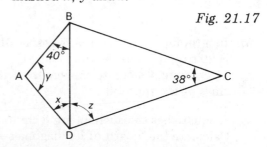

Fig. 21.17

2. In Fig. 21.18, ABCD is a quadrilateral. Find the angles marked *x* and *y*.

Fig. 21.18

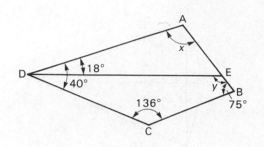

3. Fig. 21.19 shows a parallelogram. Write down the sizes of the angles marked *a* and *b*.

Fig. 21.19

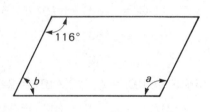

4. For the rhombus ABCD (Fig. 21.20), find the sizes of the angles DAC, BCD, ABC and DOC.

Fig. 21.20

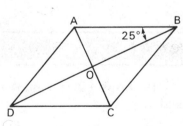

5. In a rhombus ABCD, the diagonal BD is 20 cm long and the diagonal AC is 12 cm long. Calculate the length of the sides of the rhombus.

6. A square has sides which are 8 cm long. Calculate the length of its diagonals.

7. The rectangle ABCD (Fig. 21.21) has its diagonal AC = 15 cm and its side AB = 11 cm. Calculate the length of the side BC.

Fig. 21.21

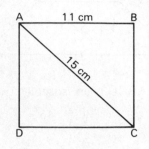

8. Fig. 21.22 shows an iron rectangular gate made from thin tubing. What length of tubing is needed to make it?

Fig. 21.22

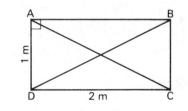

9. In Fig. 21.23, ABCD is a rhombus. BD = 10 cm, AC = 6 cm and XE is a straight line parallel to BC.

(a) What is the size of the angle AXD?

(b) What is the length of AX?

(c) What is the name of the figure EXCB?

Fig. 21.23

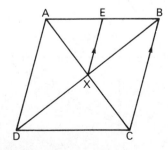

10. In Fig. 21.24, ABCD is a parallelogram.

    (a) What is the name of the quadrilateral BCDX?

    (b) Calculate the size of the angles XDC and DXB.

Fig. 21.24

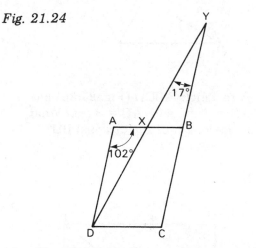

11. Fig. 21.25 shows a parallelogram ABCD. Find

    (a) The size of the angle marked x.

    (b) The size of the angle marked y.

    (c) The size of $\angle ACB$.

    (d) What type of triangle is ACD?

Fig. 21.25

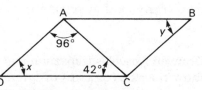

12. In a rhombus ABCD (Fig. 21.26), the diagonal AC = 24 cm and the diagonal BD = 10 cm. Find AO, BO and AB.

Fig. 21.26

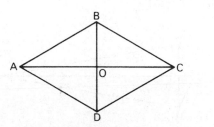

13. Name and draw two quadrilaterals which have one axis of symmetry.

14. Name a quadrilateral which has two pairs of equal sides and two axes of symmetry.

15. In Fig. 21.27, find the size of the angles marked a and p.

Fig. 21.27

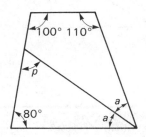

16. In Fig. 21.28, ABCD is a quadrilateral. $\angle ABC$ is bisected by BE. Find the size of the angles ABE and BED.

Fig. 21.28

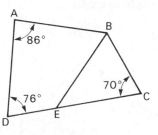

17. In Fig. 21.29, ABCD is a quadrilateral. $\angle BAD = 60°$ and the other three angles of the quadrilateral are equal. EC is drawn parallel to AB. Find the size of the angles ABC and ECD.

Fig. 21.29

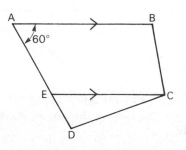

18. In Fig. 21.30, ABCD is a rhombus. Find the angle y if $x = 29°$.

Fig. 21.30

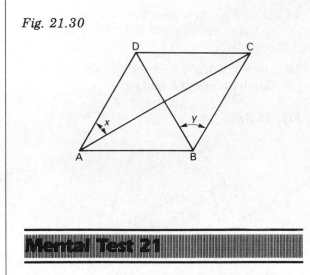

Fig. 21.33

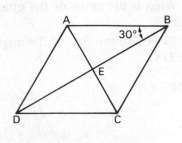

4. A rectangle ABCD (Fig. 21.34) has AB = 4 cm and AD = 3 cm. What is the length of the diagonal BD?

Fig. 21.34

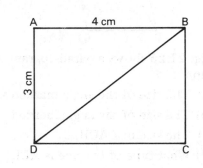

Mental Test 21

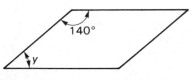

Try to answer the following questions without writing anything down except the answer.

1. Fig. 21.31 shows a quadrilateral. Find the angle marked x.

Fig. 21.31

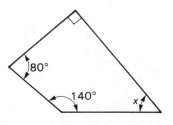

2. Find the size of the angle marked y in the parallelogram shown in Fig. 21.32.

Fig. 21.32

3. In the rhombus ABCD (Fig. 21.33) write down the size of the angles BEA and ABC.

5. How many axes of symmetry has a square?

6. Draw an isosceles trapezium and on it show the axis (or axes) of symmetry.

7. In the kite ABCD (Fig. 21.35) find the size of the angles ACD and ACB.

Fig. 21.35

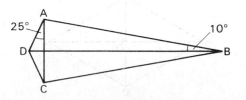

8. Fig. 21.36 shows a square ABCD surmounted by an equilateral triangle ABE. What is the size of the angle EAD?

Fig. 21.36

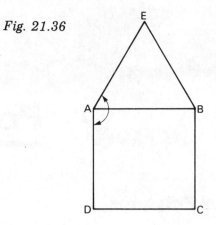

22

Polygons

Types of Polygons

A **polygon** is any plane figure bounded only by straight lines. Thus a triangle is a polygon with three sides, and a quadrilateral is a polygon with four sides.

A **convex polygon** (Fig. 22.1) has no interior angle greater than 180°.

Fig. 22.1

Exterior angle — Interior angle

Exterior angle + interior angle = 180°

A **re-entrant polygon** (Fig. 22.2) has at least one angle greater than 180°.

Fig. 22.2

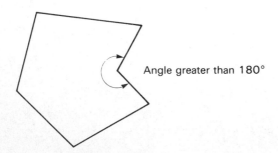

Angle greater than 180°

A **regular polygon** has all its sides and angles equal.

Sum of the Interior Angles of a Polygon

To find the sum of the interior angles of a polygon, choose any point such as O (Fig. 22.3) which lies within the polygon. Next draw lines from O to the vertices A, B, C ... to form triangles. The number of triangles is the same as the number of sides of the polygon: if the polygon has n sides then there are n triangles. Since the sum of the angles of each triangle is 180°, the sum of the angles of the triangles equals $180n$ degrees or $2n$ right angles. The sum of the angles round O equals 360° or 4 right angles. Hence *the sum of the interior angles of a polygon is $2n - 4$ right angles.*

Fig. 22.3

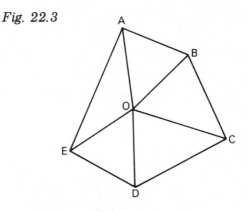

It makes no difference whether the polygon is convex or re-entrant, the sum of the interior angles is $(2n - 4)$ right angles.

Names of Polygons

Name	Number of sides
Pentagon	5
Hexagon	6
Heptagon	7
Octagon	8
Nonagon	9
Decagon	10
Hendecagon	11
Dodecagon	12

Example 1

(a) Find the sum of the interior angles of a heptagon (seven-sided polygon).

Here we have the number of sides
$$= n = 7$$

$$
\begin{aligned}
\text{Sum of the} \\
\text{interior angles} &= (2n - 4) \text{ right angles} \\
&= (2 \times 7 - 4) \text{ right angles} \\
&= 10 \text{ right angles} \\
&= 10 \times 90 \text{ degrees} \\
&= 900 \text{ degrees}
\end{aligned}
$$

(b) Find the size of each interior angle of a regular pentagon.

Since a pentagon has 5 sides, $n = 5$.

$$
\begin{aligned}
\text{Sum of the} \\
\text{interior angles} &= (2n - 4) \text{ right angles} \\
&= (2 \times 5 - 4) \text{ right angles} \\
&= 6 \text{ right angles} \\
&= 540 \text{ degrees}
\end{aligned}
$$

$$
\begin{aligned}
\text{Size of each interior angle} &= 540° \div 5 \\
&= 108°
\end{aligned}
$$

Sum of the Exterior Angles of a Convex Polygon

Fig. 22.4

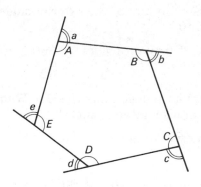

In Fig. 22.4, the interior angles of the polygon are A, B, C, D and E. The exterior angles are a, b, c, d and e. Now

$$
\begin{aligned}
A + a &= 180° \\
&= 2 \text{ right angles} \\
B + b &= 180° \\
&= 2 \text{ right angles}
\end{aligned}
$$

and so on. Therefore:

Sum of the interior angles + Sum of the exterior angles $= 2n$ right angles

$$
\begin{aligned}
\text{Sum of the} \\
\text{interior angles} &= (2n - 4) \text{ right angles}
\end{aligned}
$$

$$
\begin{aligned}
\text{Sum of the} \\
\text{exterior angles} &= [2n - (2n - 4)] \\
&\qquad \text{right angles} \\
&= 4 \text{ right angles} \\
&= 360°
\end{aligned}
$$

No matter how many sides the convex polygon has, the sum of its exterior angles is 360°.

Example 2

(a) Each interior angle of a regular polygon is 150°. How many sides has it?

Each exterior angle $= 180° - 150°$

$= 30°$

Since the sum of the exterior angles is 360°,

Number of sides $= 360 \div 30$

$= 12$

(b) A regular polygon has 15 sides. What is the size of each interior angle?

Each exterior angle $= 360° \div 15$

$= 24°$

Each interior angle $= 180° - 24°$

$= 156°$

Symmetry of Regular Polygons

Regular polygons have the number of axes of symmetry equal to the number of sides of the polygon. Thus a pentagon (Fig. 22.5) has five axes of symmetry and an octagon (Fig. 22.6) has eight axes of symmetry.

Fig. 22.5

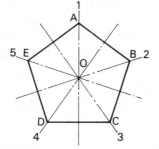

In Fig. 22.5, each of the triangles AOB, BOC, COD, DOE and EOA are congruent. The angle $AOB = 360° \div 5 = 72°$. Therefore $\angle ABO + \angle OBC = 180° - 72° = 108°$. Therefore each interior angle of a pentagon is 108°. This method gives us a third way of finding the size of the interior angle of a regular polygon.

Fig. 22.6

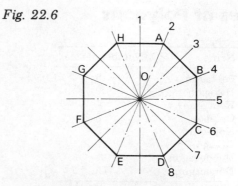

In Fig. 22.6 we note that there are 8 congruent triangles. Therefore the size of the interior angle of a regular octagon is $180° - (360° \div 8) = 180° - 45° = 135°$.

Exercise 22.1

1. Find the sum, in degrees, of the interior angles of polygons having

   (a) 6 sides   **(b)** 9 sides   **(c)** 11 sides

2. A regular polygon has 8 sides. Find the size of

   (a) each exterior angle

   (b) each interior angle

3. Draw a regular hexagon and show on it all the axes of symmetry. Hence find the size of each interior angle of the hexagon.

4. Draw a regular seven-sided polygon and on it show all the axes of symmetry. Hence find the size of each interior angle of the polygon.

5. A regular polygon has 12 sides. Find the size of each interior angle.

6. A pentagon (five-sided polygon) has interior angles of 104°, 108° and 112°. If the remaining two angles are equal, what is their size?

7. Each interior angle of a regular polygon is 120°. How many sides has it?

8. An irregular hexagon has interior angles of 110°, 90°, 130°, 140° and 135°. Find the size of its sixth angle.

9. A regular polygon has an interior angle of 160°. Calculate **(a)** the size of each exterior angle, **(b)** the number of sides possessed by the polygon.

10. Fig. 22.7 shows a regular pentagon ABCDE inscribed in a circle of centre O. Calculate the size of angles AOB and ABC.

Fig. 22.7

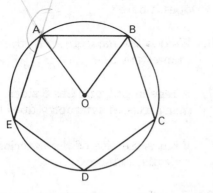

11. The exterior angle of a regular polygon is 24°. How many sides has the polygon?

12. UVWXYZ is a regular hexagon (Fig. 22.8). What is the size of the angle UVO? O is the centre of the circumscribing circle of the polygon.

Fig. 22.8

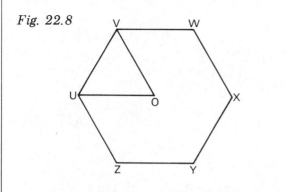

13. Calculate the size of the angles marked *x* and *y* for the regular polygon shown in Fig. 22.9. O is the centre of the circumscribing circle of the polygon.

Fig. 22.9

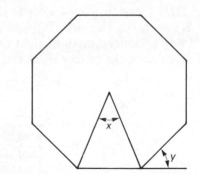

14. AB, BC and CD are three adjacent sides of a regular polygon with ten sides (Fig. 22.10). Calculate **(a)** the size of the exterior angle at B, **(b)** the size of angle ACD.

Fig. 22.10

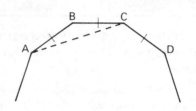

15. ABCDE is a regular pentagon and ABX is an equilateral triangle (Fig. 22.11). Calculate the angle AEX.

Fig. 22.11

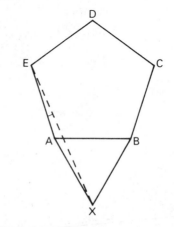

16. A polygon has *n* sides. Two of its angles are right angles and each of the remaining angles is 144°. Calculate *n*.

17. In the irregular pentagon ABCDE (Fig. 22.12) the interior angles at D and E are equal. The sides AB and DC when produced meet at right angles at F. Calculate the angles BCD and AED.

Fig. 22.12

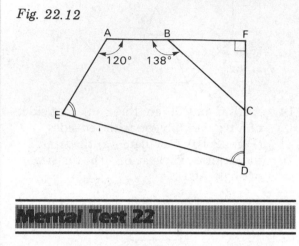

Mental Test 22

Try to answer the following questions without writing anything down except the answer.

1. The exterior angle of a polygon is 40°. What is the size of its interior angle?

2. A regular polygon has exterior angles of 60°. How many sides has the polygon?

3. A regular polygon has interior angles of 135°. How many sides has the polygon?

4. A regular polygon has 10 sides. Work out the size of its exterior angles.

5. A regular polygon has each interior angle equal to 120°. How many sides does it have?

6. What is the name given to the polygon in question 5?

7. A regular polygon has 8 sides. How many axes of symmetry does it possess?

8. What is the size of the exterior angle of a regular octagon?

9. Find the size of each interior angle of a regular octagon.

The Circle

Figure 23.1 shows the main parts of a circle.

Fig. 23.1

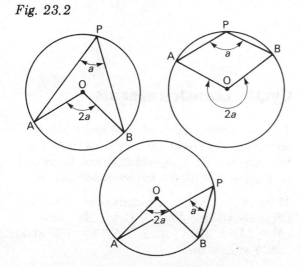

Angles in Circles

(1) *The angle which an arc of a circle subtends at the centre is twice the angle which the arc subtends at the circumference.*

Thus in Fig. 23.2, $\angle AOB = 2 \times \angle APB$.

Fig. 23.2

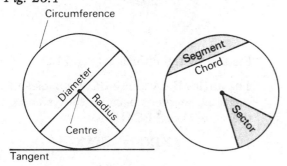

Example 1

In Fig. 23.3, O is the centre of the circle and $\angle ACB = 40°$. Find the size of $\angle AOB$.

Because $\angle AOB$ is the angle subtended by the arc AB at the centre of the circle and $\angle ACB$ is the angle subtended by the arc AB at the circumference

$$\angle AOB = 2 \times \angle ACB$$
$$= 2 \times 40°$$
$$= 80°$$

Fig. 23.3

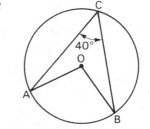

(2) *If a triangle is drawn in a circle with a diameter as the base of the triangle, then the angle opposite that diameter is a right angle.*

Because the diameter AB (Fig. 23.4) forms the angle (180°) subtended by the semicircle ABD, the angle subtended by this arc at the circumference is 90°.

Fig. 23.4

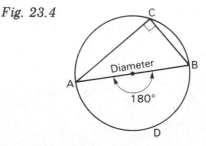

Example 2

In Fig. 23.5, calculate the diameter of the circle whose centre is O.

Fig. 23.5

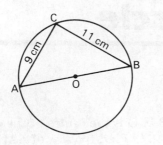

Since AB is a diameter, $\angle ACB = 90°$.

In $\triangle ABC$, using Pythagoras' theorem

$$AB^2 = AC^2 + BC^2$$
$$= 9^2 + 11^2$$
$$= 81 + 121$$
$$= 202$$
$$AB = \sqrt{202}$$
$$= 14.2$$

Hence the diameter of the circle is 14.2 cm.

(3) *Angles in the same segment of a circle are equal.*

In Fig. 23.6, the angles ACB and ADB stand on the same chord AB and so they are angles in the same segment.
Therefore $\angle ACB = \angle ADB$.

Fig. 23.6

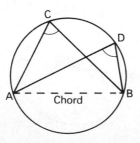

Example 3

In Fig. 23.7, find the size of the angles marked x and y.

Fig. 23.7

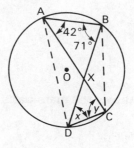

First draw the chords AD and BC.

The angles BAX and XDC are angles in the same segment because they stand on the same chord BC. Hence

$$\angle XDC = \angle BAX$$
$$= 42°$$

i.e. $x = 42°$

The angles ABX and XCD are angles in the same segment because they stand on the same chord AD. Hence

$$\angle XCD = \angle ABX$$
$$= 71°$$

i.e. $y = 71°$

Cyclic Quadrilaterals

The opposite angles of any quadrilateral inscribed in a circle are supplementary (i.e. the sum is $180°$). A quadrilateral inscribed in a circle is called a **cyclic quadrilateral**.

Thus the cyclic quadrilateral ABCD (Fig. 23.8) has $\angle A + \angle C = 180°$ and $\angle B + \angle D = 180°$. Also, $\angle CDX = \angle B$ and $\angle BCY = \angle A$ etc.

Fig. 23.8

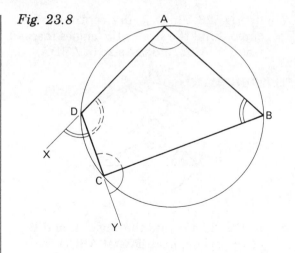

Example 4

ABCD (Fig. 23.9) is a cyclic quadrilateral with ∠A = 100° and ∠B and ∠D equal. Find the angles C and D.

Fig. 23.9

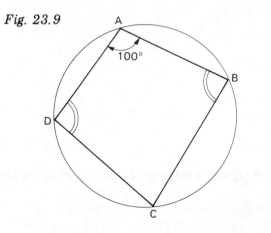

Since ABCD is a cyclic quadrilateral

$$\angle A + \angle C = 180°$$

$$\angle C = 180° - \angle A$$

$$= 180° - 100°$$

$$= 80°$$

Also

$$\angle B + \angle D = 180°$$

$$\angle B = \angle D$$

Therefore ∠D = 90°

Exercise 23.1

1. In Fig. 23.10, AC is the diameter of the circle. Find the angles ACB, ADB and BDC.

Fig. 23.10

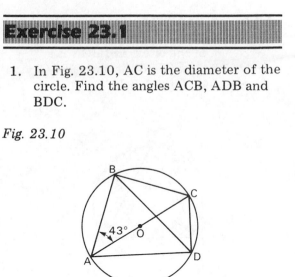

2. In Fig. 23.11, AC is the diameter of the circle. Calculate the length of AC.

Fig. 23.11

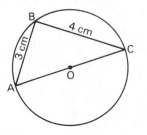

3. A circle centre O (Fig. 23.12) has ∠AOB = 40°. Find the size of the angles ABO and OAC.

Fig. 23.12

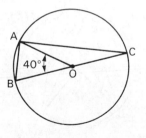

4. In Fig. 23.13, ABC is an isosceles triangle with AB = AC. O is the centre of the circle and BOY is a straight line. Find the angles BAC, BYC, ACY and YBC.

Fig. 23.13

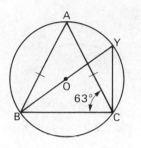

5. In Fig. 23.14, K is the centre of the circle, AC is a diameter and BD is parallel to AC. Find the angles CBA, ACB and CAD.

Fig. 23.14

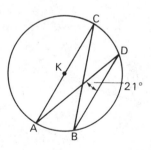

6. In Fig. 23.15, O is the centre of the circle. Find the size of the angles AOB and ADB.

Fig. 23.15

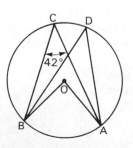

7. In Fig. 23.16, O is the centre of the circle. Find the size of the angles marked x and y. Also find the size of $\angle$BDA.

Fig. 23.16

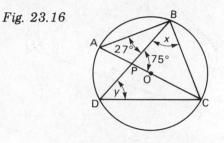

8. In Fig. 23.17, find the angles C and D of the cyclic quadrilateral ABCD. What is the size of $\angle$BCX?

Fig. 23.17

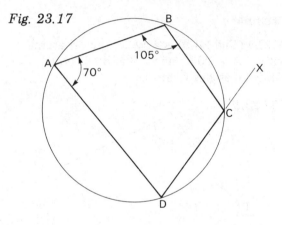

Tangent Properties of a Circle

A **tangent** is a line which touches a circle at one point. This point is called the **point of tangency** (Fig. 23.18).

Fig. 23.18

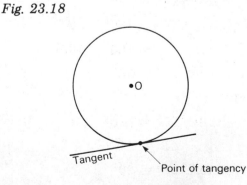

(1) *A tangent to a circle lies at right angles to a radius drawn to the point of tangency.*

Thus in Fig. 23.19, $\angle OTS$ is a right angle.

Fig. 23.19

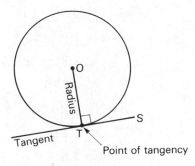

Example 5

In Fig. 23.20, TS is a tangent, T being the point of tangency and O the centre of the circle. If the radius of the circle is 5 cm and OS = 7 cm, find the distance TS.

Fig. 23.20

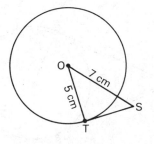

Since OT is a radius drawn to the point of tangency, $\angle OTS = 90°$.
By Pythagoras' theorem

$$TS^2 = OS^2 - OT^2$$
$$= 7^2 - 5^2$$
$$= 49 - 25$$
$$= 24$$
$$TS = \sqrt{24}$$
$$= 4.90$$

Hence the distance TS is 4.90 cm.

(2) *If, from a point outside a circle, tangents are drawn to the circle, then the two tangents are equal in length.*

Thus in Fig. 23.21, PX = PY. The quadrilateral OXPY is a kite since OX = OY (radii) and it is symmetrical about the line OP. Hence $\angle XPY$ is bisected and $\angle KXP = \angle KYP$.

Fig. 23.21

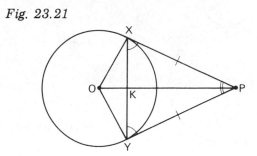

Example 6

In Fig. 23.22, calculate the size of the angles marked a and b.

Fig. 23.22

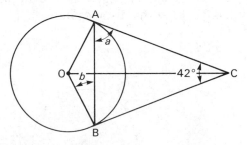

The two tangents AC and BC meet at C. Hence AC = BC and ABC is isosceles. Therefore

$$a = \tfrac{1}{2} \times (180° - 42°)$$
$$= \tfrac{1}{2} \times 138°$$
$$= 69°$$

Since OBC is the angle between a tangent and a radius, it equals $90°$. Therefore

$$b = \angle OBC - \angle ABC$$
$$= 90° - 69°$$
$$= 21°$$

(3) *If two circles touch internally or exter-
nally then the line which passes through
their centres also passes through the
point of tangency* (Fig. 23.23).

Fig. 23.23

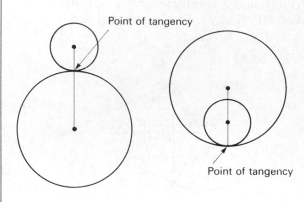

Point of tangency

Point of tangency

Example 7

Three circles are arranged as shown in
Fig. 23.24. Find the distance h.

Fig. 23.24

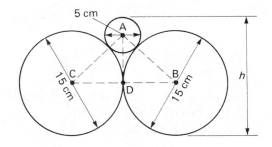

5 cm

15 cm

15 cm

Because the circles are tangential to
each other

$$AC = 7.5 + 2.5$$

$$= 10\,cm$$

$$AB = 7.5 + 2.5$$

$$= 10\,cm$$

$$BC = 7.5 + 7.5$$

$$= 15\,cm$$

Therefore triangle ABC is isosceles, and
hence

$$CD = \tfrac{1}{2} \text{ of } 15$$

$$= 7.5\,cm$$

In $\triangle$ACD, using Pythagoras' theorem

$$AD^2 = AC^2 - CD^2$$

$$= 10^2 - 7.5^2$$

$$= 100 - 56.25$$

$$= 43.75$$

$$AD = \sqrt{43.75}$$

$$= 6.61$$

$$h = 7.5 + 6.61 + 2.5$$

$$= 16.61\,cm$$

Exercise 23.2

1. In Fig. 23.25, calculate the distance OP
   given that OA is a radius and PA and
   PB are tangents meeting at P.

Fig. 23.25

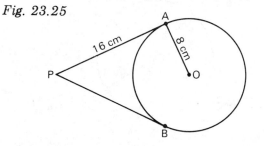

16 cm

8 cm

2. In Fig. 23.26, OA is a radius and AP
   and BP are tangents meeting at P. Find
   the size of the angles marked a and b.

Fig. 23.26

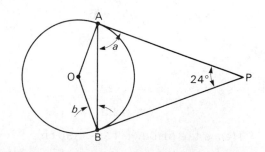

24°

3. In Fig. 23.27, OA is the radius of the circle and CA and CB are tangents meeting at C. Find the length of CB.

Fig. 23.27

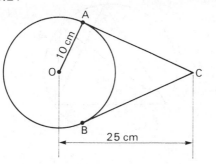

6. Fig. 23.30 shows two circles which are touching. By using Pythagoras' theorem, calculate the distance x.

Fig. 23.30

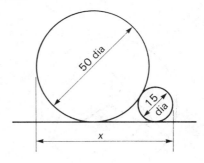

4. In Fig. 23.28, AB and AC are tangents to the circle whose centre is O. AOD is a straight line. Find the size of the angles AOB, OBD and CBD.

Fig. 23.28

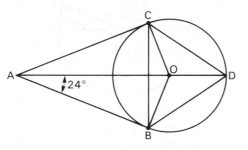

7. In Fig. 23.31, find the distance h.

Fig. 23.31

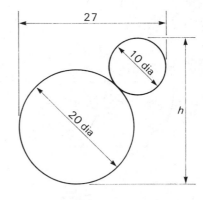

5. In Fig. 23.29, C is the centre of the circle and AP is a tangent to the circle. Calculate the radius of the circle.

Fig. 23.29

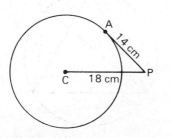

8. Three circles are arranged as shown in Fig. 23.32. Find the distance h.

Fig. 23.32

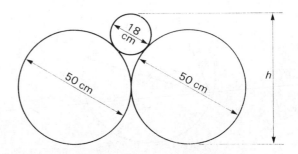

Mental Test 23

1. Find the angle marked *a* in Fig. 23.33. O is the centre of the circle.

Fig. 23.33

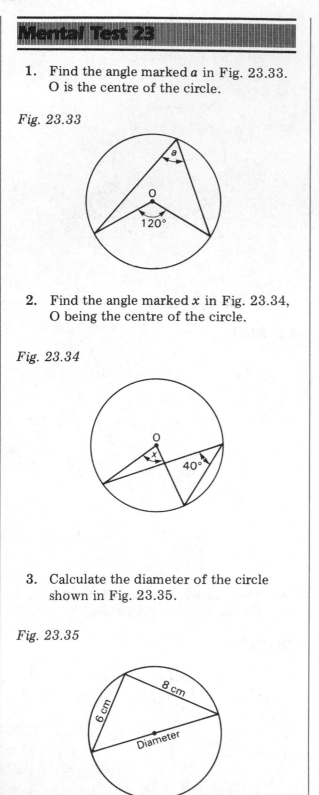

2. Find the angle marked *x* in Fig. 23.34, O being the centre of the circle.

Fig. 23.34

3. Calculate the diameter of the circle shown in Fig. 23.35.

Fig. 23.35

4. What is the size of the angle marked *p* in Fig. 23.36?

Fig. 23.36

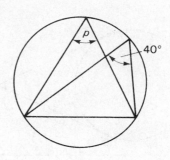

5. In Fig. 23.37, what is the size of ∠ABO, O being the centre of the circle?

Fig. 23.37

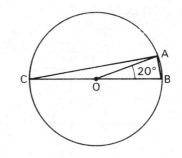

6. In Fig. 23.37, what is the size of ∠CAO?

7. In Fig. 23.38, ABC is an equilateral triangle. What is the size of ∠BYC?

Fig. 23.38

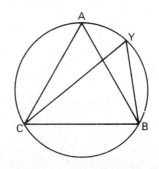

8. In Fig. 23.39, TS is a tangent and OT is a radius. Find the length of OS.

Fig. 23.39

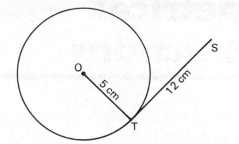

9. In Fig. 23.40, find the angle marked x. O is the centre of the circle and AT and BT are tangents.

Fig. 23.40

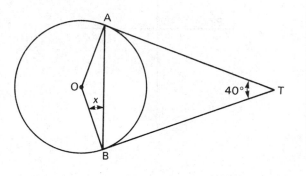

10. In Fig. 23.41, the three circles touch. Find the lengths AB, AC and BC if A, B and C are the centres of the circles.

Fig. 23.41

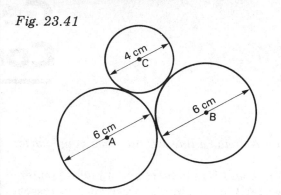

Geometrical Constructions

(1) *To divide a line AB into two equal parts:*

With A and B as centres and a radius greater than $\frac{1}{2}$AB, draw circular arcs which intersect at X and Y (Fig. 24.1). Join XY. The line XY divides AB into two equal parts and is also perpendicular to AB.

Fig. 24.1

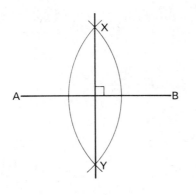

(2) *To draw a perpendicular from a given point A on a straight line:*

With centre A and any radius, draw a circle to cut the straight line at points P and Q (Fig. 24.2). With centres P and Q and a radius greater than AP (or AQ), draw circular arcs to intersect at X and Y. Join XY. This line will pass through A and is perpendicular to the given line.

Fig. 24.2

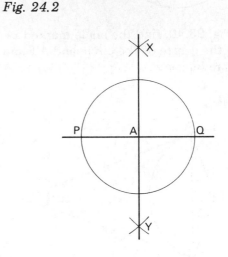

(3) *To draw a perpendicular from a point A at the end of a line:*

From any point O outside the line and with radius OA, draw a circle to cut the line at B (Fig. 24.3). Draw the diameter BC and join AC. AC is perpendicular to the straight line (because the angle in a semicircle is 90°).

Fig. 24.3

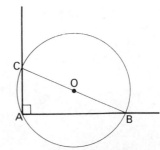

(4) *To draw the perpendicular to a line AB from a given point P which is not on the line:*

With P as the centre, draw a circular arc to cut AB at points C and D. With C and D as centres and a radius greater than $\frac{1}{2}$CD, draw circular arcs to intersect at E. Join PE. The line PE is the required perpendicular (Fig. 24.4).

Fig. 24.4

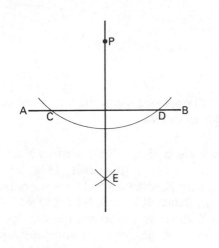

(5) *To construct an angle of 60°:*

Draw a line AB. With A as centre, and any radius, draw a circular arc to cut AB at D. With D as centre and the *same* radius, draw a second arc to cut the first arc at C. Join AC. The angle CAD is then 60° (Fig. 24.5).

Fig. 24.5

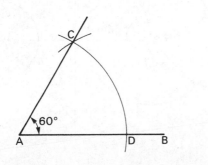

(6) *To bisect a given angle BAC:*

With centre A and any radius, draw an arc to cut AB at D and AC at E. With centres D and E and a radius greater than $\frac{1}{2}$DE, draw arcs to intersect at F. Join AF, then AF bisects ∠BAC (Fig. 24.6). Note that by bisecting an angle of 60°, an angle of 30° is obtained. An angle of 45° is obtained by bisecting a right angle.

Fig. 24.6

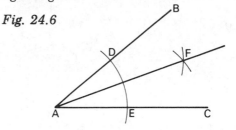

(7) *To construct an angle equal to a given angle BAC:*

With centre A and any radius, draw an arc to cut AB at D and AC at E. Draw a line XY. With centre X and the same radius, draw an arc to cut XY at W. With centre W and radius equal to DE draw an arc to cut the first arc at V. Join VX, then ∠VXW = ∠BAC (Fig. 24.7).

Fig. 24.7

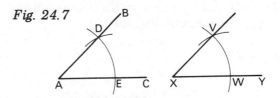

(8) *To draw through point P a line parallel to a given line AB:*

Mark off any two points X and Y on AB. With centre P and radius XY, draw an arc. With centre Y and radius XP, draw a second arc to cut the first arc at Q. Join PQ, then PQ is parallel to AB (Fig. 24.8).

Fig. 24.8

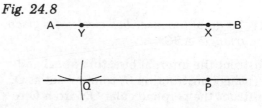

(9) *To divide a straight line AB into a number of equal parts:*

Suppose that AB has to be divided into four equal parts. Draw AC at an angle to AB. Mark off on AC four equal parts AP, PQ, QR, RS of any convenient length. Join SB. Draw RV (using construction 8), QW and PX each parallel to SB. Then AX = XW = WV = VB (Fig. 24.9).

Fig. 24.9

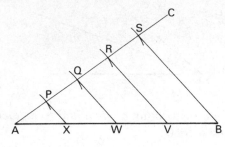

(10) *To draw the circumscribed circle of a given triangle ABC:*

Construct the perpendicular bisectors of the sides AB and AC (using construction 1) so that they intersect at O. With centre O and radius AO, draw a circle which is the required circumscribed circle (Fig. 24.10).

Fig. 24.10

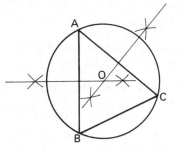

(11) *To draw the inscribed circle of a given triangle ABC:*

Construct the internal bisectors of ∠B and ∠C (using construction 6) to intersect at O. Construct the perpendicular ON from O to

BC (using construction 4). With centre O, and radius ON, draw the inscribed circle of the triangle ABC (Fig. 24.11).

Fig. 24.11

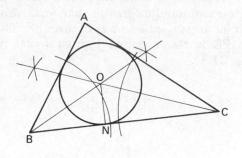

(12) *To construct a common tangent to two given circles:*

The two given circles have centres X and Y and radii x and y, respectively (Fig. 24.12). With centre X, draw a circle whose radius is $(x - y)$. Bisect XY at Z. With radius XZ and centre Z, draw an arc to cut the previously drawn circle at M. Join XM and produce to P at the circumference of the circle. Draw YQ parallel to XP (using construction 8), Q being at the circumference of the circle. Join PQ, which is the required tangent.

Fig. 24.12

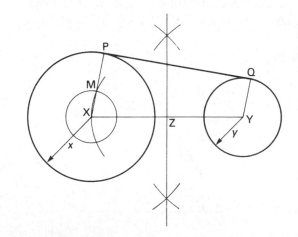

(13) *To construct a pair of tangents from an external point to a given circle:*

It is required to draw a pair of tangents from the point P to the circle, centre O (Fig. 24.13). Join OP. Bisect OP at X. With OX as radius and centre X draw a circular arc to cut the given circle at points A and B. Join PA and PB, which are the required pair of tangents.

Fig. 24.13

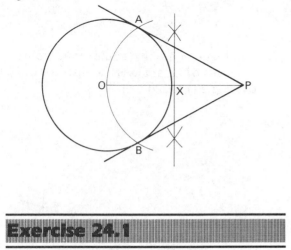

![Exercise 24.1]

In this exercise, only a rule and compasses should be used:

1. Draw a line PQ 8 cm long and construct its perpendicular bisector.

2. Draw a line XY 9 cm long and mark off a point A such that AX = 4 cm. Through the point A draw a line perpendicular to XY.

3. Draw a line PQ 7 cm long, and through P draw a line perpendicular to PQ.

4. Draw a horizontal line AB which is 9 cm long. Then mark off a point P such that P is 6 cm from A and 5 cm from B. Now construct a line perpendicular to AB which also passes through point P.

5. (a) Construct an angle of $60°$.

   (b) By bisecting your $60°$ angle, construct an angle of $30°$.

6. Draw a line AB 8 cm long and construct a line perpendicular to AB which passes through the point A. By bisecting the right angle so formed, construct an angle of $45°$.

7. Draw a line 9 cm long and divide it into 5 equal parts without measuring each division.

8. (a) Construct the triangle ABC given that AB = 7 cm, BC = 6 cm and AC = 5 cm.

   (b) Draw the circumscribed circle for triangle ABC and measure its diameter.

9. (a) Construct triangle XYZ given XY = 5 cm, YZ = 7 cm and $\angle XYZ = 50°$.

   (b) Draw the inscribed circle for $\triangle XYZ$ and state its diameter.

10. (a) Draw a triangle ABC given that AB = 6.3 cm, AC = 7.7 cm and BC = 8.4 cm.

    (b) Draw the circle which passes through A, B and C. Measure the radius of this circle.

11. Construct the quadrilateral ABCD from the following data: AB = 10 cm, BC = 6 cm, AC = 14 cm, $\angle DAC = 60°$ and $\angle DCA = 45°$. Measure and record the length DB.

12. Draw a line AB 8 cm long. Now construct an equilateral triangle on the line segment AB.

13. Draw two circles whose centres are 9 cm apart and whose diameters are 5 cm and 7 cm. Draw the common tangent to these circles.

14. Draw a circle, centre O, whose radius is 5 cm. Mark off any point P so that OP = 9 cm. Construct a pair of tangents from P to the circle.

Loci

A **locus** is a set of points traced out by a point which moves according to some given law. For instance the locus of a point which moves so that it is always 3 cm from a given fixed point is a circle having a radius of 3 cm.

It often helps if we mark off a few points according to the given law. By doing this we may gain some idea of what the locus will be. Sometimes three or four points will be sufficient but sometimes ten or more points may be required before the locus can be recognised.

Example 1

Given a straight line of length 6 cm, find the locus of a point P so that APB is always a right angle.

By drawing a number of points P_1, P_2, P_3, ..., etc., so that the angles AP_1B, AP_2B, AP_3B, ..., etc. are all right angles, it appears that the locus is a circle with AB as diameter as shown in Fig. 24.14.

Fig. 24.14

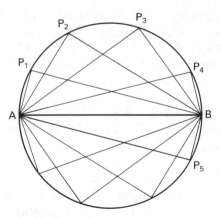

Standard Loci

The following standard loci should be remembered:

(1) The locus of a point equidistant from two given points, A and B, is the per-pendicular bisector of the line AB (Fig. 24.15).

Fig. 24.15

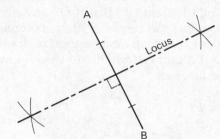

(2) The locus of a point equidistant from the arms of an angle is the bisector of the angle (Fig. 24.16).

Fig. 24.16

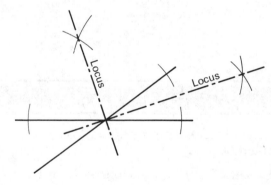

Intersecting Loci

Sometimes two pieces of information are given about the position of a point. Each piece of information is then dealt with separately because any attempt to comply with the two conditions at the same time will lead to a trial and error method which is not acceptable. Each piece of information will partially locate the point and the inter-section of the two loci will determine the required position of the point.

Example 2

A point P lies 3 cm from a given straight line and it is also equidistant from two fixed points not on the line and not perpendicular to it. Find the two possible positions of P.

> *Condition 1.* The point P lies 3 cm from the given straight line (AB in Fig. 24.17). To meet this condition draw two straight lines, X_1Y_1 and X_2Y_2, parallel to AB and on either side of it.
>
> *Condition 2.* P is equidistant from the two fixed points R and S in Fig. 24.17. To meet this condition we draw the perpendicular bisector of RS because $\triangle$RSP must be isosceles. The intersections of the two loci give the required positions of the point P. These are the points P_1 and P_2 in the diagram.

Fig. 24.17

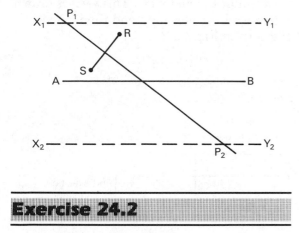

Exercise 24.2

1. Find the locus of a point P which is always 5 cm from a fixed straight line of infinite length.

2. Find the locus of a point P which moves so that it is always 5 cm from a given straight line AB which is 8 cm long.

3. XYZ is a triangle whose base XY is fixed. If XY = 5 cm and the area of XYZ = 10 cm², find the locus of Z.

4. Given a square of 10 cm side, find all the points which are 8 cm from two of the vertices. How many points are there?

5. Find all the points which are 5 cm from each of two intersecting straight lines inclined at an angle of 45°. How many points are there?

6. AB is a fixed line of length 8 cm and P is a variable point. The distance of P from the mid-point of AB is 5 cm and the distance of P from AB is 4 cm. Construct a point P so that both of these conditions are satisfied. State the number of possible positions for P.

7. AB is a fixed line of length 4 cm and R is a point such that the area of triangle ABR is 5 cm². S is the mid-point of AR. State the locus of R and the locus of S.

8. Draw a circle centre O, having a radius of 4 cm. Construct the locus of the mid-points of all the chords of this circle which are 6.5 cm long.

Orthographic Projection

In **orthographic projection** a full view of each of the faces of an object is shown in turn. Usually, but not always, three of these views are enough to show all the details of the object.

There are two versions of orthographic projection, first-angle projection and third-angle projection. Only first-angle projection will be described in this book.

First-Angle Projection

Figure 25.1 shows a pictorial drawing of a wooden block. To represent this in first-angle projection, we look at the front of the block in the direction of arrow A. The view we see is called the **elevation.**

Fig. 25.1

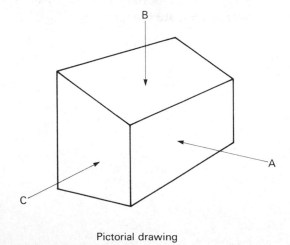

Pictorial drawing

Next we look at the object in the direction of arrow B. This view is called the **plan view.** The plan view is drawn directly underneath the elevation, as shown in Fig. 25.2. Finally we look at the object in the direction of arrow C. The view we see is called the **end elevation** or the **end view.** This view is drawn in line with the elevation and the plan view, as shown in Fig. 25.2.

Fig. 25.2

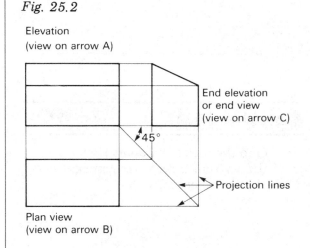

Elevation
(view on arrow A)

End elevation
or end view
(view on arrow C)

45°

Projection lines

Plan view
(view on arrow B)

Projection Lines

Projection lines are used in orthographic projection to make sure that the views are correctly positioned relative to each other. In Fig. 25.2, the faint projection lines are shown and these lines ensure that the plan view, elevation and end view line up properly.

Hidden Details

Hidden details are shown by means of dotted lines. In drawing the object shown in Fig. 25.3 we cannot see the face AB when looking at the top of the object. Hence this face is represented by dotted lines in the plan view.

Fig. 25.3

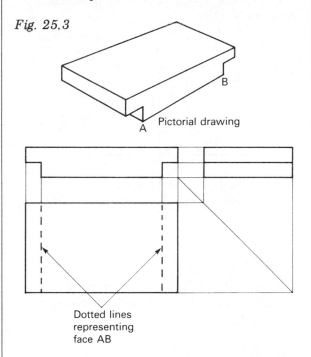

Pictorial drawing

Dotted lines representing face AB

Two-View Drawings

A two-view drawing is often enough to represent an object fully. Thus the object shown in Fig. 25.4 may be fully represented by an elevation and an end view.

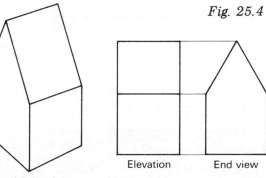

Fig. 25.4

Elevation   End view

Pictorial drawing

Make three-view drawings, in correct projection, of each of the objects shown pictorially below:

1.
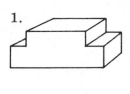

2.

3.
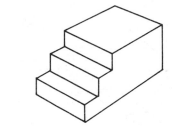

4.
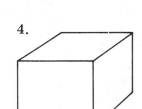

5.
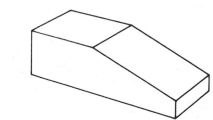

6.

7.
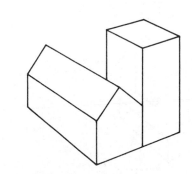

8. 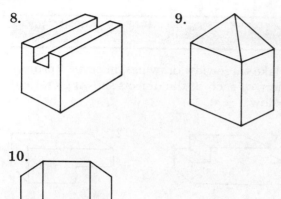   **9.**

10.

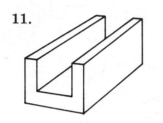

Make two-view drawings of the objects shown below:

11.

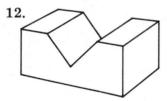

12.

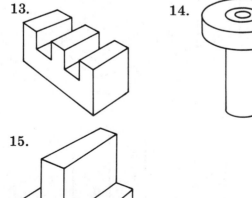

13.   **14.**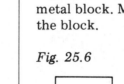

15.

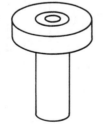

Isometric Projection

Isometric projection is a way of representing an object by means of a pictorial drawing. It is often used to make a pictorial drawing from two- and three-view drawings.

Fig. 25.5 shows the **isometric axes** OX, OY and OZ from which an isometric drawing is made.

Fig. 25.5

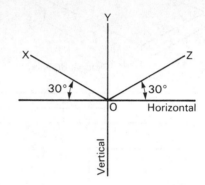

Example 1

Fig. 25.6 shows a two-view drawing of a metal block. Make an isometric drawing of the block.

Fig. 25.6

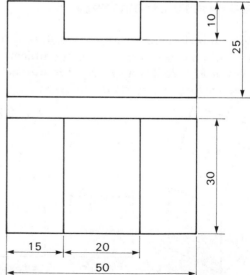

The first step is to draw the isometric axes OX, OY and OZ (Fig. 25.7). Vertical edges of the block are represented by vertical lines in the isometric drawing. For rectangular-type objects, all the lines making up

the isometric drawing either lie along the isometric axes or are parallel to them, as shown in Fig. 25.7. Note that when making an isometric drawing, all measurements are made full size.

Fig. 25.7

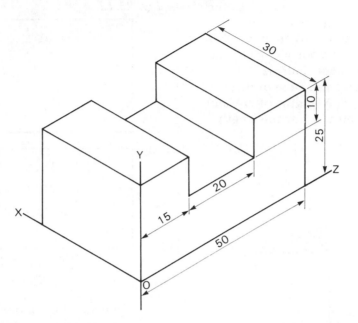

Isometric Representation of Circles

Representing circles in an isometric drawing sometimes causes difficulty. Fig. 25.8 gives the clue necessary to overcome this difficulty. Fig. 25.8(a) shows a true circle inscribed in a square. At the points marked X the circumference of the circle and the sides of the square touch. This must be the same in the isometric view shown in Fig. 25.8(b).

Fig. 25.8

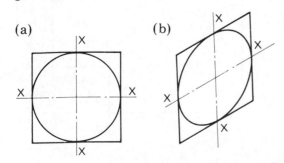

Fig. 25.9 shows the method of making an isometric drawing of a cylinder. The rectangular shape shown by the faint lines is needed to obtain the shape of the circular ends of the cylinder.

Fig. 25.9

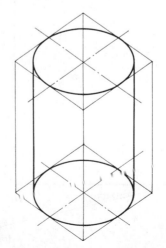

Oblique Projection

Oblique projection is a second way of representing objects pictorially. Figure 25.10 shows the axes used. OA is horizontal, OB is vertical whilst OC is drawn at 45° to the horizontal. Measurements made along OA and OB (or parallel to them) are made full-size but measurements made along OC (or parallel to it) are made half-size.
Fig. 25.11 shows a 50 mm cube drawn in oblique projection. The lengths of OA and OB are both made 50 mm but the length OC is 25 mm.

Fig. 25.10

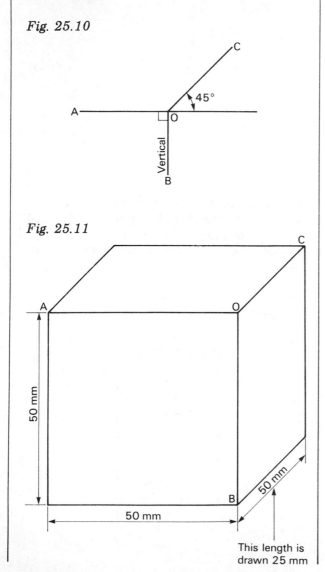

Fig. 25.11

This length is drawn 25 mm

Example 2

Figure 25.12 shows a two-view orthographic drawing of a metal block. Draw an oblique projection of this block.

Fig. 25.12

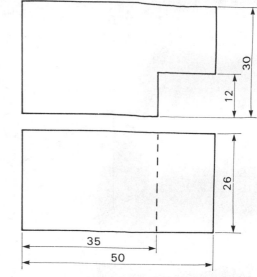

To make the oblique drawing, first draw the three axes OA, OB and OC. The front face of the block in oblique projection is the same as the front elevation of the block. Thus in Fig. 25.13 the front face of the block is drawn using the axes OA and OB. The length OC is made 13 mm (i.e. half of 26 mm).

Fig. 25.13

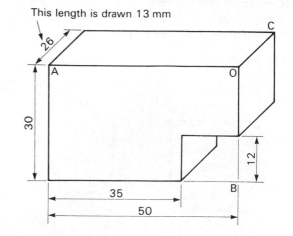

This length is drawn 13 mm

It is easier to represent circles in oblique projection than in isometric projection. Figure 25.14 is an example. It will be noticed that the circles are drawn as true circles at the front face of the oblique projection.

Fig. 25.14

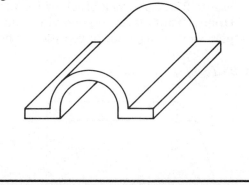

Exercise 25.2

Make **(a)** isometric drawings, **(b)** oblique drawings of each of the objects shown in orthographic projection below. All dimensions are in millimetres:

1.

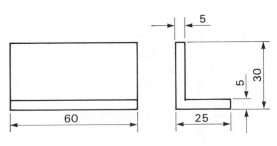

2.

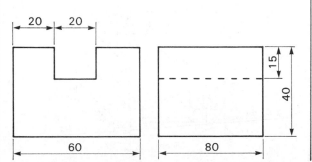

3.

4.

5.

6.

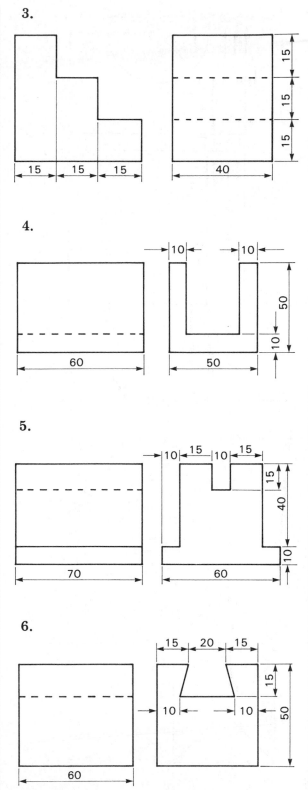

7.

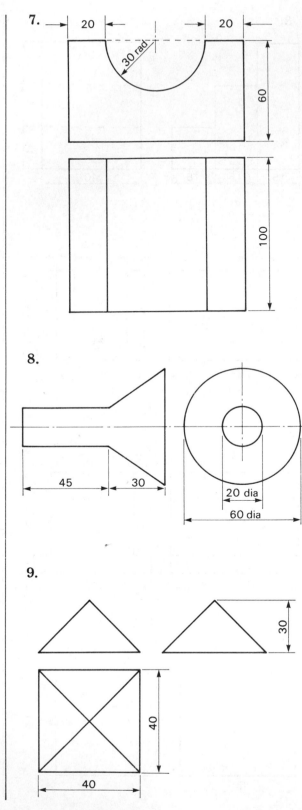

8.

9.

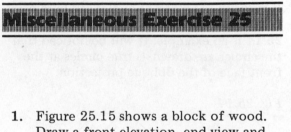

1. Figure 25.15 shows a block of wood. Draw a front elevation, end view and plan view using 1 cm to represent 5 cm.

Fig. 25.15

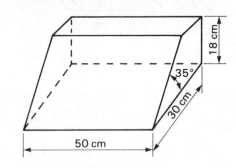

2. Figure 25.16 illustrates a blackboard cleaner consisting of a piece of felt inserted into a wooden holder. Draw, full-size, the plan, front elevation and end elevation of the cleaner. The three views must be correctly positioned to each other.

Fig. 25.16

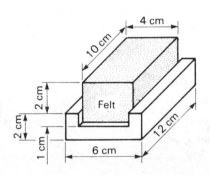

3. Make an isometric drawing of the object shown in Fig. 25.17.

Fig. 25.17

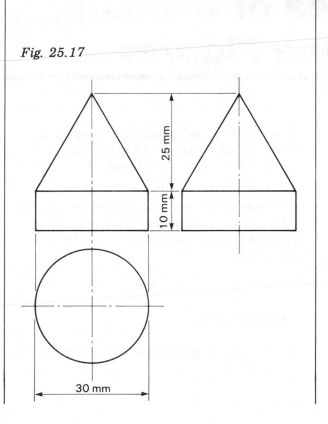

30 mm

4. A pyramid with a square base stands on a horizontal table (Fig. 25.18). Make a three-view drawing of the pyramid with the views correctly positioned to each other.

Fig. 25.18

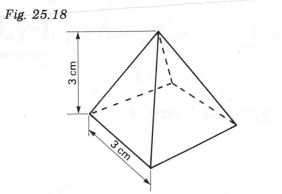

3 cm

3 cm

5. Make an isometric drawing of the object shown in Fig. 25.19.

Fig. 25.19

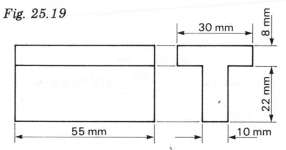

30 mm

8 mm

55 mm

22 mm

10 mm

26 Area of Plane Figures

Types of Plane Figures

A plane figure (or shape) is bounded by lines which are called the **sides** of the figure.

Triangles (Fig. 26.1) have three sides.

Fig. 26.1

Quadrilaterals (Fig. 26.2) have four sides.

Fig. 26.2

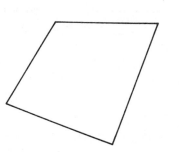

Polygons (Fig. 26.3) are plane figures bounded by straight lines. Thus a triangle is a polygon with three sides and a quadrilateral is a polygon with four sides.

Fig. 26.3

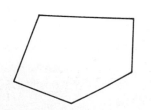

A **regular polygon** has all its sides equal in length and all its angles equal in size. An equilateral triangle and a square are examples of regular polygons.

A **pentagon** has five sides, a **hexagon** has six sides and an **octagon** has eight sides. The names of these polygons should be memorised.

Perimeters

The distance round a plane figure is called its **perimeter**.

Example 1

(a) An equilateral triangle has all its sides 5 cm in length. Calculate its perimeter.

$$\text{Perimeter} = (5 + 5 + 5)\,\text{cm}$$
$$= 15\,\text{cm}$$

(b) The longer side of a rectangle is 8 cm in length and its shorter side is 5 cm long. What is the perimeter of the rectangle?

$$\text{Perimeter} = (8 + 5 + 8 + 5)\,\text{cm}$$
$$= 26\,\text{cm}$$

(c) A regular hexagon has each of its sides 7 cm long. Calculate the perimeter of the hexagon. Since a regular hexagon has six equal sides

$$\text{Perimeter} = (6 \times 7)\,\text{cm}$$
$$= 42\,\text{cm}$$

Exercise 26.1

1. Each side of a square is 8 cm long. Find its perimeter.

2. A rhombus has sides 5 cm long. Calculate its perimeter.

3. A regular pentagon has each of its sides 6 cm long. What is its perimeter?

4. Figure 26.4 shows a scalene triangle. Find its perimeter.

Fig. 26.4

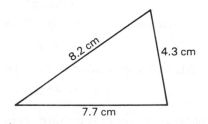

8.2 cm   4.3 cm

7.7 cm

5. Fig. 26.5 shows a trapezium. Calculate its perimeter.

Fig. 26.5

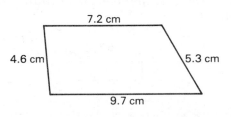

7.2 cm

4.6 cm   5.3 cm

9.7 cm

6. A rectangle has a long side 8.2 cm in length and a short side 7.5 cm in length. Calculate the perimeter of the rectangle.

7. An isosceles triangle has a base 7.8 cm long and equal sides 8.2 cm long. What is the perimeter of the triangle?

8. A kite has a long side 7.3 cm in length and a short side 3.8 cm long. Find its perimeter.

Units of Area

The amount of surface that a shape possesses is called its **area**.

The area of a shape is measured by counting the number of equal squares contained in the shape.

Example 2

Look at the shapes shown in Fig. 26.6. By counting the number of equal squares contained in each plane figure, find out which figure has the greatest area.

Fig. 26.6

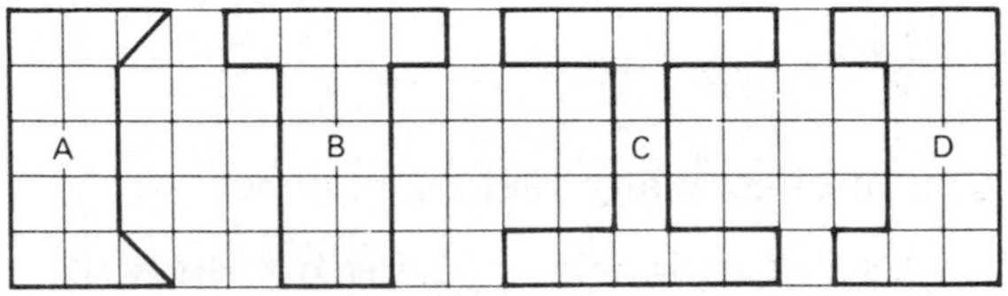

Figure A contains 11 equal squares
Figure B contains 12 equal squares
Figure C contains 13 equal squares
Figure D contains 12 equal squares

Therefore Figure C has the greatest area because it contains the largest number of equal squares.

In practice the squares used for measuring area have sides of 1 metre or 1 centimetre or 1 millimetre. The area inside a square having a side of 1 metre is 1 square metre; 1 square centimetre is the area inside a square with a side of 1 centimetre; 1 square millimetre is the area inside a square having a side of 1 millimetre.

The standard abbreviations for the units of area are:

$$\text{square metre} = \text{m}^2$$

$$\text{square centimetre} = \text{cm}^2$$

$$\text{square millimetre} = \text{mm}^2$$

In the imperial system, areas are measured in square inches (in^2), square feet (ft^2) and square yards (yd^2).

Area of a Rectangle

The rectangle shown in Fig. 26.7 contains $4 \times 2 = 8$ equal squares, each of which has an area of 1 square centimetre. Hence the area of the rectangle is $8\,\text{cm}^2$. What we have done is to multiply the length by the breadth in order to find the area of the rectangle.

Fig. 26.7

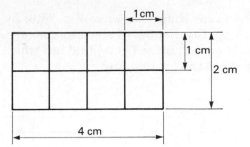

This applies to any rectangle, hence

    Area of rectangle = Length × Breadth

If we let A = area of the rectangle, l = length of the rectangle and b = breadth of the rectangle then

$$A = lb$$

When using this formula the units of l and b must be the same. That is, they both must be stated in metres, or centimetres, or millimetres, etc.

Example 3

A rectangular carpet measures 6 m by 5 m. Calculate the area of the carpet.

    Here $l = 6\,\text{m}$ and $b = 5\,\text{m}$.

$$A = lb$$
$$= (6 \times 5)\,\text{m}^2$$
$$= 30\,\text{m}^2$$

Example 4

A room 9 m long and 7 m wide is to be carpeted so as to leave a surround 50 cm wide as shown in Fig. 26.8. Find the area of the surround.

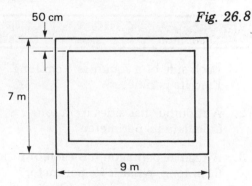

Fig. 26.8

One way of solving this problem is to calculate the area of the room, and from it subtract the area of the carpet.

$$\text{Area of room} = (7 \times 9)\,\text{m}^2$$
$$= 63\,\text{m}^2$$
$$\text{Area of carpet} = (6 \times 8)\,\text{m}^2$$
$$= 48\,\text{m}^2$$
$$\text{Area of surround} = (63 - 48)\,\text{m}^2$$
$$= 15\,\text{m}^2$$

Many shapes can be split up into rectangles. Then by calculating the area of each rectangle the area of the shape can be found by simple addition or subtraction.

Example 5

Find the area of the steel girder shown in Fig. 26.9.

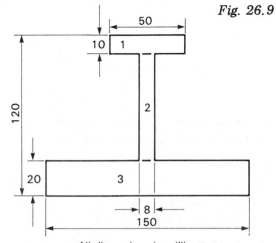

Fig. 26.9

All dimensions in millimetres

The shape can be divided up into three rectangles as shown in the diagram.

Area of shape = Area of rectangle 1
+ Area of rectangle 2
+ Area of rectangle 3

$$= (50 \times 10)\,\text{mm}^2$$
$$+ (90 \times 8)\,\text{mm}^2$$
$$+ (150 \times 20)\,\text{mm}^2$$
$$= 500\,\text{mm}^2 + 720\,\text{mm}^2$$
$$+ 3000\,\text{mm}^2$$
$$= 4220\,\text{mm}^2$$

Example 6

A piece of wood is 85 cm long and 30 cm wide. What is its area in square metres?

In problems of this kind it is best to express each dimension in metres before attempting to find the area. Thus 85 cm = 0.85 m, and 30 cm = 0.3 m.

Area of wood = $(0.85 \times 0.3)\,\text{m}^2$

$$= 0.255\,\text{m}^2$$

Sometimes we are given the area of a rectangle and one dimension (such as the width) and we have to find the other dimension (such as the length). The formula $A = lb$ can be transposed to give

$$l = \frac{A}{b} \quad \text{or} \quad b = \frac{A}{l}$$

Note carefully that, when using these transposed formulae, the units for each term in the formula must be of the same kind. Thus if the area is given in square metres, the given dimension (length or width) must be in metres and the remaining dimension will then also be in metres.

Example 7

The area of a rectangle is 63 in². If its width is 7 in, find the length of the rectangle.

We are given that $A = 63\,\text{in}^2$ and $b = 7\,\text{in}$.

$$l = \frac{A}{b}$$

$$= \frac{63}{7}\,\text{in}$$

$$= 9\,\text{in}$$

Hence the length of the rectangle is 9 in.

Area of a Square

Since a square is a rectangle with all sides equal in length

Area of square = Side × Side

$$= \text{Side}^2$$

If A = area of the square and a = length of side

$$A = a^2$$

Example 8

(a) A square has sides 8 cm long. Calculate its area.

We are given that $a = 8$ cm, hence

$$A = a^2$$

$$= 8^2\,\text{cm}^2$$

$$= 64\,\text{cm}^2$$

(b) A square plate has an area of 25 ft². Calculate the length of its sides.

We are given $A = 25\,\text{ft}^2$ and we have to find a.

$$a^2 = A$$

$$a = \sqrt{A}$$

$$= \sqrt{25}$$

$$= 5$$

Hence the plate has sides which are 5 ft long.

Exercise 26.2

Find the areas of the following rectangles:

1. Length 7 cm, width 5 cm.

2. Length 20 mm, width 11 mm.

3. Length 35 ft, width 8 ft.

4. Length 8.3 cm, width 5.2 cm.

5. Length 91 mm, width 20 mm.

6. A piece of wood is 3.7 m long and 30 cm wide. Find its area in square metres.

7. A rectangular metal plate is 120 cm long and 80 cm wide. Calculate its area in square metres.

8. A rectangular floor 5.8 ft long and 4.9 ft wide is to be covered with vinyl. What area of vinyl is needed?

9. What is the total area of the walls of a room 6 m long, 5 m wide and 2 m high?

10. A rectangular lawn is 32 yd long and 23 yd wide. A path 1.5 yd wide is made around the lawn. What is the area of the path?

11. A room 8.3 m long and 6.3 m wide is to be carpeted to leave a surround 60 cm wide round the carpet. Work out (a) the area of the room, (b) the area of the carpet, (c) the area of the surround.

12. Find the areas of the steel sections shown in Fig. 26.10.

Fig. 26.10

(a)

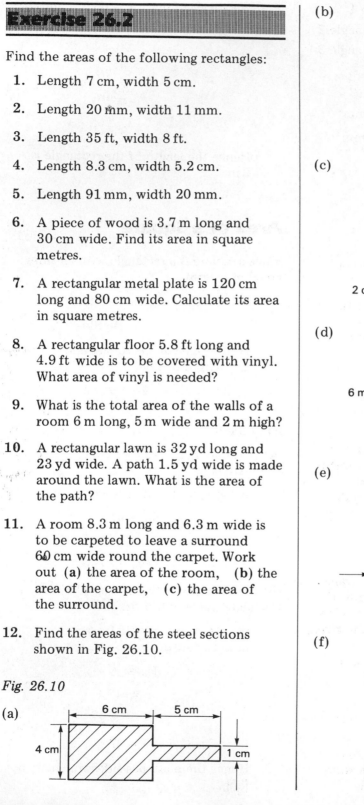

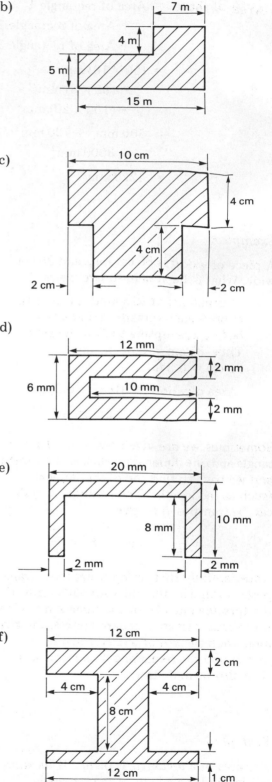

(g)

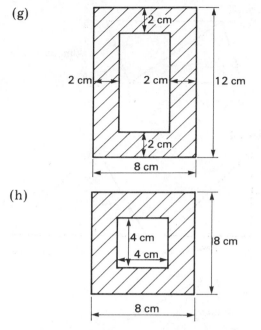

(h)

13. The area of a rectangle is $72\,\text{ft}^2$. If its length is 12 ft, what is its width?

14. The floor area of a room is $84\,\text{m}^2$. If its width is 7 m, calculate its length.

15. A square metal plate has an area of $36\,\text{cm}^2$. What is the length of its sides?

16. A tile is square and its sides are 8 cm long. Calculate its area.

17. A room is 5.4 m long and 4.2 m wide. It takes 1575 square tiles to cover the floor. Calculate the area of each tile and then find the length of its sides.

Area of a Parallelogram

A parallelogram is in effect a rectangle pushed out of square as shown in Fig. 26.11, where the equivalent rectangle is shown in dotted outline. Hence

Area of parallelogram = Length of base
× Vertical height

If A is the area, b the length of the base and h the vertical height (often called the altitude) then

$$A = bh$$

Fig. 26.11

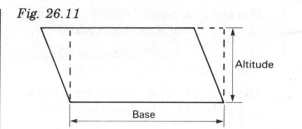

Example 9

(a) Find the area of a parallelogram whose base is 15 cm long and whose altitude is 8 cm.

We are given $b = 15$ and $h = 8$ and we have to find A.

$$A = bh$$
$$= 15 \times 8$$
$$= 120$$

Hence the area of the parallelogram is $120\,\text{cm}^2$.

(b) A parallelogram has an area of $36\,\text{in}^2$. If its base is 9 in long, find its altitude.

Since $A = bh$

$$h = \frac{A}{b}$$

We are given $A = 36$ and $b = 9$ and we have to find h.

$$h = \frac{36}{9}$$
$$= 4$$

Therefore the altitude is 4 in.

Exercise 26.3

Find the areas of the following parallelograms:

1. Base 7 cm, altitude 5 cm.

2. Base 15 mm, altitude 12 mm.

3. Base 5 ft, altitude 3 ft.

4. The area of a parallelogram is $64 \, \text{m}^2$. If its base is $16 \, \text{m}$ long, find its altitude.

5. The area of a parallelogram is $56 \, \text{in}^2$. If its altitude is $7 \, \text{in}$, find its base.

6. Figure 26.12 shows a steel section. Calculate its area in square centimetres.

Fig. 26.12

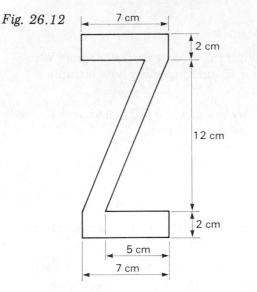

7. Determine the length of the side of a square whose area is equal to that of a parallelogram with a base of $6 \, \text{m}$ and an altitude of $3 \, \text{m}$.

8. The perimeter of a square is $36 \, \text{cm}$.

   (a) Calculate the area of the square.

   (b) Find the altitude of a parallelogram with the same area as the square, if its base is $6 \, \text{cm}$ long.

Area of a Triangle

The diagonal of a parallelogram (Fig. 26.13) splits the parallelogram into two equal triangles. Hence

Area of triangle $= \frac{1}{2} \times$ Base $\times$ Altitude

As a formula the statement becomes

$$A = \tfrac{1}{2}bh$$

where A is the area of the triangle, b the length of the base and h the altitude (or vertical height); see Fig. 26.14.

Fig. 26.13

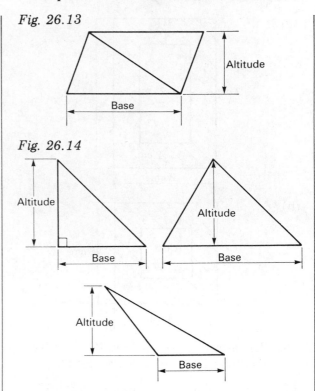

Fig. 26.14

Example 10

(a) A triangle has a base $8 \, \text{cm}$ long and an altitude of $5 \, \text{cm}$. Calculate its area.

$$A = \tfrac{1}{2}bh$$
$$= (\tfrac{1}{2} \times 8 \times 5) \, \text{cm}^2$$
$$= 20 \, \text{cm}^2$$

(b) A triangle has an area of $30 \, \text{in}^2$ and a base $6 \, \text{in}$ long. What is its altitude?

Since $A = \tfrac{1}{2}bh$,

$$h = \frac{2A}{b}$$

We are given $A = 30$ and $b = 6$ and we have to find h.

$$h = \frac{2A}{b}$$
$$= \frac{2 \times 30}{6}$$
$$= 10$$

Hence the altitude of the triangle is $10 \, \text{in}$.

When we are given the length of the three sides of a triangle, the area of the triangle may be found by using the formula

$$A = \sqrt{s(s-a)(s-b)(s-c)}$$

where s stands for half the perimeter of the triangle and a, b and c are the lengths of the sides of the triangle.

Example 11

The lengths of the sides of a triangle are 7 cm, 8 cm and 13 cm. Calculate the area of the triangle.

$$s = \frac{a + b + c}{2}$$

$$= \frac{7 + 8 + 13}{2}$$

$$= \frac{28}{2}$$

$$= 14$$

$$A = \sqrt{14 \times (14 - 7) \times (14 - 8) \times (14 - 13)}$$

$$= \sqrt{14 \times 7 \times 6 \times 1}$$

$$= \sqrt{588}$$

$$= 24.2$$

Hence the area of the triangle is 24.2 cm².

Exercise 26.4

Calculate the area of the following triangles:

1. Base 6 cm, altitude 3 cm.
2. Base 8 cm, altitude 7 cm.
3. Base 11 in, altitude 16 in.
4. Base 3.2 m, vertical height 4.3 m.
5. Base $4\frac{1}{2}$ cm, altitude $1\frac{1}{3}$ cm.
6. Fig. 26.15 shows several right-angled triangles. Work out the area of each.
7. A triangle has an area of 60 yd² and a base 12 yd long. Calculate its altitude.

Fig. 26.15

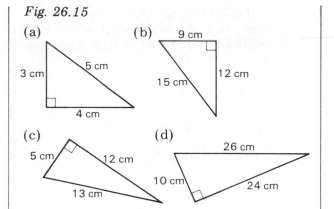

(a)   (b)

(c)   (d)

8. A triangle has an area of 80 mm² and a vertical height of 5 mm. Find the base of the triangle.

9. A triangle has sides which are 4 cm, 5 cm and 7 cm long. Calculate its area.

10. The sides of a triangle are 3 ft, 5 ft and 6 ft long. Find the area of the triangle.

11. Find the area of the right-angled triangle shown in Fig. 26.16.

Fig. 26.16

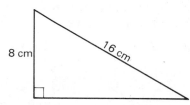

12. Calculate the area of the quadrilateral shown in Fig. 26.17.

Fig. 26.17

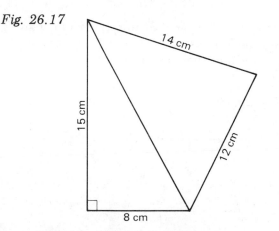

13. In Fig. 26.18, ABCD is a rhombus. The length of diagonal BD is 10 cm and the length of diagonal AC is 6 cm. Calculate the area of ABCD.

Fig. 26.18

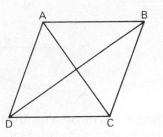

14. A triangle has sides 37 mm, 52 mm and 63 mm long. Calculate the area of the triangle in square centimetres.

Area of a Trapezium

A trapezium is a quadrilateral with one pair of parallel sides (Fig. 26.19).

Area of trapezium = $\frac{1}{2}$ × Sum of parallel
sides
× Distance between
parallel sides

As a formula this becomes

$$A = \tfrac{1}{2}(a + b)h$$

where A is the area of the trapezium, a and b are the lengths of the parallel sides and h is the distance between the parallel sides.

Fig. 26.19

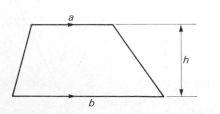

Example 12

(a) The parallel sides of a trapezium are 12 cm and 16 cm long. The distance between them is 9 cm. Calculate the area of the trapezium.

We are given $a = 12$, $b = 16$ and $h = 9$. We have to find A.

$$A = \tfrac{1}{2}(a + b)h$$
$$= \tfrac{1}{2} \times (12 + 16) \times 9$$
$$= \tfrac{1}{2} \times 28 \times 9$$
$$= 126$$

Hence the area of the trapezium is 126 cm².

(b) The area of a trapezium is 220 cm². The parallel sides are 26 cm and 14 cm long. Find the distance between the parallel sides.

We are given $A = 220$, $a = 26$ and $b = 14$. We have to find h.

$$220 = \tfrac{1}{2} \times (26 + 14) \times h$$
$$220 = \tfrac{1}{2} \times 40 \times h$$
$$220 = 20 \times h$$
$$h = \frac{220}{20}$$
$$= 11$$

Hence the distance between the parallel sides is 11 cm.

Exercise 26.5

1. A trapezium has parallel sides 12 in and 8 in long. If the distance between the parallel sides is 6 in, calculate the area of the trapezium.

2. Find the area of a trapezium whose parallel sides are 7 cm and 9 cm long if the distance between them is 5 cm.

3. The area of a trapezium is 300 cm². Its parallel sides are 35 cm and 65 cm long. What is the distance between the parallel sides?

4. Find the area of a trapezium whose parallel sides are 25 in and 35 in, the distance between them being 20 in.

5. Find the area of the trapezium shown in Fig. 26.20.

Fig. 26.20

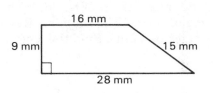

Similar Shapes

Two shapes are similar if they are equiangular and the ratio of corresponding sides is the same. The parallelograms shown in Fig. 26.21 are similar because they are equiangular and $AD/WZ = CD/YZ = \frac{1}{3}$.

Fig. 26.21

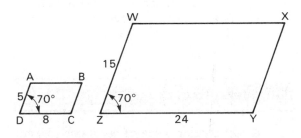

The ratio of the areas of similar shapes is equal to the ratio of the squares on corresponding sides. For the parallelograms of Fig. 26.21, AD and WZ are corresponding sides because they lie opposite to equal angles. Therefore

$$\frac{\text{Area of WXYZ}}{\text{Area of ABCD}} = \frac{WZ^2}{AD^2}$$

$$= \frac{15^2}{5^2}$$

$$= \frac{9}{1}$$

Example 13

(a) Find the area of triangle XYZ, given that the area of triangle ABC is $12\,\text{cm}^2$ (Fig. 26.22).

Fig. 26.22

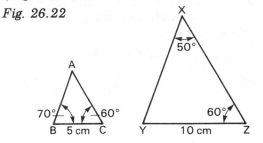

In $\triangle XYZ$, $Y = 70°$ and in $\triangle ABC$, $A = 50°$. Hence triangles XYZ and ABC are equiangular, i.e. similar. YZ and BC are corresponding sides because they lie opposite to the equal angles A and X.

Hence

$$\frac{\text{Area of XYZ}}{\text{Area of ABC}} = \frac{YZ^2}{BC^2}$$

$$\frac{\text{Area of XYZ}}{12} = \frac{10^2}{5^2}$$

$$= \frac{4}{1}$$

$$\text{Area of XYZ} = 4 \times 12$$

$$= 48$$

Hence the area of triangle XYZ is 48 cm².

(b) Two similar polygons have areas of $8\,\text{cm}^2$ and $18\,\text{cm}^2$. The shortest side of the larger polygon is 4 cm long. Calculate the length of the shortest side of the smaller polygon.

Let the length of the shortest side of the smaller polygon be x cm. Then

$$\frac{8}{18} = \frac{x^2}{4^2}$$

$$x^2 = \frac{8 \times 4^2}{18}$$

$$= 7.11$$

$$x = \sqrt{7.11}$$

$$= 2.67$$

Hence the shortest side of the smaller polygon is 2.67 cm long.

Exercise 26.6

1. In Fig. 26.23, triangles ABC and EFG are similar. If the area of triangle ABC is $8\,m^2$, calculate the area of triangle EFG.

Fig. 26.23

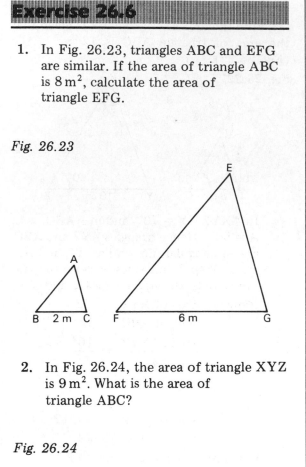

2. In Fig. 26.24, the area of triangle XYZ is $9\,m^2$. What is the area of triangle ABC?

Fig. 26.24

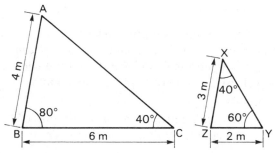

3. A rectangle ABCD has a length of 30 in and an area of $600\,in^2$. A similar rectangle WXYZ has a length of 20 in. What is the area of WXYZ?

4. A trapezium ABCD (Fig. 26.25) has an area of $12\,300\,mm^2$. Calculate the area of the trapezium PQRS.

Fig. 26.25

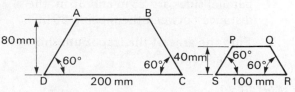

5. A rhombus ABCD (Fig. 26.26) has an area of $18\,m^2$. The rhombus EFGH has an area of $72\,m^2$. Find the length of EF.

Fig. 26.26

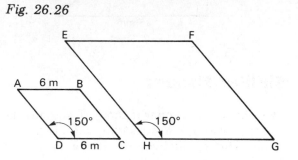

6. Two similar polygons have areas of $160\,in^2$ and $360\,in^2$. The shortest side of the larger polygon is 18 in long. Calculate the length of the shortest side of the smaller polygon.

Miscellaneous Exercise 26

1. A rectangular piece of cardboard 10 cm by 6 cm has equal squares of side 2 cm cut from its corners. The final shape is shown in Fig. 26.27. Calculate **(a)** the perimeter, **(b)** the area of the shape.

Fig. 26.27

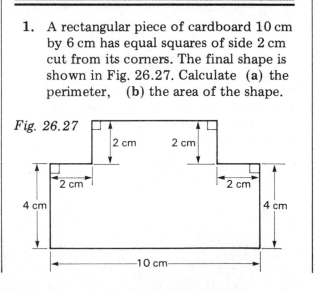

2. A square has a perimeter of 20 in. What is its area?

3. Fig. 26.28 shows a trapezium ABCD. Calculate its area.

Fig. 26.28

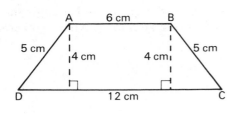

4. Refer to Fig. 26.29. Calculate
   (a) the length of AB
   (b) the area of the triangle AED

Fig. 26.29

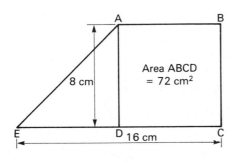

5. Find the area of the parallelogram shown in Fig. 26.30.

Fig. 26.30

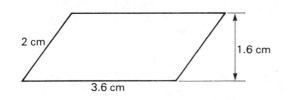

6. A rectangular table-cloth is 150 cm by 200 cm. What is its area in square metres?

7. In Fig. 26.31, ABCD is a rectangle 12 cm by 8 cm. The shaded portions are cut off. Find the area of DLMNPC.

Fig. 26.31

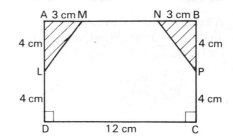

8. A square has an area of 81 cm². Calculate its perimeter.

9. Express the area of a rectangle 10 in long and 6 in wide as a fraction of the area of a square with sides 8 in long.

10. Find the area of the kite shown in Fig. 26.32.

Fig. 26.32

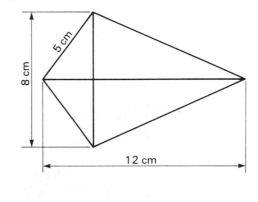

Mental Test 26

1. What is the area of a rectangle whose length is 7 cm and whose width is 3 cm?

2. A square has an area of 25 in². What is the length of its sides?

3. A parallelogram has a base 8 cm long and a vertical height of 5 cm. Find its area.

4. A parallelogram has an area of 80 cm². Its base is 10 cm long. What is its vertical height?

5. A triangle has a base 8 ft long and an altitude of 4 ft. What is its area?

6. A triangle has an area of 36 cm². Its altitude is 9 cm. What is the length of its base?

7. A trapezium has parallel sides 12 cm and 8 cm long. The distance between these sides is 5 cm. What is the area of the trapezium?

8. An isosceles right-angled triangle has equal sides 4 cm long. What is its area?

27 Mensuration of the Circle

The names of the main parts of a circle are shown in Fig. 27.1.

Fig. 27.1

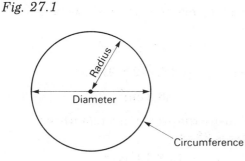

Circumference of a Circle

In a circle the value of

$$\frac{\text{Circumference}}{\text{Diameter}} = 3.141\,59\ldots$$

It has been proved that the exact value cannot be worked out, but for most problems a value of 3.142 is sufficiently accurate when working in decimals. When working in fractions a value of $\frac{22}{7}$ may be used. When making mental calculations, a value of $\frac{25}{8}$ is acceptable.

The value circumference/diameter is so important that it has been given the special symbol π (the Greek letter pi).

We take the value of π as being 3.142 or $\frac{22}{7}$.

Since Circumference/Diameter $= \pi$ it follows that

$$\text{Circumference} = \pi \times \text{Diameter}$$

or $\text{Circumference} = 2 \times \pi \times \text{Radius}$

or, as a formula

$$C = 2\pi r$$
$$= \pi d$$

where C is the circumference, r the radius and d the diameter.

Example 1

(a) The diameter of a circle is 30 cm. Calculate its circumference.

$$C = \pi d$$
$$= 3.142 \times 30$$
$$= 94.3$$

Hence the circumference is 94.3 cm.

(b) The radius of a circle is 14 in. Calculate its circumference.

$$C = 2\pi r$$
$$= \frac{2}{1} \times \frac{22}{\cancel{7}_{1}} \times \frac{\cancel{14}^{2}}{1}$$
$$= 2 \times 22 \times 2$$
$$= 88$$

Hence the circumference is 88 in.

The formulae $C = \pi d$ and $C = 2\pi r$ can be transposed to give

$$d = \frac{C}{\pi} \qquad \text{and} \qquad r = \frac{C}{2\pi}$$

Example 2

Find the radius of a circle whose circumference is 93.8 cm.

$$r = \frac{C}{2\pi}$$

$$= \frac{93.8}{2 \times 3.142}$$

$$= 14.9$$

This calculation is best done using a calculator:

Input	Display
93.8	93.8
÷	93.8
2	46.9
÷	46.9
3.142	3.142
=	14.9 ...

Hence the radius is 14.9 cm.

Exercise 27.1

Taking $\pi = \frac{22}{7}$, find the circumference of each of the following circles:

1. Radius 21 cm.   2. Radius 350 cm.

3. Radius 14 ft.   4. Diameter 28 cm.

5. Diameter 560 ft.

Taking $\pi = 3.142$, find the circumference of each of the following circles:

6. Radius 43 cm.   7. Radius 3.16 m.

8. Diameter 85 in.

9. Diameter 42.7 mm.

10. Diameter 86.15 ft.

11. A circular flower bed has a diameter of 64 m. What is its circumference?

12. A circular pond has a circumference of 12.62 m. Calculate its radius.

13. Find the diameter of a circle whose circumference is 110 cm.

14. A garden roller has a diameter of 80 cm. If it is 1.5 m wide, what area of lawn does it roll in 70 revolutions?

15. A wheel has a diameter of 220 in. What is the distance around its rim?

Area of a Circle

It can be shown that

$$\text{Area of a circle} = \pi \times \text{Radius}^2$$

or

$$A = \pi r^2$$

Example 3

(a) Find the area of a circle whose radius is 3 cm.

$$A = (\pi \times 3^2)\,\text{cm}^2$$

$$= (3.142 \times 9)\,\text{cm}^2$$

$$= 28.3\,\text{cm}^2$$

(b) Calculate the area of a circle whose diameter is 28 cm.

$$A = (\pi \times 14^2)\,\text{cm}$$

$$= \frac{22}{7_1} \times \frac{\cancel{14}^2}{1} \times \frac{14}{1}\,\text{cm}^2$$

$$= (22 \times 2 \times 14)\,\text{cm}^2$$

$$= 616\,\text{cm}^2$$

(c) The area of a circle is 88 in². Find its radius.

$$A = \pi r^2$$

$$88 = \frac{22}{7}r^2$$

$$r^2 = \frac{\cancel{88}^4}{1} \times \frac{7}{\cancel{22}_1}$$

$$= 4 \times 7$$

$$= 28$$

$$r = \sqrt{28}$$

$$= 5.29$$

The circle has a radius of 5.3 in.

Area of an Annulus

An annulus consists of two concentric circles, as shown in Fig. 27.2.

Area of annulus = Area of outer circle
                  − Area of inner circle

$$= \pi R^2 - \pi r^2$$
$$= \pi(R^2 - r^2)$$

Fig. 27.2

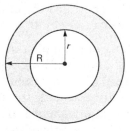

Example 4

Find the area of an annulus which has an outer diameter of 20 cm and an inner diameter of 12 cm.

We are given that $R = \frac{20}{2} = 10$ and $r = \frac{12}{2} = 6$.

$$
\begin{aligned}
A &= \pi(R^2 - r^2) \\
  &= 3.142 \times (10^2 - 6^2) \\
  &= 3.142 \times (100 - 36) \\
  &= 3.142 \times 64 \\
  &= 201
\end{aligned}
$$

Hence the area of the annulus is 201 cm².

Exercise 27.2

Taking $\pi = \frac{22}{7}$, find the area of each of the following circles:

1. Radius 7 cm.
2. Radius 140 mm.
3. Radius 28 in.
4. Diameter 42 m.
5. Diameter 35 ft.

Taking $\pi = 3.142$, find the area of each of the following circles:

6. Radius 4 cm.
7. Radius 70 in.
8. Diameter 14 m.
9. Diameter 50 ft.
10. Diameter 130 mm.
11. An annulus has an inner diameter of 12 in and an outer diameter of 18 in. Calculate its area.
12. A copper pipe has a bore of 32 mm and an outside diameter of 42 mm. Calculate the cross-sectional area of the pipe.
13. A pond with a diameter of 36 m has a path 1 m wide around its circumference. Calculate the area of the path.
14. A hollow shaft has an outside diameter of 3.25 in and an inside diameter of 2.50 in. Calculate the cross-sectional area of the shaft.
15. A hollow shaft has a cross-sectional area of 8.68 cm² and a bore of 0.75 cm diameter. Find the outside diameter of the shaft.

Sector of a Circle

The area and the length of arc of a sector of a circle depend upon the angle that the sector subtends at the centre of the circle (Fig. 27.3).

Fig. 27.3

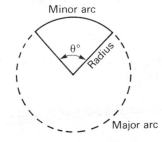

Referring to Fig. 27.3, where θ is measured in degrees:

Length of minor
arc of the sector $= 2 \times \pi \times \text{Radius} \times \dfrac{\theta}{360}$

$$= \pi \times \text{Diameter} \times \dfrac{\theta}{360}$$

or, as a formula

$$l = 2\pi r \times \dfrac{\theta}{360}$$

or $\qquad l = \pi d \times \dfrac{\theta}{360}$

Area of sector $= \pi \times \text{Radius}^2 \times \dfrac{\theta}{360}$

or $\qquad A = \pi r^2 \times \dfrac{\theta}{360}$

Example 5

Find the length of the minor arc and the area
of the sector of a circle which subtends an
angle of $120°$ at the centre if its radius is
8 cm.

The sector is shown in Fig. 27.4.

Fig. 27.4

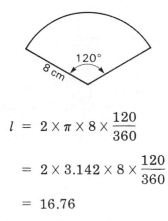

$$l = 2 \times \pi \times 8 \times \dfrac{120}{360}$$

$$= 2 \times 3.142 \times 8 \times \dfrac{120}{360}$$

$$= 16.76$$

Hence the length of the minor arc is
16.76 cm.

$$A = 3.142 \times 8^2 \times \dfrac{120}{360}$$

$$= 67.0$$

Therefore the area is 67.0 cm².

Exercise 27.3

Find the length of arc and the area for each
of the following sectors of a circle:

1. Radius 4 cm; sector angle 45°.

2. Radius 10 cm; sector angle 90°.

3. Radius 3 in; sector angle 120°.

4. If an arc 7 cm long subtends an angle of
   45° at the centre of the circle, what is
   the radius of the circle?

5. In marking out the plan of part of a
   building, a line 8 m long is pegged down
   at one end. Then with the line held
   horizontally and taut, the free end is
   swung through an angle of 57°. Calculate the distance moved by the free end
   of the line and find the area swept out.

6. A bay window is semicircular in plan. It
   is to be covered with lead. If the radius
   of the bay is 2.4 yd, calculate the area
   of lead required.

7. Calculate the area of the cross-section
   shown in Fig. 27.5.

Fig. 27.5

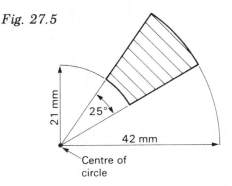

Areas of Composite Shapes

Many shapes are composed of straight lines
and arcs of circles. The areas of such shapes
are found by splitting up the shape into
figures such as rectangles, triangles and
sectors of circles.

Example 6

Find the cross-sectional area of the steel section shown in Fig. 27.6.

Fig. 27.6

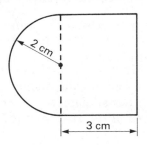

The shape consists of a semicircle and a rectangle.

$$\text{Area of semicircle} = \tfrac{1}{2} \times \pi \times \text{Radius}^2$$

$$= \tfrac{1}{2} \times 3.142 \times 2^2$$

$$= 6.28$$

$$\text{Area of rectangle} = \text{Length} \times \text{Width}$$

$$= 4 \times 3$$

$$= 12$$

$$\text{Total area of cross-section} = 6.28\,\text{cm}^2 + 12\,\text{cm}^2$$

$$= 18.28\,\text{cm}^2$$

Example 7

Find the shaded area in Fig. 27.7.

Fig. 27.7

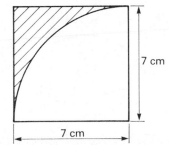

The shaded area is calculated by subtracting the area of the quarter-circle from the area of the square.

$$\text{Area of square} = 7^2$$

$$= 49$$

$$\text{Area of quarter-circle} = \tfrac{1}{4} \times \pi \times \text{Radius}^2$$

$$= \tfrac{1}{4} \times 3.142 \times 7^2$$

$$= 38.5$$

$$\text{Shaded area} = 49\,\text{cm}^2 - 38.5\,\text{cm}^2$$

$$= 10.5\,\text{cm}^2$$

Exercise 27.4

Find the areas of the shapes drawn in Fig. 27.8. They are all composed of squares or rectangles and parts of circles.

Fig. 27.8

1.

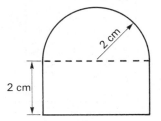

2.

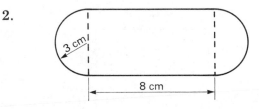

3.

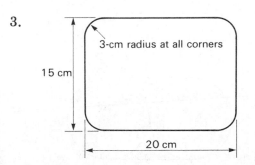

Find the areas of the shaded portions of the shapes shown in Fig. 27.9.

Fig. 27.9

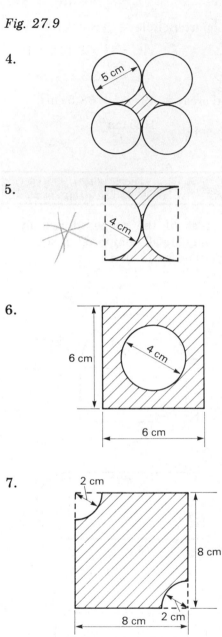

4.

5.

6.

7.

8.

1. A wheel has a radius of 35 cm. What is the distance round its rim? (Take $\pi = \frac{22}{7}$)

2. A circular flower bed measures 7 yd across its centre. How far is it round its edge? (Take $\pi = \frac{22}{7}$)

3. A pipe has an outside diameter of 14 cm and an internal diameter of 9 cm. What is the area of its cross-section? (Take $\pi = 3.14$)

4. Find the area of a circle whose radius is 21 in. (Take $\pi = \frac{22}{7}$)

5. A steel bar has a diameter of 6.5 cm. Find the area of its cross-section. Take $\pi = 3.14$ and round off the answer to the nearest tenth.

6. Calculate the area of a circle with a diameter of 6 cm. (Take $\pi = 3.14$)

7. The circumference of a circle is 88 in. Taking $\pi = \frac{22}{7}$, find the radius of the circle.

8. (a) Calculate the area of the sector shown in Fig. 27.10, stating your answer correct to 2 significant figures. (Take $\pi = 3.14$)

    (b) What is the length of arc, correct to 2 significant figures?

Fig. 27.10

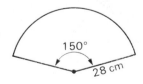

9. Calculate the area of a semicircle which has a radius of 14 cm. Take $\pi = \frac{22}{7}$.

10. A circle has a circumference of 176 in. Taking $\pi = \frac{22}{7}$, find its radius.

11. Calculate the shaded area of the shape shown in Fig. 27.11.

Fig. 27.11

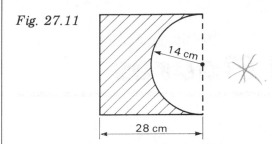

12. The circumference of a circle is 44 cm. Find the length of an arc of this circle formed by an angle of 45° at the centre.

13. Calculate the radius of a circle which has an area of 616 cm².

14. A pipe has an internal diameter of 8 in and an external diameter of 12 in. Calculate the cross-sectional area of the pipe.

15. Fig. 27.12 shows a metal framework made of thin wire. ABCD is a square and A, B, C and D are the centres of four circles which have equal radii. Taking $\pi = 3.142$ and given that the radius of each circle is 3 cm, calculate

   (a) the circumference of one of the circles

   (b) the area of one of the circles

   (c) the total length of the straight pieces of metal in the framework

   (d) the total length of metal used

Fig. 27.12

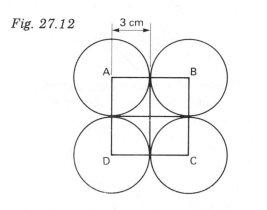

Mental Test 27

1. Find the circumference of a circle whose radius is 7 cm.

2. Find the circumference of a circle whose diameter is 28 in.

3. Find the area of a circle whose radius is 7 cm.

4. A circle has a circumference of 24 in. What is the length of an arc of this circle which subtends an angle of 60° at the centre of this circle?

5. A circle has an area of 48 cm². Find the area of a sector of this circle which subtends an angle of 120° at the centre.

6. Find the area of an annulus whose outside radius is 8 cm and whose inside radius is 4 cm. Give the answer as a multiple of π.

7. A circle has a radius of 8 cm. What is the length of an arc of this circle which subtends an angle of 45° at the centre of the circle? Leave your answer as a multiple of π.

8. A circle has a radius of 6 cm. Find the area of a sector of this circle which subtends an angle of 120° at the centre. Leave your answer as a multiple of π.

9. Taking $\pi = \frac{25}{8}$, find the area of a circle whose radius is 4 cm.

10. Taking $\pi = \frac{25}{8}$, calculate the circumference of a circle whose diameter is 32 cm.

Solid Figures

Introduction

The plane figures discussed in Chapter 26, such as rectangles and parallelograms, have two dimensions, namely length and breadth. They have no height (or thickness).

Solid figures have three dimensions, namely length, breadth and height (or thickness).

Types of Solid Figures

(1) A **sphere** is a circular solid. Examples are a football and a ball bearing (Fig. 28.1).

Fig. 28.1

(2) **Prisms** are solid figures which have a uniform cross-section. There are various types of prisms:

(a) **The cuboid** is a rectangular solid (Fig. 28.2). That is, it has a cross-section which is a rectangle. An example of a cuboid is a box.

(b) **The cube** (Fig. 28.3) has all its edges of equal length, and each of its faces is a square. A sugar lump is an example of a cube.

Fig. 28.2   *Fig. 28.3*

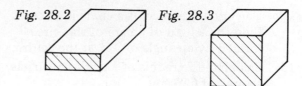

(c) **The cylinder** has a constant cross-section which is a circle. It could be called a circular prism. Most tins are cylindrical in shape (Fig. 28.4).

Fig. 28.4

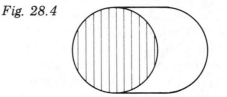

(d) A **triangular prism** has a uniform cross-section which is a triangle (Fig. 28.5). A ridge tent is an example of a triangular prism.

Fig. 28.5

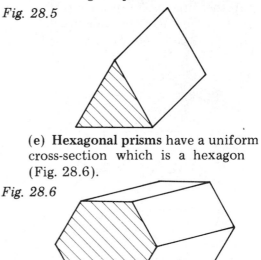

(e) **Hexagonal prisms** have a uniform cross-section which is a hexagon (Fig. 28.6).

Fig. 28.6

(f) Irregular prisms have a cross-section with no well-known shape (Fig. 28.7). Some of the steel sections used in the construction industry fall into this category (Fig. 28.8).

Fig. 28.7

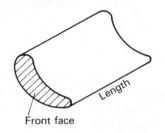

Fig. 28.8

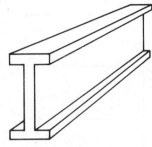

(3) A **pyramid** is a solid figure which stands on a flat base which may be triangular, square, rectangular or polygonal. It tapers to a point which means that each of its sides is a triangle. Fig. 28.9 shows a triangular pyramid and a rectangular pyramid.

Fig. 28.9

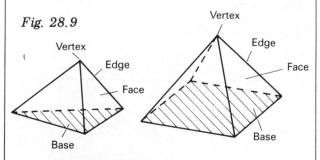

(4) A **regular tetrahedron** is a triangular pyramid with all its faces equilateral triangles (Fig. 28.10). A tetrahedron has therefore all its edges of equal length.

Fig. 28.10

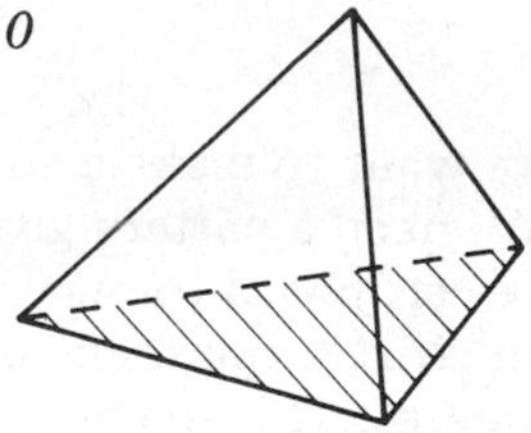

(5) A **cone** is a solid figure with a circular base which tapers to a point (Fig. 28.11).

Fig. 28.11

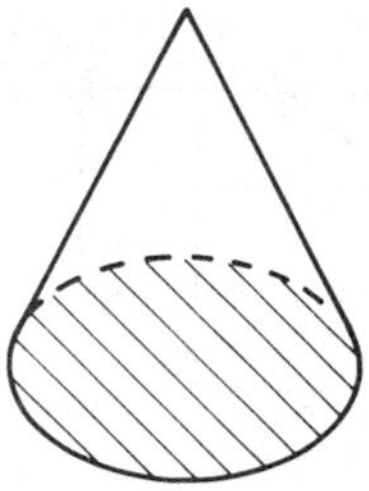

Exercise 28.1

Write down the number of faces possessed by the following solid figures:

1. Cuboid.

2. Hexagonal prism.

3. Tetrahedron.

4. Cube.

5. Triangular prism.

6. Square pyramid.

Write down the number of edges on each of the following solid figures:

7. Cube.

8. Rectangular pyramid.

9. Triangular prism.

10. Tetrahedron.

11. Draw three different types of prism and label them carefully.

12. Draw three different types of pyramid and label them.

Nets

Suppose that we want to make a cube out of cardboard. We need a pattern giving us the shape of the cardboard to make the cube. As shown in Fig. 28.12, the shape of the pattern is six squares. This shape can be folded to make the cube.

Fig. 28.12

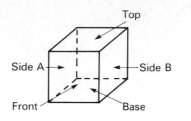

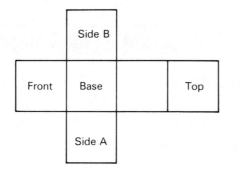

It is possible for there to be more than one net for a solid object. For instance the cube in Fig. 28.12 can also be made from the net shown in Fig. 28.13.

Fig. 28.13

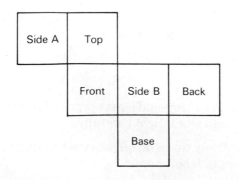

Example 1

Sketch the net of the triangular prism shown in Fig. 28.14.

Fig. 28.14

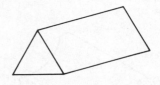

The net is sketched in Fig. 28.15. It consists of three rectangles representing the base and two sides and two triangles representing the two ends.

Fig. 28.15

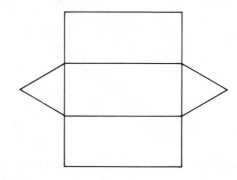

Nets of Curved Surfaces

The net for a cylinder without a top and a bottom is shown in Fig. 28.16. The length of the net is equal to the circumference of the cylinder.

Fig. 28.16

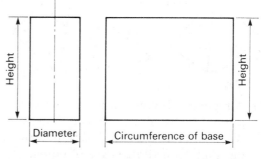

Circumference = π × diameter

Example 2

A cone has a base which is a circle 10 cm in diameter and a height of 15 cm. Draw a net for this cone.

The net is shown in Fig. 28.17 and it is seen to be the sector of a circle. Note that the arms of the sector are equal to the slant height of the cone and that the length of the arc is equal to the circumference of the base.

Fig. 28.17

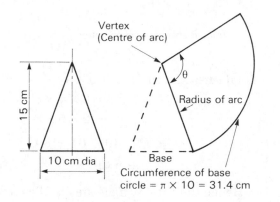

If the slant height is l, then by Pythagoras

$$l^2 = 5^2 + 15^2$$
$$= 25 + 225$$
$$= 250$$
$$l = \sqrt{250}$$
$$= 15.8$$

If the length of arc is C then

$$C = 2 \times \pi \times r$$
$$= 2 \times 3.142 \times 5$$
$$= 31.4$$

If the sector angle is θ then

$$C = 2 \times \pi \times l \times \frac{\theta}{360}$$

$$31.4 = 2 \times 3.142 \times 15.8 \times \frac{\theta}{360}$$

$$31.4 = 99.3 \times \frac{\theta}{360}$$

$$\theta = \frac{31.4 \times 360}{99.3}$$

$$= 114°$$

Exercise 28.2

Sketch the nets of the following solid figures:

1. A cuboid 8 cm long, 3 cm wide and 4 cm high.

2. A triangular prism whose ends are right-angled triangles of base 3 cm and height 4 cm and whose length is 6 cm.

3. A pyramid with a square base of side 5 cm and a vertical height of 8 cm.

4. A cube with an edge of 4 cm.

5. A cylinder with a height of 5 cm and a diameter of 3 cm.

6. A cone with a vertical height of 8 cm and a base diameter of 7 cm. Calculate the slant height of this cone and mark it on your diagram.

7. The diagrams in Fig. 28.18 show the nets of various solids. Name the solids.

Fig. 28.18

(a)

(b)

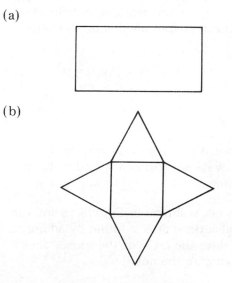

(c)

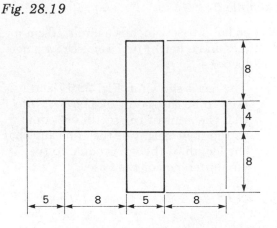

Fig. 28.19

All dimensions in centimetres

Total surface area $=\ 2 \times 5 \times 4$

$+\ 2 \times 4 \times 8$

$+\ 2 \times 5 \times 8$

$=\ 40 + 64 + 80$

$=\ 184$

Hence the total surface area is 184 cm^2.

We could rewrite the terms comprising the total surface area as follows (see Fig. 28.20):

Total surface area $=\ (2 \times 4 + 2 \times 5) \times 8$

$+\ 2 \times (5 \times 4)$

$=\ \text{Perimeter of end}$

$\times \text{ Length}$

$+\ \text{Area of ends}$

$=\ \text{Lateral surface area}$

$+\ \text{Area of ends}$

Surface Area

We often need to find the surface area of solid figures such as cylinders and rectangular blocks. The total surface area of such figures is composed of the lateral surface area plus the area of the ends. For a cylinder the lateral surface is the curved surface and for a rectangular block it is the top, bottom and the two sides.

The surface area can be found by drawing a net of the object and working out the areas of the various shapes making up the net.

Fig. 28.20

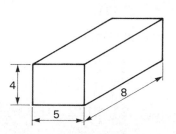

Example 3

A cuboid is 8 cm long, 5 cm wide and 4 cm high. Draw its net and hence find its total surface area.

The net is shown in Fig. 28.19 and the total surface area is found by adding together the areas of the six rectangles making up the net.

The lateral surface area of any solid figure with a constant cross-section (i.e. cuboid, prism and cylinder) can be found by multiplying the perimeter of the cross-section by the length of the solid:

Lateral
surface area = Perimeter of cross-section
$\times$ Length of solid

Example 4

(a) Find the total surface area of a cylinder whose diameter is 28 cm and whose height is 20 cm.

Perimeter of cross-section = Circumference
$= \pi \times$ Diameter
$= \frac{22}{7} \times 28$ cm
$= 88$ cm

Lateral surface area $= (88 \times 20)$ cm^2
$= 1760$ cm^2

Area of one end $= \pi \times$ Radius^2
$= \frac{22}{7} \times 14^2$
$= 616$ cm^2

Total surface area $= 1760$ cm^2
$+ (2 \times 616)$ cm^2
$= 1760$ cm^2
$+ 1232$ cm^2
$= 2992$ cm^2

(b) Fig. 28.21 shows a triangular prism. Find its lateral surface area.

Fig. 28.21

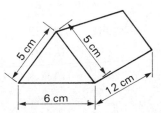

Perimeter of cross-section $= (6 + 5 + 5)$ cm
$= 16$ cm

Lateral
surface area = Perimeter of cross-section
$\times$ Length
$= (16 \times 12)$ cm^2
$= 192$ cm^2

Exercise 28.3

1. A room is 5 m long, 4 m wide and 3 m high. Calculate the total surface area of the walls.

2. A rectangular block of wood is 5 cm long, 6 cm wide and 4 cm high. Calculate its total surface area.

3. A cylinder has a diameter of 28 cm and a height of 30 cm. Calculate its total surface area.

4. A triangular prism is shown in Fig. 28.22. Calculate
   (a) its lateral surface area
   (b) the total surface area

Fig. 28.22

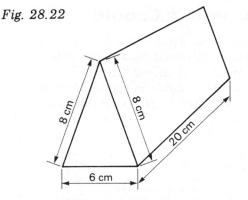

5. Find the curved surface area of a cylinder whose diameter is 7 cm and whose height is 8 cm.

6. Calculate the total surface area of a cube whose edge is 6 cm long.

7. A closed water tank has a cross-section in the form of a rectangle with semicircular ends as shown in Fig. 28.23. The tank is 7 m high. Find its total surface area.

Fig. 28.23

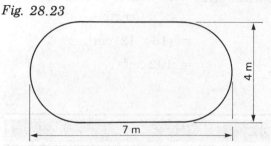

8. A tin is in the form of a cuboid and it is made from sheet metal. It is 9 cm long, 6 cm wide and 5 cm high. If it has no top, calculate the area of sheet metal needed to make it.

Units of Volume

Volume is measured by seeing how many cubic units the solid contains. The cubic centimetre (abbreviation cm³) is the volume contained inside a cube of side 1 cm.

Similarly a cubic metre (abbreviation m³) is the volume contained inside a cube of side 1 m.

In the imperial system, volumes are measured in cubic inches (in³) and cubic feet (ft³).

Volume of a Cuboid

The cuboid shown in Fig. 28.24 has been divided into three layers of small cubes, each having a volume of 1 cm³.
There are 5 × 4 cubes in each layer and therefore the total number of small cubes is 5 × 4 × 3 = 60.
Therefore the cuboid has a volume of 60 cm³.

Fig. 28.24

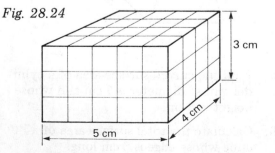

What we have done to find the volume of the cuboid is to multiply the length by the breadth by the height. This rule applies to any cuboid, and hence

Volume of cuboid = Length × Breadth × Height

Since the area of the end of a cuboid is breadth by height we can also write

Volume of cuboid = Area of the end × Length

This statement is true for any solid which has the same cross-section throughout its length. As a formula we write

$$V = Al$$

where V is the volume of the solid, A the area of the cross-section and l the length of the solid.

Example 5

(a) Find the volume of a cuboid which is 15 in long, 8 in wide and 4 in high.

$$\text{Volume of a cuboid} = (15 \times 8 \times 4)\,\text{in}^3$$
$$= 480\,\text{in}^3$$

(b) A tank is in the form of a cuboid. It is 120 cm long, 80 cm wide and 50 cm high. Calculate the volume of the tank in cubic metres.

In problems of this kind it is best to convert all the dimensions to metres before attempting to find the volume. Thus

$$\text{Length} = 120\,\text{cm}$$
$$= 1.20\,\text{m}$$
$$\text{Breadth} = 80\,\text{cm}$$
$$= 0.8\,\text{m}$$
$$\text{Height} = 50\,\text{cm}$$
$$= 0.5\,\text{m}$$
$$\text{Volume of tank} = (1.2 \times 0.8 \times 0.5)\,\text{m}^3$$
$$= 0.48\,\text{m}^3$$

(c) A block of wood has the cross-section shown in Fig. 28.25. If it is 9 m long find its volume, stating the answer in cubic metres.

Fig. 28.25

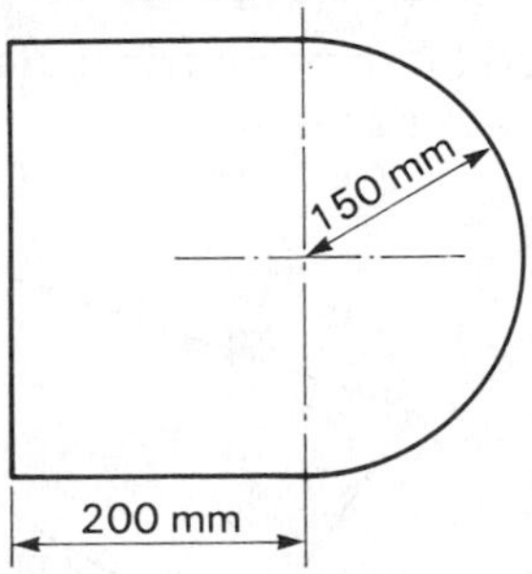

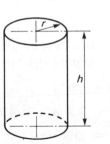

150 mm

200 mm

Because the length is given in metres we first convert all the other dimensions into metres.

$$\text{Area of cross-section} = (\tfrac{1}{2} \times \pi \times 0.15^2)\,\text{m}^2$$
$$+ (0.2 \times 0.3)\,\text{m}^2$$
$$= (0.035 + 0.06)\,\text{m}^2$$
$$= 0.095\,\text{m}^2$$

$$\text{Volume} = \text{Area of cross-section} \times \text{Length}$$
$$= (0.095 \times 9)\,\text{m}^3$$
$$= 0.855\,\text{m}^3$$

Volume of a Cylinder

Since a cylinder has a constant cross-section which is a circle (Fig. 28.26)

$$\text{Volume of cylinder} = \text{Area of cross-section} \times \text{Height}$$
$$= \pi \times \text{Radius}^2 \times \text{Height}$$

As a formula

$$V = \pi r^2 h$$

Fig. 28.26

r

h

Example 6

(a) Find the volume of a cylinder which has a radius of 14 in and a height of 8 in.

$$V = \pi r^2 h$$
$$= (\tfrac{22}{7} \times 14^2 \times 8)\,\text{cm}^3$$
$$= 4928\,\text{in}^3$$

(b) A pipe has the dimensions shown in Fig. 28.27. Calculate the volume of metal in the pipe.

Fig. 28.27

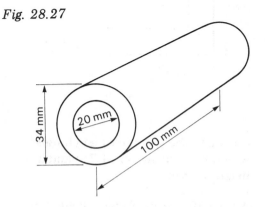

34 mm

20 mm

100 mm

Since the cross-section is an annulus

$$\text{Area of cross-section} = \pi(R^2 - r^2)$$
$$= 3.142 \times (17^2 - 10^2)\,\text{mm}^2$$
$$= 3.142 \times (289 - 100)\,\text{mm}^2$$
$$= (3.142 \times 189)\,\text{mm}^2$$
$$= 594\,\text{mm}^2$$

$$\text{Volume} = \text{Area of cross-section} \times \text{Length}$$
$$= (594 \times 100)\,\text{mm}^3$$
$$= 59\,400\,\text{mm}^3$$

Exercise 28.4

1. Find the volume of a rectangular block of wood which is 8 in long, 5 in wide and 4 in high.

2. Fig. 28.28 shows the cross-section of a steel beam which is 8 m long. Find its volume in cubic metres.

Fig. 28.28

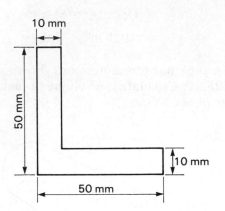

3. Calculate the volume of a cylinder which has a diameter of 7 m and a height of 8 m.

4. A hole 40 mm diameter is drilled in a plate which is 50 mm thick. Find the volume of metal removed in drilling the hole.

5. A rectangular tank is 500 cm long, 300 cm wide and 400 cm high. Calculate its volume in cubic metres.

6. A block of wood has the cross-section shown in Fig. 28.29. If it is 8 cm long, find its volume in cubic centimetres.

Fig. 28.29

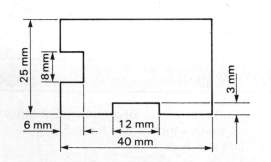

7. Calculate the volume of a metal tube which has an inside diameter of 14 in and an outside diameter of 28 in, and a length of 30 in.

8. Fig. 28.30 shows a triangular prism. Calculate its volume.

Fig. 28.30

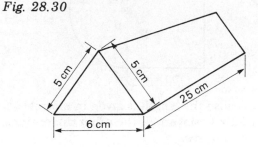

Volume of a Cone

The volume of a cone (Fig. 28.31) is one-third of the volume of an equivalent cylinder (i.e. a cylinder with the same radius as the base radius of the cone and having the same height). That is

Volume of cone $= \frac{1}{3} \times \pi \times \text{Radius}^2 \times \text{Height}$

or $\qquad V = \frac{1}{3}\pi r^2 h$

Fig. 28.31

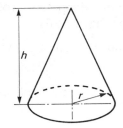

Example 7

Find the volume of a cone which has a base radius of 6 cm and a height of 7 cm.

$$V = \frac{1}{3}\pi r^2 h$$
$$= \frac{1}{3} \times \frac{22}{7} \times 6^2 \times 7 \text{ cm}^3$$
$$= 264 \text{ cm}^3$$

Volume of a Pyramid

The volume of a pyramid (Fig. 28.32) is one-third of the volume of an equivalent prism. Hence

Volume of a pyramid $= \frac{1}{3} \times$ Area of base $\times$ Height

or $\qquad V = \frac{1}{3}Ah$

Fig. 28.32

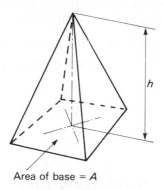

Area of base $= A$

Example 8

Find the volume of a pyramid whose base is a square of 8 in side and whose vertical height is 9 in.

$$\text{Area of base} = 8 \times 8 \, \text{in}^2$$
$$= 64 \, \text{in}^2$$
$$\text{Volume} = \frac{1}{3} \times 64 \times 9 \, \text{in}^3$$
$$= 192 \, \text{in}^3$$

The Sphere

It can be shown that

Volume of a sphere $= \frac{4}{3} \times \pi \times \text{Radius}^3$

or $\qquad V = \frac{4}{3}\pi r^3$

Surface area of a sphere $= 4 \times \pi \times \text{Radius}^2$

or $\qquad A = 4\pi r^2$

Example 9

A ball has a diameter of 14 cm. Calculate the volume and the surface area of the ball.

$$V = \frac{4}{3}\pi r^3$$
$$= \frac{4}{3} \times \frac{22}{7} \times 7^3 \, \text{cm}^3$$
$$= 1437 \, \text{cm}^3$$
$$A = 4\pi r^2$$
$$= 4 \times \frac{22}{7} \times 7^2 \, \text{cm}^2$$
$$= 616 \, \text{cm}^2$$

Exercise 28.5

1. Find the volume of a cone whose base radius is 28 cm and whose height is 10 cm.

2. A cone has a base diameter of 10 in and height of 8 in. Calculate its volume.

3. A pyramid has a rectangular base 4 cm by 3 cm and a vertical height of 6 cm. Calculate its volume.

4. A pyramid has a square base of side 8 ft and an altitude of 5 ft. What is its volume?

5. A sphere has a radius of $3\frac{1}{2}$ cm. Calculate its volume and its surface area.

6. A concrete dome is in the shape of a hemisphere. Its internal and external diameters are 4.8 m and 5 m, respectively. Calculate the volume of concrete used in its construction.

7. A small roof is in the form of a pyramid. The base is a regular pentagon whose area is 15 m², and its vertical height is 4 metres. Calculate the volume enclosed by the roof.

8. Fig. 28.33 shows a flask which may be considered as a sphere with a cylindrical neck. Calculate the volume of the flask.

Fig. 28.33

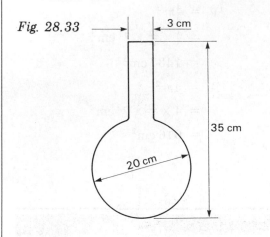

Capacity

The capacity of a container is the volume of liquid that it will hold. Capacity is sometimes measured in the same units as volume (i.e. cubic centimetres or cubic metres) but more often it is measured in litres (abbreviation ℓ) such that

$$1 \text{ litre} = 1000 \text{ cm}^3$$

Small capacities are often measured in centilitres ($c\ell$) or millilitres ($m\ell$).

$$100 \text{ c}\ell = 1000 \text{ m}\ell$$
$$= 1 \ell$$
$$1 \text{ m}\ell = 1 \text{ cm}^3$$

In the imperial system, capacity is measured in fluid ounces, pints and gallons.

$$20 \text{ fluid ounces (fl oz)} = 1 \text{ pint (pt)}$$
$$8 \text{ pints (pt)} = 1 \text{ gallon (gal)}$$

Example 10

(a) How many 5-millilitre doses of medicine can be obtained from a cylindrical medicine bottle which has a diameter of 5 cm and a height of 12 cm?

Volume of bottle $= \pi \times \text{Radius}^2 \times \text{Height}$
$$= 3.142 \times 2.5^2 \times 12 \text{ cm}^3$$
$$= 236 \text{ cm}^3$$
$$= 236 \text{ m}\ell$$
Number of doses $= 236 \div 5$
$$= 47.2$$
Number of full doses $= 47$

(b) A rectangular water tank is 4 m long, 2 m wide and 3 m high. When it is full how many litres of water does it hold?

Volume of tank $= (400 \times 200 \times 300) \text{ cm}^3$
$$= 24\,000\,000 \text{ cm}^3$$

Capacity
of the tank $= (24\,000\,000 \div 1000) \ell$
$$= 24\,000 \ell$$

Theory of Dimensions

Perimeters are always measured in **linear** units, for instance in centimetres (cm), inches (in), etc.

The circumference of a circle of radius r is given by the formula $2\pi r$. 2 and π are numbers and the variable, r, is raised to the first power. Therefore the circumference of a circle is measured in linear units.

The perimeter of a rectangle of length a and breadth b is given by the formula $2(a + b)$. a and b are both measured in linear units and therefore the variable $(a + b)$ is also measured in linear units. Hence the perimeter of a rectangle is measured in linear units.

Area is measured in square units, for instance square inches (in^2), square centimetres (cm^2), etc.

The area of a circle of radius r is given by the formula πr^2. π is merely a number but the variable r is raised to the second power and hence the area of a circle is measured in square units.

The surface area of cuboid of width a, breadth b and length l is given by the formula $2l(a + b)$. In this case l and $(a + b)$ are both variables raised to the first power. The index of the formula is $1 + 1 = 2$ and hence the surface area of a cuboid is measured in square units.

Volume is measured in cubic units, for instance cubic centimetres (cm³), cubic inches (in³), etc.

The volume of a cylinder of radius r and height h is given by the formula $\pi r^2 h$. π is merely a number but the variable r is raised to the second power and the variable h is raised to the first power. Therefore the index of the formula is $2 + 1 = 3$. Hence the volume of a cylinder is measured in cubic units.

The volume of a cuboid of width a, breadth b and length l is given by the formula abl. Each of the variables a, b and l is raised to the first power. Hence the index of the formula is $1 + 1 + 1 = 3$. Hence the volume of a cuboid is measured in cubic units.

When a formula consists of more than one term it is essential that all of the terms have the same units. For instance if

$$A = \pi r^3 + 2\pi r^2$$

the first term has a cubic units whilst the second term has square units and therefore this formula cannot be correct. However if

$$V = \pi r^3 + \pi r^2 l$$

both the terms have cubic units and hence the formula could be correct.

Example 11

The formulae for the surface area and the volume of the container shown in Fig. 28.34 are contained in the following list:

(1) $\pi d + 2h$ (4) $\frac{1}{6}\pi d^3 + \frac{1}{4}\pi d^2 h$
(2) $\pi d^2 + \frac{1}{6}\pi d^2 h$ (5) $\frac{1}{6}\pi d^3 + \pi dh$
(3) $\pi d^2 + \pi dh$

Formula (1) cannot be correct for surface area or volume because the index of both πd and $2h$ is 1 and surface area must be measured in square units and volume in cubic units.

Fig. 28.34

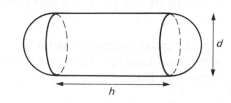

Formula (2) cannot be correct because the index of πd^2 is 2 and the index of $\frac{1}{6}\pi d^2 h$ is 3 and hence this formula contains a mixture of square and cubic units.

Formula (3) could be correct for the surface area since the index of both πd^2 and πdh is 2. Neither formulae (4) and (5) have an index of 2 and so formula (3) is correct for the surface area of the container.

In formula (4) both $\frac{1}{6}\pi d^3$ and $\frac{1}{4}\pi d^2 h$ have an index of 3 and this formula is correct for the volume because formula (5) contains a mixture of cubic and square units.

Therefore

$$\text{Surface area} = \pi d^2 + \pi dh$$
$$\text{Volume} = \frac{1}{6}\pi d^3 + \frac{1}{4}\pi d^2 h$$

Exercise 28.6

1. A small can in the form of a cylinder has a radius of 7 cm and a height of 8 cm. How many litres will the can hold?

2. A rectangular tank is 3 m long, 1 m wide and 0.5 m tall. How many litres of liquid will it hold when full?

3. A basin may be considered to be a hemisphere of diameter 28 cm. How many litres of water will it hold when full?

4. A cylindrical garden pool has a radius of 3 m. How many litres of water will it take to fill the pool to a depth of $1\frac{1}{2}$ m?

5. A rectangular medicine bottle is 8 cm wide, 4 cm long and 12 cm high. How many centilitres will it hold when full?

6. A conical wine glass is 4.5 cm in diameter and 6 cm high. What is its capacity in millilitres? How many such glasses would be filled from a bottle of wine containing 70 centilitres?

7. An ice cream carton is cylindrical in shape. It is 6 cm in diameter and 8 cm tall. How many litres of ice cream are needed to fill 50 of these cartons?

8. A petrol storage tank is rectangular in shape and measures 3 m by 2 m by 1.5 m. Calculate the capacity of the tank in litres.

9. A water tank with vertical sides has a horizontal base as shown in Fig. 28.35. The tank has a vertical height of h metres. The formula for the volume of the container is contained in the following list:

    (1) $2\pi rh + 2kh$
    (2) $\pi r^2 h + 2kh$
    (3) $\pi r^2 h + 2rkh$
    (4) $2rkh + 2\pi rh$

    Write down the correct formula for the volume of the tank.

Fig. 28.35

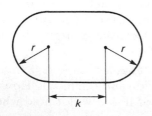

10. Fig. 28.36 represents a bird cage in the form of a cylinder surmounted by a cone. The formulae for the total surface area and the volume are given in the following list:

    (1) $\pi r^2 h + \pi r^2 + \pi rl$
    (2) $2\pi rh + \pi r^2 + \pi rl$
    (3) $2\pi rl + \pi r^2 h$
    (4) $\pi r^2 h + \pi r^2 + \frac{1}{3}\pi r^2 k$
    (5) $2\pi rh + \pi rl + \pi r^2 h$
    (6) $\pi r^2 h + \frac{1}{3}\pi r^2 k$

Write down the correct formulae for the total surface area and the volume.

Fig. 28.36

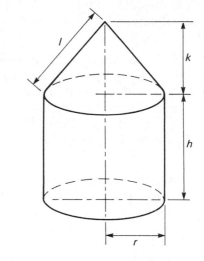

Similar Solids

For solids to be similar they must be wide, long and tall in the same proportions.

For instance, the diagram (Fig. 28.37) shows two child's building bricks. Looking at the diagram you might say that the larger brick is three times bigger than the smaller brick, and indeed all the lengths are three times bigger. However, 9 small bricks would sit on top of the larger brick and it would take 27 small bricks to make a cube as big as the larger brick.

Fig. 28.37

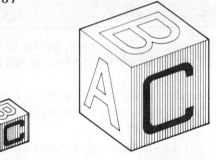

Notice that the length is 3 times bigger.

The surface area is $3 \times 3 = 9$ times bigger.

The volume and weight is $3 \times 3 \times 3 = 27$ times bigger.

From this example we see that:

(1) The volume of similar solids vary as the cube of the corresponding dimensions.

(2) The surface areas of similar solids vary as the square of the corresponding dimensions.

Fig. 28.38 shows two cylinders. If the ratio of corresponding dimensions is the same then the cylinders are similar. In the diagram

$$\frac{\text{Height of cylinder A}}{\text{Height of cylinder B}} = \frac{\text{Diameter of cylinder A}}{\text{Diameter of cylinder B}}$$

and hence cylinders A and B are similar.

Fig. 28.38

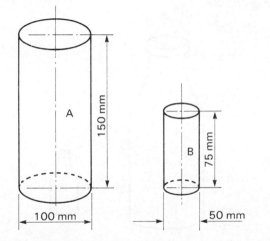

It follows that

$$\frac{\text{Volume of cylinder A}}{\text{Volume of cylinder B}}$$

$$= \left(\frac{\text{Height of cylinder A}}{\text{Height of cylinder B}}\right)^3$$

$$= \left(\frac{\text{Diameter of cylinder A}}{\text{Diameter of cylinder B}}\right)^3$$

$$\frac{\text{Surface area of cylinder A}}{\text{Surface area of cylinder B}}$$

$$= \left(\frac{\text{Height of cylinder A}}{\text{Height of cylinder B}}\right)^2$$

$$= \left(\frac{\text{Diameter of cylinder A}}{\text{Diameter of cylinder B}}\right)^2$$

Example 12

(a) The volume of a cone whose height is 135 mm is 1090 mm^3 and its surface area is 680 mm^2. Find the volume and surface area of a similar cone which has a height of 72 mm.

$$\frac{\text{Volume of smaller cone}}{\text{Volume of larger cone}}$$

$$= \left(\frac{72}{135}\right)^3$$

$$= 0.1517$$

Volume of smaller cone

$$= 0.1517 \times \text{Volume of larger cone}$$

$$= 0.1517 \times 1090 \text{ cm}^3$$

$$= 165.4 \text{ cm}^3$$

$$\frac{\text{Surface area of smaller cone}}{\text{Surface area of larger cone}}$$

$$= \left(\frac{72}{135}\right)^2$$

$$= 0.2844$$

Surface area of smaller cone

$\quad = 0.2844 \times$ Surface area of larger cone

$\quad = 0.2844 \times 680\ \text{mm}^2$

$\quad = 193\ \text{mm}^2$

(b) The model of a lorry is made on a scale of 1 in 10.

   (i) The windscreen on the model has an area of $100\ \text{cm}^2$. Calculate the area, in square centimetres, of the windscreen of the lorry.

  (ii) The fuel tank of the lorry, when full, holds 100 litres. Calculate the capacity of the fuel tank on the model.

The lorry and its model are similar solids. Each dimension on the lorry is 10 times larger than the corresponding dimension on the model.

(i) $\dfrac{\text{Area of windscreen on the lorry}}{\text{Area of windscreen on the model}}$

$\quad = 10^2$

$\quad = 100$

Area of windscreen on the lorry

$\quad = 100 \times$ Area of windscreen on the model

$\quad = 100 \times 100\ \text{cm}^2$

$\quad = 10\,000\ \text{cm}^2$

(ii) $\dfrac{\text{Capacity of fuel tank on model}}{\text{Capacity of fuel tank on lorry}}$

$\quad = \left(\dfrac{1}{10}\right)^3$

$\quad = \dfrac{1}{1000}$

Capacity of fuel tank on model

$\quad = \dfrac{1}{1000} \times$ Capacity of fuel tank on lorry

$\quad = \dfrac{1}{1000} \times 100\ \text{litres}$

$\quad = 0.1\ \text{litre}$

Exercise 28.7

1. Two tins of beans are similar in shape but the larger tin is twice as tall as the smaller tin.

   (a) If the surface area of the larger tin is $200\ \text{cm}^2$, what is the surface area of the smaller tin?

   (b) The volume of the smaller tin is $35\ \text{cm}^3$. What is the volume of the larger tin?

2. Two cubes are made of the same type of wood. The larger cube is 8 times heavier than the smaller cube. If the smaller cube has an edge 3 in long, what is the length of the edge on the larger cube?

3. A hemisphere has a radius of 2 cm and a volume of $6\ \text{cm}^3$. Work out the volume of a hemisphere with a radius of 3 cm.

4. The volume of a cone of height 14.2 in is $210\ \text{in}^3$. Find the volume of a similar cone whose height is 9.4 in.

5. Fig. 28.39 shows two bottles which are similar in shape. Bottle A is $2\frac{1}{2}$ times as tall as bottle B and holds 27 pints of liquid. How many pints does bottle B hold?

Fig. 28.39

6. A model aircraft, similar to the full-size aircraft, is made to a scale of 1 in 20. The model has a tail 0.15 m high, a wing area of 0.28 m² and a cabin volume of 0.016 m³. Find the corresponding figures for the full size aircraft.

7. A sphere has a radius of 3 ft. Calculate its surface area and its volume. Hence find the surface area and the volume of a sphere with a radius of 5 ft.

8. The curved surface of a cone has an area of 20.5 cm². Calculate the curved surface area of a similar cone whose height is $1\frac{1}{2}$ times greater than that of the first cone.

Miscellaneous Exercise 28

1. A piece of alloy is melted down to form 48 toy soldiers. how many similar soldiers of twice the height could be made from the same piece of metal?

2. Fig. 28.40 shows a large greenhouse. Calculate (a) the area of one end, (b) the volume of air contained in the greenhouse, (c) the perimeter of one end, (d) the total area of glass comprising both ends, both sides and the roof.

Fig. 28.40

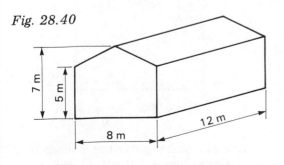

3. Fig. 28.41 shows two solid wooden blocks taken from a child's building outfit. Each block is a right prism.

Fig. 28.41

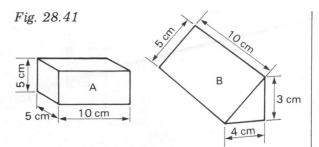

Prism A is a rectangular block 10 cm long, its uniform cross-section being a square of side 5 cm. Prism B is a triangular prism 10 m long, its uniform cross-section being a right-angled triangle of sides 3, 4 and 5 cm. Calculate

(a) the volume of prism A

(b) the total surface area of prism A

(c) the total surface area of prism B

4. Which of the diagrams in Fig. 28.42 are nets of

(a) a cube   (b) a tetrahedron

Fig. 28.42

(a)

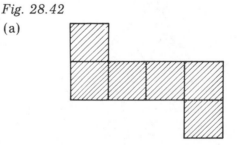

(b)

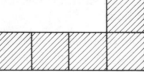

(c)

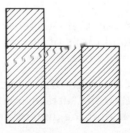

(d)

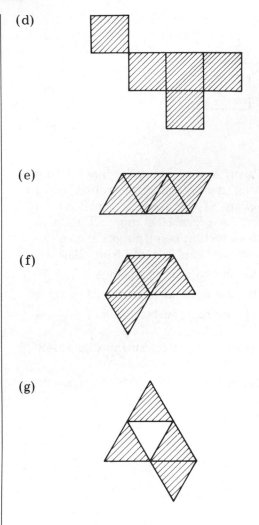

(e)

(f)

(g)

Fig. 28.43

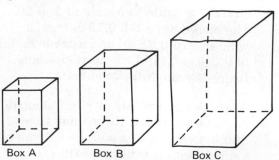

Box A   Box B   Box C

5. Fig. 28.43 shows three similar rectangular boxes.

(a) The dimensions of box A are 1.6 in by 3.5 in by 5.3 in. Calculate its volume.

(b) In box C the dimensions are double the dimensions of box A. Calculate the volume of box C.

(c) If the ratio of the surface area of box B to the surface area of box A is 9 to 4, find the dimensions of box B.

6. (a) A rectangular tank with a base 72 cm long and 45 cm wide contains, when full, 81 litres of water. Calculate the volume of the tank in cubic centimetres.

(b) Find the height of the tank.

7. A pipe 20 m long and with 70 cm outside diameter is to be laid under a road. A trench 1.2 m wide and 1.5 m deep is dug across the road, and the pipe is to be laid in concrete as shown in Fig. 28.44. Calculate, to the nearest cubic metre, the volume of concrete that will be needed for the job.

Fig. 28.44

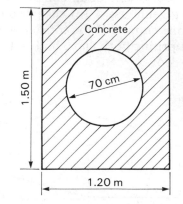

Concrete

1.50 m

70 cm

1.20 m

8. Plastic rain-water guttering is made in rectangular sections (type A) and semi-circular sections (type B) in 2 m lengths and with the other dimensions shown in Fig. 28.45.

Fig. 28.45

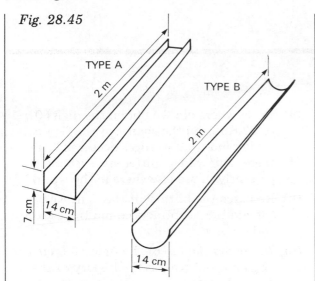

TYPE A

2 m

TYPE B

2 m

7 cm

14 cm

14 cm

(a) Calculate the area in square centimetres of the flat piece of plastic needed to make a length of type A guttering.

(b) Find the area in square centimetres of the flat piece of plastic needed to make a length of type B guttering.

(c) If the ends of each section were blocked up and the gutters were filled with water, find the volume of water, in litres, in each type of guttering.

9. Which is the better buy: a packet of soap powder which is 28 cm high costing 48p, or a geometrically similar packet of the same powder which is 35 cm high and costs 90p?

Mental Test 28

1. Find the volume of a cuboid whose length is 5 in, whose width is 4 in and whose height is 2 in.

2. A triangular prism has a constant cross-section whose area is $5 \, \text{cm}^2$. If the prism is 20 cm long, find its volume.

3. A match box is 7 cm wide, 3 cm high and 10 cm long. What is its volume?

4. A block of wood has a cross-section which is a square of 5 in side. If the block is 10 in long, calculate its volume.

5. Find the lateral surface area for the block of wood in question 4.

6. A cylinder has a cross-sectional area of $12 \, \text{cm}^2$. If it is 8 cm high, determine its volume.

7. A cylinder has a diameter of 7 cm and a height of 5 cm. Find its curved surface area.

8. A cylinder has a base radius of 5 cm and a height 4 cm. Find its volume, stating the answer as a multiple of π.

9. A cone has a base area of $21 \, \text{cm}^2$ and a height of 4 cm. What is its volume?

10. A pyramid has a base $36 \, \text{cm}^2$ in area and a height of 10 cm. What is the volume of the pyramid?

11. A sphere has a radius of 3 cm. Find its surface area as a multiple of π.

12. The ratio of the volumes of two similar cylinders is $1 : 8$. What is the ratio of their base diameters?

Coursework 2

Assignment 11   Ma 4

Wages for the staff at six stores are delivered by a security van from one bank. The stores are lettered A to F and the position of the bank is shown by X.

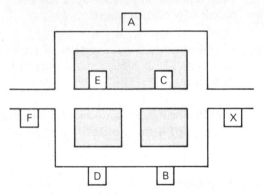

The van must proceed forward at all times and must not turn completely round in the road. The shaded portions are office blocks.

The first delivery must be to one of the stores A to E but the van must not pass in front of a store without stopping to make a delivery.

The last delivery is always to store F and each store is visited on a single journey each Thursday.

This week, the order was XBEACDF. The security firm is anxious to vary the route each week. How many different routes can you find ending at F?

Assignment 12   Ma 2, 3

For this assignment you will need to carry out some historical research by consulting various reference books. Here are some suggestions which are worth investigating:

(a) From where did we obtain the digits 0 to 9? How did the shapes of the figures evolve and what changes occurred? Are there any other different systems presently in use elsewhere in the world?

(b) Research the history of the signs we use for addition, subtraction, multiplication and division.

(c) Where did the use of the figure 0 (zero) originate and what is the history of the decimal point?

(d) What do we know about the mathematical symbol π (pi) and its value?

(e) How and when did algebra originate?

(f) Examine the history of time measurement and the calendar.

(g) How have the units of length and weight varied in the last two thousand years?

Assignment 13   Ma 2, 3, 5

(a) We use simple rules for checking whether a number is divisible by 2, 3 or 5. Work out your own sets of rules for testing the divisibility of numbers by 4, 6, 8, 9, 15 and 25. Try and make your rules as simple and short as possible. Consult reference books to find a method which tests for divisibility by 11.

(b) On the next page is a flow chart which can be used to test for divisibility by a particular number:

To pin it down, you will need to introduce a range of carefully selected numbers into the flow chart and tabulate the results for each circuit around the loops. How does the routine identify divisibility by this number? Computer enthusiasts might consider writing a program based on the flow chart.

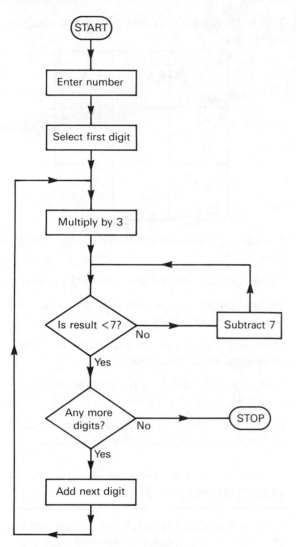

Substitute a range of positive integers for n, starting with $n = 1$, and record the results. At what point does the formula fail? Why does it fail? Can you make an adjustment to the formula which will improve its efficiency?

Assignment 14   Ma 4

The firm Awkward Shapes Ltd produces plastic blocks of various shapes and sizes. Here are four of the shapes, each of which is 1 cm thick:

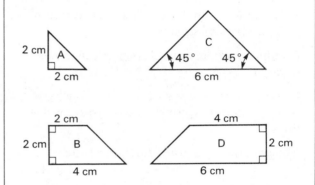

The shapes are sold in the following sets:

Set 1: 2 of A, 2 of B, 4 of C, 2 of D
Set 2: 2 of A, 2 of B, 2 of C, 2 of D
Set 3: 1 of A, 1 of B, 2 of C, 1 of D
Set 4: 3 of A, 3 of B, 2 of C, 3 of D

Investigate the various packing boxes which could be used to hold these sets of blocks.

(c) Under what circumstances will the LCM of three different numbers be the same as their product? What will the HCF of the numbers be in this case?

(d) What conditions must be met for the LCM of three different numbers p, q and r to be exactly four times their HCF? How does the sum of the three numbers, $p + q + r$, relate to part (b) of this assignment? Can you show why it happens using algebra?

(e) Here is an interesting formula:

$$A = n^2 - n + 11$$

Assignment 15   Ma 4, 5

Using four lines, we can draw one square, but if five lines are used, we can produce two squares:

Using six lines, these diagrams can be constructed:

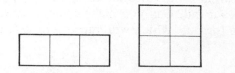

There are three squares in the first diagram. In the second diagram there are four small squares and one large square giving a total of five squares.

Investigate the use of lines to draw squares.

Is there anything significant about the maximum number of squares each time?

Make tables to include different formations and numbers of squares.

Comment on patterns shown in the tables of numbers.

Continue the investigation until you have reached patterns constructed from twelve lines.

Is it possible to predict further results?

Assignment 16   Ma 4

There are seven different wall tiles available at the local DIY store:

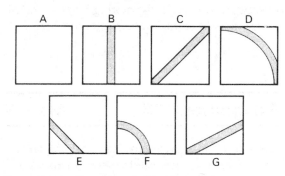

Investigate the use of the tiles:

(a) to form closed shapes

(b) to produce symmetrical patterns for bathroom or kitchen.

This is an example of a closed shape using eight tiles:

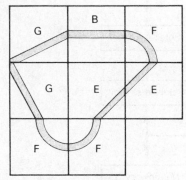

Is a closed shape possible with only three tiles?

Investigate the use of more than three tiles.

Include patterns in which tiles do not necessarily have a common edge.

Does any tile present a particular problem compared with other tiles?

Design a tile which will produce better results in the investigation.

Construct symmetrical patterns to fit a rectangle measuring 15 tiles by 12 tiles.

Assignment 17   Ma 4

A fly was allowed to settle on X, one vertex of a cube. It was then free to walk along the edges. The routes it creates are fly-paths.

Fly-paths are identified by the number of edges the fly uses in one journey. A fly-path does not use any edge of the cube more than once.

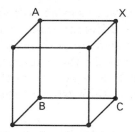

From X to A is called a 1-path.
From X to A and then to B is a 2-path.

Investigate fly-paths on the cube and other solids.

First construct a cube. One candidate used lengths of wire held together at each vertex by a blob of plasticine. Availability of materials will influence the choice of construction.

Label each vertex with a letter and record the fly-paths in a list:

1-path: XA, XC, etc.
2-path: XAB, XCB, etc.

All paths must start at X and no edge must appear twice in one path. There are 12 edges. Is there a possible 12-path? If not, what is the maximum fly-path? Is there anything particularly significant about any of the paths?

At this stage give some thought to constructing a cube using a single length of wire. It will obviously be necessary to run a double wire on one or more edges. Here is one attempt:

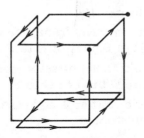

Does this method use the minimum possible length of wire? If the wire is allowed to follow diagonals across faces, is it possible to form a cube without duplicating any edges? Does that allow for the use of less wire? Produce diagrams to illustrate your conclusions.

Investigate these solids:

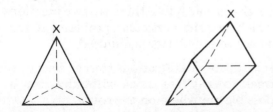

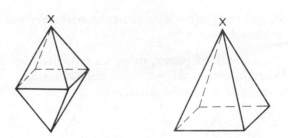

Examine fly-paths from point X. Are the numbers of faces and vertices in a given solid at all significant? Does any solid present a unique problem? Is the position of point X important?

Assignment 18   Ma 4

Construct two circles which have:

(a) four common tangents

(b) only three common tangents

(c) only two common tangents

(d) only one common tangent

(e) no common tangents.

Assignment 19   Ma 4

A puzzle is to be constructed in which five solids will be assembled to make a cube. These notes explain how to make four of the solids, but you must work out how to produce the fifth solid yourself.

Use this net to construct four pyramids as shown:

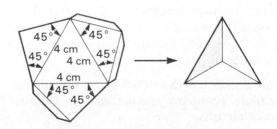

If adhesive tape is used, tabs may not be required.

On a piece of paper, draw an equilateral triangle with a side of 8 cm. Glue the four pyramids on to the triangle as shown:

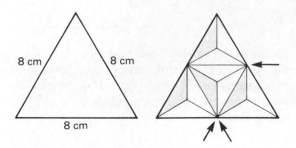

Fold the shape as indicated by the three arrows to form a cube. There will be a gap in the centre.

Decide what shape is required to fill the centre gap perfectly and construct the fifth solid according to the measurements from your cube. Finally, cut the triangular paper base to release your first four pyramids and you will then have a puzzle for constructing a cube from five other solids.

Assignment 20   Ma 4

A train set contains three types of rail.

Straight   Curve   Cross

Curves, which can be used to turn left or right, are each one-eighth of a circle. This means that a single curve gives a rotation of 45°. Curves will join on to any line or on to each other.

Investigate the use of all types of rail and explore problems encountered in various track circuits.

Firstly, examine the various combinations:

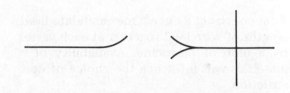

Draw and explain these and other combinations which are possible with the rails.

Investigate a typical circuit which incorporates all the different types of rail.

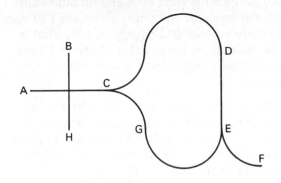

Explain why

(a) a train can travel from A to C and also from H to B but not H to C

(b) we could travel directly from A to D and A to G but not G to F

(c) we could start at A and travel back to A without stopping

(d) the train starting at F would need to stop twice if it wished to return to F.

The part of the line CGEDC is a closed circuit. Investigate the use of straights and curves in different closed circuits.

A figure-of-eight route is a particular type of closed circuit. Invent different circuits of that kind using all types of rail.

On the circuit A to F there are two junctions which present particular problems if free access to each letter is required.

Investigate modifications to each part of the track so that free travel without a stop is possible anywhere on the route.

Try to construct circuits for solving problems similar to the one presented in the next diagram.

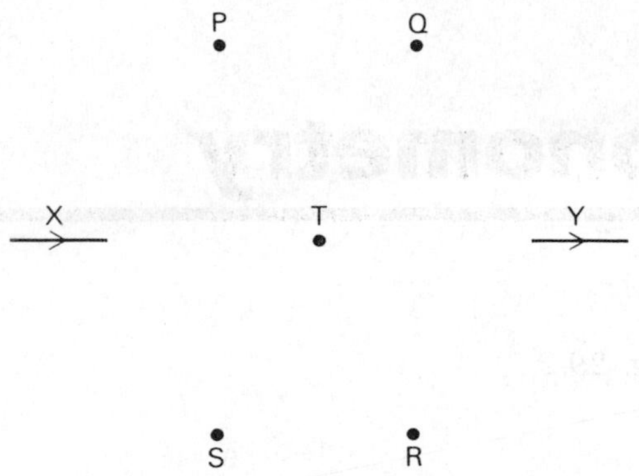

P   Q
•   •

X   T   Y
→   •   →

S   R
•   •

Stations P, Q, R, S and T are to be connected by rail to the main line entering at X and leaving at Y. A continuous route is required so that any or all of the five stations may be visited on a journey from X to Y without stopping or reversing. It may, of course, be necessary to pass through some points when travelling to other stations.

Create other circuits to include engine-storage sheds and loop lines. Show two or more routes linking a pair of points. Suggest new types of rail which might usefully be included in the train set.

Trigonometry

Notation for a Right-angled Triangle

Look at the right-angled triangle in Fig. 29.1 and note the special names given to its sides.

Fig. 29.1

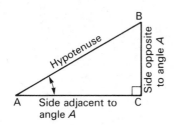

For the angle *A*:

> **Hypotenuse:** the side AB opposite to the right angle.
>
> **Opposite:** the side BC opposite to the angle *A*.
>
> **Adjacent:** the side AC adjacent to the angle *A* as it forms the angle *A* with the hypotenuse.

For the angle *B* (Fig. 29.2):

> **Hypotenuse:** the side AB opposite to the right angle.
>
> **Opposite:** the side AC opposite the angle *B*.
>
> **Adjacent:** the side BC adjacent to the angle *B* as it forms the angle *B* with the hypotenuse.

Fig. 29.2

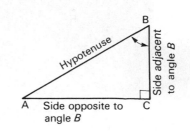

The Trigonometrical Ratios

In any right-angled triangle the following three ratios between the sides of the triangle can be formed:

$$\frac{\text{Opposite}}{\text{Hypotenuse}} \quad \text{called the sine of the angle}$$

$$\frac{\text{Adjacent}}{\text{Hypotenuse}} \quad \text{called the cosine of the angle}$$

$$\frac{\text{Opposite}}{\text{Adjacent}} \quad \text{called the tangent of the angle}$$

These three ratios are called the **trigonometrical ratios.** For any angle in a triangle which is not a right angle, each of these ratios has a definite value. These values can be found by using a calculator.

The abbreviations **sin, cos** and **tan** are usually used for sine, cosine and tangent respectively.

The Sine of an Angle

In Figs. 29.1 and 29.2

$$\sin A = \frac{\text{Side opposite } A}{\text{Hypotenuse}}$$

$$= \frac{BC}{AB}$$

$$\sin B = \frac{\text{Side opposite } B}{\text{Hypotenuse}}$$

$$= \frac{AC}{AB}$$

Example 1

Find, by drawing, the value of $\sin 30°$.

Draw the lines AX and AY which intersect at A so that $\angle YAX = 30°$, as shown in Fig. 29.3. Along AX measure off AC equal to 1 unit (say 10 cm) and from C draw CB perpendicular to AY. Measure BC.

BC will be found to be 0.5 unit (5 cm in this case).

$$\sin 30° = \frac{BC}{AC} = \frac{0.5}{1} = 0.5$$

Fig. 29.3

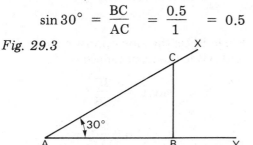

It is possible to find the sine of any angle by the drawing method shown in Example 1, but this is inconvenient and not very accurate.

Finding the Sine of an Angle

Before using a calculator to find the sine of an angle it must be converted to degrees and decimals of a degree.

Example 2

Find the sine of $27°53'41''$.

The first step is to convert the given angle to degrees and a decimal of a degree. Using one of the methods given on page 155 we find

$$27°53'41'' = 27.8947$$

Input	Display
27.8947	27.8947
sin	0.4678484

So $\sin 27°53'41'' = 0.4678$ correct to 4 d.p.

Example 3

Find the angle whose sine is 0.1711.

Input	Display
0.1711	0.1711
INV	0.1711
$\sin^{-1}$	9.8517815
INV	9.8517815
° ' ''	9°51'6.41''

So the angle whose sine is 0.1711 is $9°51'6''$ to the nearest second.

Exercise 29.1

1. Find, by drawing, the sines of
   (a) $45°$ (b) $60°$

2. Convert to degrees and decimals of a degree correct to 4 d.p.
   (a) $18°24'$ (b) $38°42'15''$
   (c) $79°54'53''$

3. Convert to degrees, minutes and seconds
   (a) $43.2°$ (b) $54.752°$
   (c) $43.196°$

4. Find the sines of the following angles correct to 4 d.p.
   (a) $14°$ (b) $73°$
   (c) $25.2°$ (d) $59.8°$
   (e) $54°12'18''$ (f) $79°42'39''$
   (g) $19°53'54''$ (h) $76°14'19''$

5. Use a calculator to find correct to the nearest second the angles whose sines are

    (a) 0.9135   (b) 0.9816
    (c) 0.7349   (d) 0.9943
    (e) 0.8112   (f) 0.1452

Calculations on the Sine of an Angle

Example 4

(a) Find the length of the side BC in Fig. 29.4.

Fig. 29.4

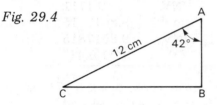

Since BC is the side opposite to the angle A and AC is the hypotenuse

$$\frac{BC}{AC} = \sin 42°$$

$$BC = AC \times \sin 42°$$

$$= 12 \times 0.6691 \text{ cm}$$

$$= 8.03 \text{ cm}$$

(b) Find the length of AC in Fig. 29.5.

Fig. 29.5

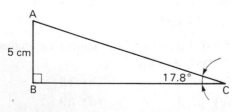

AB is the side opposite $\angle C$ and AC is the hypotenuse.

$$\frac{AB}{AC} = \sin C$$

$$AC = \frac{AB}{\sin C}$$

$$= \frac{5}{\sin 17.8°}$$

$$= \frac{5}{0.3057}$$

$$= 16.4 \text{ cm}$$

(c) Find the angles A and B in Fig. 29.6

Fig. 29.6

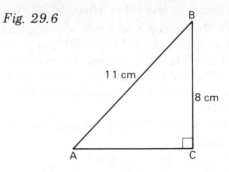

Since BC is the side opposite angle A and AB is the hypotenuse

$$\sin A = \frac{BC}{AB}$$

$$= \frac{8}{11}$$

$$= 0.7273$$

Using a calculator

$$\angle A = 46.7°$$

$$\angle B = 90° - 46.7°$$

$$= 43.3°$$

Exercise 29.2

1. Find the length of the sides marked x in Fig. 29.7.

Fig. 29.7

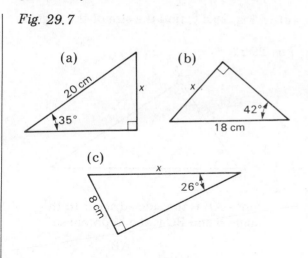

(a)

(b)

(c)

2. Find the size of the angles marked *y* in Fig. 29.8

Fig. 29.8

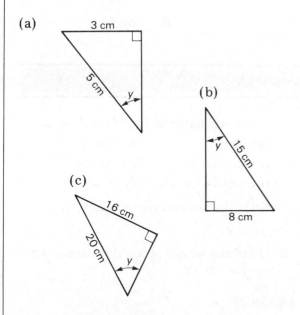

(a)

(b)

(c)

3. In the triangle ABC, $\angle B = 90°$, $\angle C = 26°21'$ and $b = 13.4$ cm. Calculate the length of the side *c* of the triangle.

4. In triangle ABC, $\angle C = 90°$, $\angle A = 69.3°$ and $a = 3.4$ cm. Calculate the length of the side *c* of the triangle.

5. An equilateral triangle has a vertical height of 20 cm. Calculate the length of the equal sides.

6. Calculate the length of the equal sides of an isosceles triangle whose altitude is 15 cm and whose equal angles are $48°36'$.

7. The equal sides of an isosceles triangle are each 12 cm long. If the altitude of the triangle is 8 cm, find the size of the three angles of the triangle.

8. In the triangle ABC, $\angle A = 90°$, $b = 10.8$ cm and $a = 12.3$ cm. Find the size of the angles *B* and *C*.

The Cosine of an Angle

In Figs. 29.1 and 29.2

$$\cos A = \frac{\text{Side adjacent to } \angle A}{\text{Hypotenuse}}$$

$$= \frac{\text{AC}}{\text{AB}}$$

$$\cos B = \frac{\text{Side adjacent to } \angle B}{\text{Hypotenuse}}$$

$$= \frac{\text{BC}}{\text{AB}}$$

The cosine of an angle may be found by drawing, the construction being similar to Fig. 29.3. However, cosines can be found by using a calculator.

Example 5

(a) Find the length of the side BC in Fig. 29.9

Fig. 29.9

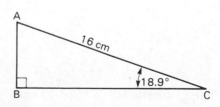

Since BC is the side adjacent to $\angle C$ and AC is the hypotenuse

$$\frac{BC}{AC} = \cos C$$

$$BC = AC \cos C$$

$$= 16 \times \cos 18.9°$$

$$= 16 \times 0.9461$$

$$= 15.14$$

Therefore BC = 15.14 cm.

(b) In triangle ABC, $\angle A = 90°$, $\angle C = 25°$ and $b = 21$ cm. Find the length of the side a.

The triangle is drawn in Fig. 29.10 where it will be seen that b is the side adjacent to the angle C whilst a is the hypotenuse.

Fig. 29.10

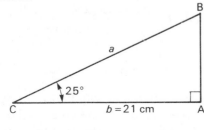

$$\frac{b}{a} = \cos C$$

$$a = \frac{b}{\cos C}$$

$$= \frac{21}{\cos 25°}$$

$$= \frac{21}{0.9063}$$

$$= 23.17$$

Therefore the side a is 23.17 cm.

(c) In Fig. 29.11, find the size of the angle B.

Fig. 29.11

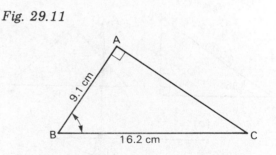

Since AB is the side adjacent to the angle B and BC is the hypotenuse

$$\cos B = \frac{AB}{BC}$$

$$= \frac{9.1}{16.2}$$

$$= 0.5617$$

$$\angle B = 55.8°$$

Exercise 29.3

1. Use a calculator to find the values of
    (a) $\cos 18°$ **(b)** $\cos 76°$
    (c) $\cos 37.3°$ **(d)** $\cos 81.6°$
    (e) $\cos 54°25'18''$ **(f)** $8°52'29''$

    State the answers correct to 4 d.p.

2. Find the length of the sides marked p in Fig. 29.12.

Fig. 29.12

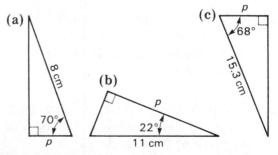

8. In Fig. 29.20, calculate the distance marked *q*.

Fig. 29.20

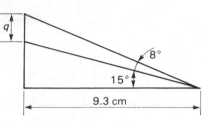

q

8°

15°

9.3 cm

Angles of Elevation and Depression

If you look upwards at an object, the angle formed between the horizontal and your line of sight is called the **angle of elevation** (Fig. 29.21).

Fig. 29.21

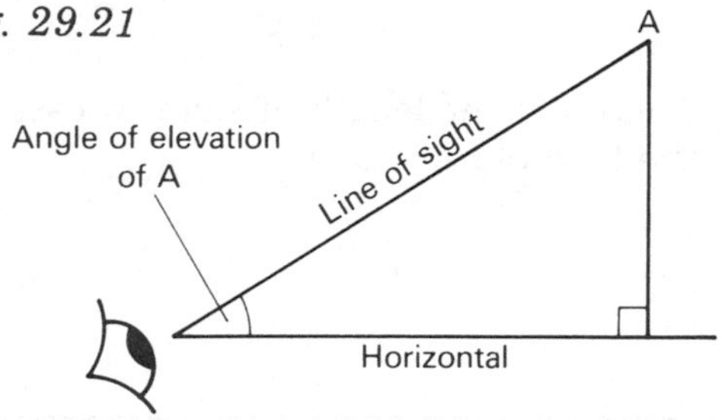

Angle of elevation of A

Line of sight

Horizontal

A

Example 7

A boy 1.5 m tall is 25 m away from a tower as shown in Fig. 29.22. The angle of elevation of the tower is 18°. How high is the tower?

Fig. 29.22

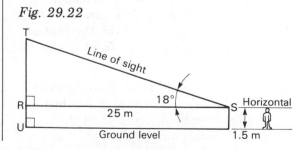

T

Line of sight

18°

R

25 m

S Horizontal

U

Ground level

1.5 m

In the diagram, ∠TSR is the angle of elevation to the top of the tower. TU is the tower and SR = 25 m.

$$\frac{TR}{SR} = \tan \angle TSR$$

$$TR = SR \times \tan \angle TSR$$

$$= 25 \times \tan 18°$$

$$= 25 \times 0.3249$$

$$= 8.12$$

$$UT = 8.12\,m + 1.5\,m$$

$$= 9.62\,m$$

If you look downwards at an object, the angle formed between the horizontal and your line of sight is called the **angle of depression** (Fig. 29.23).

Fig. 29.23

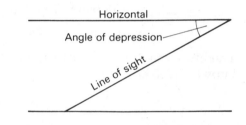

Horizontal

Angle of depression

Line of sight

Example 8

A man standing on top of a cliff 50 m high is in line with two buoys whose angles of depression are 18° and 20°. Calculate the distance between the buoys.

The problem is illustrated in Fig. 29.24 where the buoys are C and D and the observer is A.

Fig. 29.24

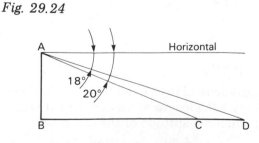

A

Horizontal

18°

20°

B

C

D

In triangle ABC,
∠BAC = 90° − 20° = 70°

$$\frac{BC}{AB} = \tan\angle BAC$$

$$BC = AB\tan\angle BAC$$

$$= 50 \times \tan 70°$$

$$= 137.4\,\text{m}$$

In triangle ABD,
∠BAD = 90° − 18° = 72°

$$\frac{BD}{AB} = \tan\angle BAD$$

$$BD = AB\tan\angle BAD$$

$$= 50 \times \tan 72°$$

$$= 153.9\,\text{m}$$

$$CD = BD - BC$$

$$= 153.9\,\text{m} - 137.4\,\text{m}$$

$$= 16.5\,\text{m}$$

Therefore the distance between the buoys is 16.5 m.

Altitude of the Sun

The **altitude of the sun** is the angle of elevation of the sun (Fig. 29.25).

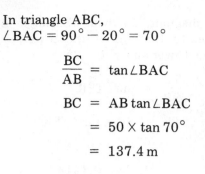

Fig. 29.25

Example 9

A flagpole is 11 m tall and it stands vertically. Calculate the length of shadow that it will cast when the altitude of the sun is 43°.

Fig. 29.26

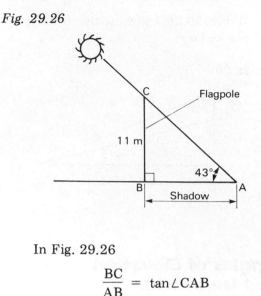

In Fig. 29.26

$$\frac{BC}{AB} = \tan\angle CAB$$

$$AB = \frac{BC}{\tan\angle CAB}$$

$$= \frac{11}{\tan 43°}$$

$$= 11.8$$

Therefore the length of shadow cast by the flagpole is 11.8 m.

Exercise 29.5

1. From a point on level ground the angle of elevation of a post is 15°. If the post is 18 m from the point of observation, calculate the height of the post.

2. A man 1.6 m tall observes the angle of elevation of the top of a spire to be 43°. If he is standing 28 m from the foot of the spire, calculate the height of the spire.

3. A man 1.8 m tall stands 20 m from the foot of a tower which is 19.8 m high. Find the angle of elevation of the top of the tower from his eye.

4. A man lying down on top of a cliff 35 m high observes the angle of depression of a buoy to be 20°. If he is in line with the buoy, calculate the distance between the buoy and the foot of the cliff, which may be assumed to be vertical.

5. A flagpole is 17 m high. Calculate the length of shadow it casts when the altitude of the sun is 64°.

6. A pole casts a shadow 20 m long when the altitude of the sun is 49°. Calculate the height of the pole.

7. A man standing on top of a mountain observes the angle of depression of a steeple to be 41°. If the mountain is 500 m high, how far is the steeple from the mountain?

8. In Fig. 29.27, a vertical cliff is 50 m high. The cliff is observed from a boat which is 60 m from the foot of the cliff. Calculate the angle of elevation of the top of the cliff from the boat.

Fig. 29.27

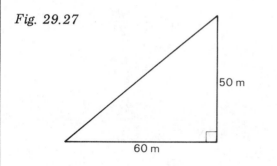

9. A person standing 30 m away from the foot of a tower observes the angles of elevation to the top and bottom of a flagpole standing on the tower to be 66° and 64° respectively. Calculate the height of the flagpole.

10. A surveyor stands 120 m from the foot of a tower on which an aerial stands. He measures the angles of elevation to the top and bottom of the aerial to be 59° and 57° respectively. What is the height of the aerial?

Area of a Triangle

Three formulae are commonly used for finding the area of a triangle.

(1) Given the base and the altitude (vertical height, Fig. 29.28):

$$\text{Area} = \tfrac{1}{2}bh$$

See also Chapter 26.

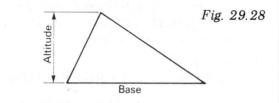

Fig. 29.28

(2) Given any two sides and the included angle (Fig. 29.29):

$$\text{Area} = \tfrac{1}{2}bc \sin A$$
$$\text{Area} = \tfrac{1}{2}ac \sin B$$
$$\text{Area} = \tfrac{1}{2}ab \sin C$$

Fig. 29.29

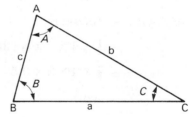

(3) Given all three sides (see also Chapter 26):

$$\text{Area} = \sqrt{s(s-a)(s-b)(s-c)}$$

where $s = \tfrac{1}{2}$ the perimeter
$$= \tfrac{1}{2}(a + b + c)$$

Example 10

(a) Find the areas of the triangles shown in Fig. 29.30.

Fig. 29.30

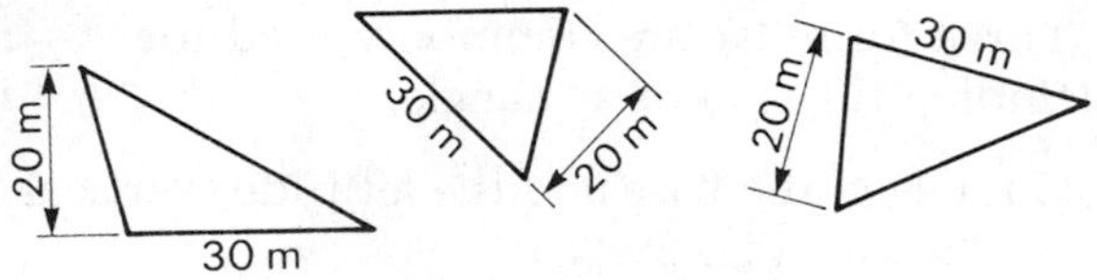

In each of the triangles shown in the diagram, the base is taken as being the side whose length is given, and the vertical height (altitude) is measured perpendicular to this side.

Area of
each triangle $= \frac{1}{2} \times$ Base $\times$ Altitude

$= \frac{1}{2} \times 30 \times 20$ m^2

$= 300$ m^2

(b) Find the area of the triangle shown in Fig. 29.31.

Fig. 29.31

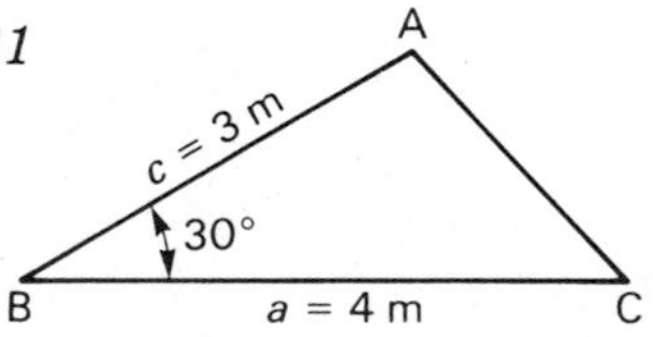

Area of $\triangle ABC = \frac{1}{2}ac \sin B$

$= \frac{1}{2} \times 4 \times 3 \times \sin 30°$

$= \frac{1}{2} \times 4 \times 3 \times 0.5$

$= 3$ m^2

(c) A triangle has sides which are 3 m, 5 m and 6 m long. Find its area.

$A = \sqrt{s(s-a)(s-b)(s-c)}$

$s = \frac{1}{2}(a+b+c)$

$= \frac{1}{2} \times (3+5+6)$

$= 7$ cm

$A = \sqrt{7 \times (7-3) \times (7-5) \times (7-6)}$

$= \sqrt{7 \times 4 \times 2 \times 1}$

$= \sqrt{56}$

$= 7.48$ m^2

Area of a Polygon

The area of a regular polygon may be found by dividing the polygon into a number of equal triangles (the number of triangles equals the number of sides). Then by finding the area of one of these triangles the area of the polygon is determined.

Example 11

The base of a steeple is a regular octagon (eight-sided figure) of side 4 m. Find its area.

In Fig. 29.32, the octagon is divided into eight equal triangles. AOB is one of these triangles. In $\triangle AOB$

$$\angle AOB = \frac{360°}{8} = 45°$$

Since $\triangle AOB$ is isosceles

$$\angle AON = \frac{45°}{2}$$

$$= 22.5°$$

$$AN = \frac{AB}{2}$$

$$= \frac{4}{2}$$

$$= 2 \text{ m}$$

In $\triangle AON$

$$\frac{AN}{ON} = \tan \angle AON$$

Fig. 29.32

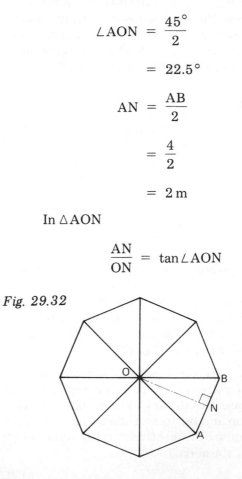

$$ON = \frac{AN}{\tan\angle AON}$$

$$= \frac{2}{\tan 22.5°}$$

$$= 4.828\,m$$

$$\text{Area of } \triangle AOB = \tfrac{1}{2} \times 4 \times 4.828$$

$$= 9.656\,m^2$$

$$\text{Area of octagon} = 8 \times 9.656\,m^2$$

$$= 77.25\,m^2$$

Exercise 29.6

1. Find the area of a triangle whose base is 7.5 cm and whose vertical height is 5.9 cm.

2. Find the area of an isosceles triangle whose equal sides are 8.2 cm long and whose base is 9.6 cm.

3. An equilateral triangle has an area of 15.6 cm². Calculate the lengths of its sides.

4. A triangle has sides of length 39.3 cm and 41.5 cm. If the angle between these sides is 41°30′, find the area of the triangle.

5. Fig. 29.33 shows a playground. Find its area.

Fig. 29.33

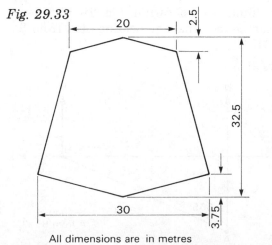

All dimensions are in metres

6. Calculate the area of triangle ABC if $a = 3$ m, $c = 6$ m and $\angle B = 63°44′$.

7. A triangle has sides 4 m, 7 m and 8 m long. Find its area.

8. Find the area of the plot of land with dimensions shown in Fig. 29.34.

Fig. 29.34

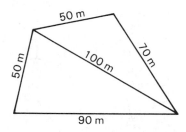

9. Find the area of a regular hexagon (six-sided figure) whose perpendicular distance between opposite sides is 4 cm.

10. A regular octagon (eight-sided figure) has sides 6 m long. Calculate its area.

Bearings

The four cardinal directions are north, south, east and west (Fig. 29.35). Directions between the cardinal points are called **bearings**.

Fig. 29.35

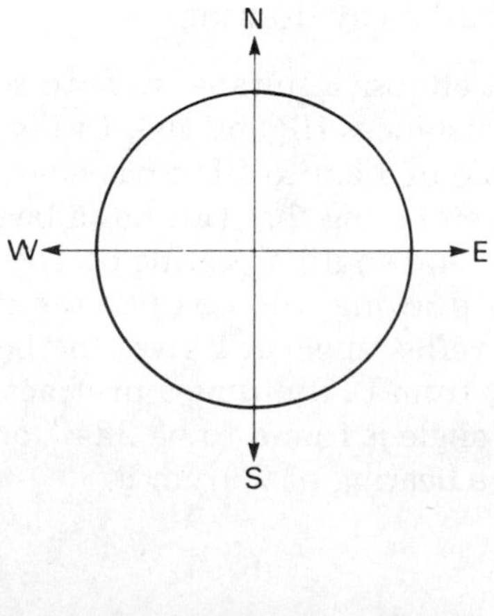

The usual way of stating a bearing is to measure the angle from north in a clockwise direction. Three figures are always stated, north being 000°. The figure 005° is written instead of 5°, 027° instead of 27°. East is 090°, south is 180° and west is 270°.

Some typical bearings are shown in Fig. 29.36.

Fig. 29.36

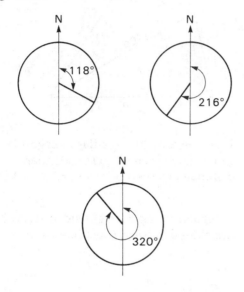

Example 12

P is a point due east of a harbour H and Q is a point on the coast 10 km due south of H. If the distance PQ is 14 km, find the bearing of Q from P.

   Method 1 (by drawing)

   First choose a suitable scale to represent the distances HP and HQ. In Fig. 29.37, a scale of 1 cm to 2 km has been used. Remembering that east has a bearing of 090° and south a bearing of 180°, the scale drawing will look like the diagram. The reflex angle at P gives the bearing of Q from P. By using a protractor this angle is found to be 224°, and this is the bearing of Q from P.

Fig. 29.37

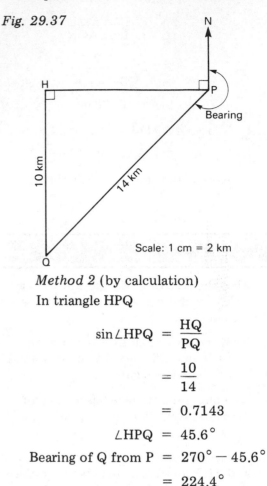

Scale: 1 cm = 2 km

Method 2 (by calculation)

In triangle HPQ

$$\sin \angle HPQ = \frac{HQ}{PQ}$$

$$= \frac{10}{14}$$

$$= 0.7143$$

$$\angle HPQ = 45.6°$$

$$\text{Bearing of Q from P} = 270° - 45.6°$$

$$= 224.4°$$

Example 13

The bearing of a point B from a point A is 065°. What is the bearing of A from B?

The situation is shown in Fig. 29.38, where it can be seen that the bearing of A from B is 245°.

Fig. 29.38

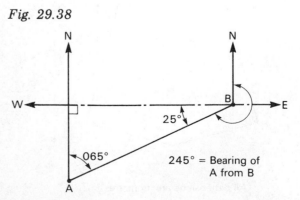

245° = Bearing of A from B

Exercise 29.7

1. Fig. 29.39 shows part of a map. Using three-digit numbers, state the bearings of
   (a) Beeson from Newtown
   (b) Dayton from Beeson

Fig. 29.39

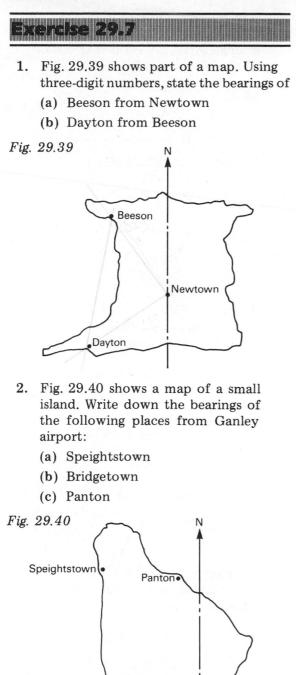

2. Fig. 29.40 shows a map of a small island. Write down the bearings of the following places from Ganley airport:
   (a) Speightstown
   (b) Bridgetown
   (c) Panton

Fig. 29.40

3. The bearing of Manley from Trestor is 155°. What is the bearing of Trestor from Manley?

4. A ship is on a bearing of 068° from a lighthouse. What is the bearing of the lighthouse from the ship?

5. B and C (Fig. 29.41) are both 100 km from A. C is on a bearing of 225° from B. Calculate, or find by scale drawing
   (a) the bearing of A from B
   (b) the size of the angle ABC

Fig. 29.41

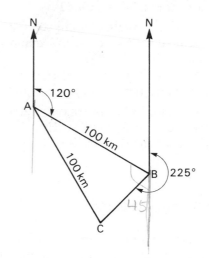

6. From Fig. 29.42, find the bearing of X from Y.

Fig. 29.42

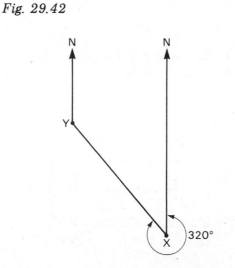

7. The distances from town A to towns B
    and C are 160 km and 340 km respec-
    tively. The bearing of B from A is 346°
    and the bearing of C from A is 035°.
    Make a scale drawing of this information
    and from it find

    (a) the distance between B and C

    (b) the bearing of C from B

8. A boat leaves a harbour H on a course
    of 120° and sails 80 km in this direction
    until it reaches a point B.

    (a) Using a scale of 1 cm to 10 km, make
    a scale drawing of the movement. Hence
    find
    (i) how far B is east of H,
    (ii) what distance B is south of H.

    (b) Find, by using trigonometry,
    (i) how far B is east of H,
    (ii) how far B is south of H.

Solid Geometry

Problems with solid figures are solved by
choosing suitable right-angled triangles in
different planes. It is essential to make a
clear three-dimensional drawing in order to
find these triangles.

Example 14

Fig. 29.43 shows a cuboid. Calculate the
length of the diagonal AG.

Fig. 29.43

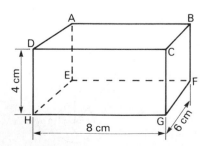

Fig. 29.44 shows that in order to find
AG we must use the right-angled
triangle AGE. GE is the diagonal of
the base rectangle EFGH.

Fig. 29.44

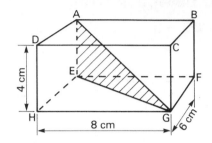

In triangle EFG, EF = 8 cm,
GF = 6 cm and $\angle EFG = 90°$.
By Pythagoras' theorem

$$EG^2 = EF^2 + GF^2$$

$$= 8^2 + 6^2$$

$$= 64 + 36$$

$$= 100$$

$$EG = \sqrt{100}$$

$$= 10$$

In triangle AGE, AE = 4 cm,
EG = 10 cm and $\angle AEG = 90°$.
By Pythagoras' theorem

$$AG^2 = AE^2 + EG^2$$

$$= 4^2 + 10^2$$

$$= 16 + 100$$

$$= 116$$

$$AG = \sqrt{116}$$

$$= 10.77 \text{ cm}$$

Example 15

Fig. 29.45 shows a pyramid with a square base. The base has sides 6 cm long and the edges of the pyramid VA, VB, VC and VD are each 10 cm long. Find the vertical height of the pyramid.

Fig. 29.45

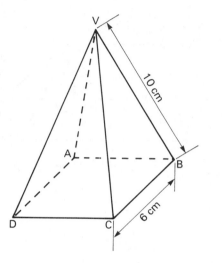

The right-angled triangle VBE (Fig. 29.46) allows the vertical height of the pyramid to be found. However, we must first find BE from the right-angled triangle BEF.

Fig. 29.46

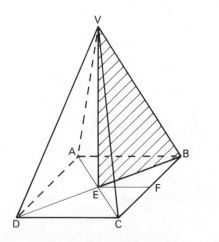

In triangle BEF, BF = EF = 3 cm and ∠BFE = 90°. By Pythagoras' theorem

$$BE^2 = BF^2 + EF^2$$
$$= 3^2 + 3^2$$
$$= 9 + 9$$
$$= 18$$
$$BE = \sqrt{18}$$
$$= 4.243 \text{ cm}$$

In triangle VBE, BE = 4.243 cm, VB = 10 cm and ∠VEB = 90°.

By Pythagoras' theorem

$$VE^2 = VB^2 - BE^2$$
$$= 10^2 - 4.243^2$$
$$= 100 - 18$$
$$= 82$$
$$VE = \sqrt{82}$$
$$= 9.055 \text{ cm}$$

Exercise 29.8

1. Fig. 29.47 shows a cuboid.

    (a) Sketch the triangle EHF and calculate the length of FH.

    (b) Sketch the triangle HBF adding known dimensions. Calculate the length of the diagonal BH.

Fig. 29.47

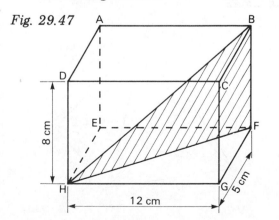

2. Fig. 29.48 shows a pyramid on a square base of side 8 cm. The altitude of the pyramid is 12 cm.

(a) Calculate EF.

(b) Draw the triangle VEF and add known dimensions.

(c) Find the angle VEF.

(d) Calculate the slant height VE.

(e) Calculate the area of triangle VAD.

(f) Calculate the complete surface area of the pyramid.

Fig. 29.48

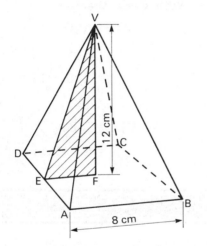

3. Fig. 29.49 shows a pyramid on a rectangular base. Calculate the length of VA.

Fig. 29.49

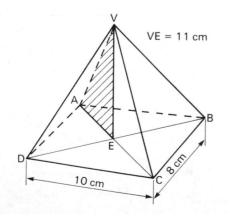

4. Fig. 29.50 shows a triangular prism with the face YDC inclined as shown in the diagram. Find the angle that the sloping face YDC makes with the base.

Fig. 29.50

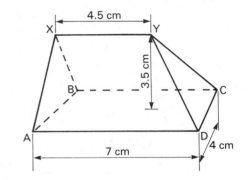

5. Fig. 29.51 shows a pyramid on a square base with VA = VB = VC = VD = 5 cm. Calculate the vertical height of the pyramid (i.e. VE).

Fig. 29.51

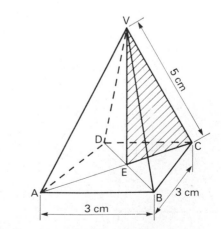

6. Fig. 29.52 shows a wooden wedge. Find the length of EA and the size of the angle marked *q*. The end faces ADF and EBC are equilateral triangles.

Fig. 29.52

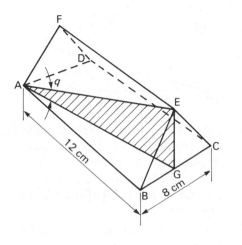

7. Fig. 29.53 shows a shed with a slanting roof ABCD. The rectangular base ABEF rests on level ground and the shed has three vertical sides.

(a) Determine the angle between the slanting roof and the ground.

(b) The roof is to be covered with roofing felt. What area of roofing felt is required?

(c) Calculate the volume of the shed in cubic metres.

Fig. 29.53

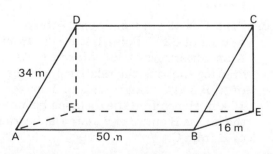

Miscellaneous Exercise 29

1. Fig. 29.54 represents a ramp. The distance AC is 5 m and the distance AB is 1 m. Use the trigonometrical tables to find **(a)** ∠ACB, **(b)** the distance BC, giving your answer correct to 3 significant figures.

Fig. 29.54

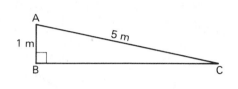

2. In the triangle ABC (Fig. 29.55) AB = AC = 10 cm, ∠ABC = 50° and AN is perpendicular to BC. Calculate **(a)** the size of ∠BAC, **(b)** the length of BC, **(c)** the length of AN.

Fig. 29.55

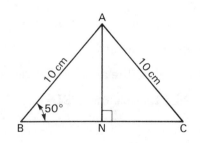

3. Given Fig. 29.56, calculate **(a)** the length of CD, **(b)** tan∠ABD, **(c)** the size of ∠ABD.

Fig. 29.56

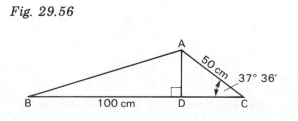

4. Fig. 29.57 represents a vertical tower TB standing on horizontal ground AB with AB = 100 m. The angle of elevation of the top of the tower T from A is 31.3°.

    (a) Calculate the height of the tower TB.

    (b) A man standing at A walks 20 m directly towards B. Calculate the angle of elevation of T from his new position.

Fig. 29.57

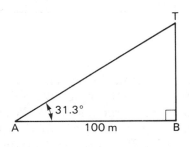

5. In Fig. 29.58, the area of the rectangle ABCD is 62 cm², AD = 8 cm, ED = 12.9 cm and ∠ADE = 90°.

    (a) Find the length of AB.

    (b) Calculate the area of triangle ADE.

    (c) Find the size of ∠AED.

Fig. 29.58

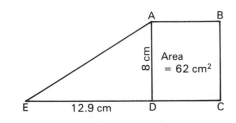

6. (a) Find x if $\sin x = 0.3891$.

    (b) For what angle is 0.4000 the cosine?

7. Fig. 29.59 represents a coastguard at C looking out to sea from the top of a cliff which is 100 m high. He sights a rowing boat at A and observes that it makes an angle of depression of 15°. If BC represents the vertical cliff, find the distance of the rowing boat from the foot of the cliff.

Fig. 29.59

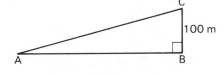

8. Fig. 29.60 shows a vertical mast PQ which stands on horizontal ground. A straight wire 20 m long runs from P at the top of the mast to a point R on the ground which is 10 m from the foot of the mast.

    (a) Calculate the angle of inclination of the wire to the ground (i.e. find ∠PRQ).

    (b) Calculate the height of the mast.

Fig. 29.60

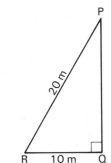

9. The bearing of point A from point B is 207°. What is the bearing of B from point A?

10. Point C is 50 km from point A on a bearing of 320°. Point B is 60 km from A on a bearing of 230°. Make a scale drawing showing the relative positions of A, B and C using a scale of 1 cm to 10 km. Hence find the distance between the points B and C and state the bearing of B from C.

Exercise 30.2

1. The plan of a school is drawn to a scale of 250:1. On the plan, a rectangular room has a length of 6 cm and a width of 5 cm. Calculate

    (a) the actual length and width of the room

    (b) the actual area of the room

2. The drawing of a house and garden is to a scale of 1 cm = 2 m. The length of the rectangular garden is shown as 8 cm and its width as 20 cm. Calculate the actual area of the garden in square metres.

3. Fig. 30.3 shows the plan of the ground floor of a house. Copy and complete the following table:

Name of room	Length (m)	Width (m)	Area (m^2)
Lounge			
Dining room			
Kitchen			
Bathroom			

The hall of the house is to be carpeted. What area of carpet is required?

Fig. 30.3

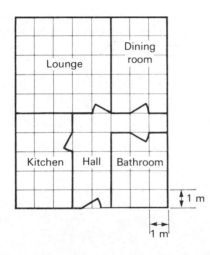

4. Fig. 30.4 represents the plan of a small apartment. Copy and complete the following table:

Name of room	Length (m)	Width (m)	Area (m^2)
Lounge			
Hall			
Bedroom			
Kitchen			
Bathroom			

Fig. 30.4

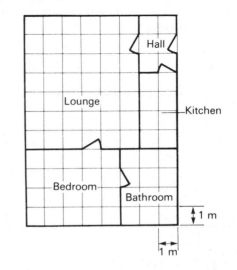

5. Fig. 30.5 shows the plan of a garage.

Fig. 30.5

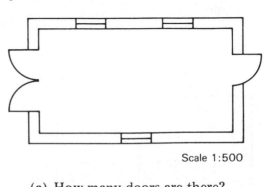

Scale 1:500

(a) How many doors are there?

(b) How many windows are there?

(c) What is the actual internal length of the garage?

(d) What is the actual internal width of the garage?

(e) What is the actual floor area of the garage?

(f) The floor is to be made of concrete to a depth of 15 cm. What volume of concrete, in cubic metres, is required?

6. Fig. 30.6 shows the plan of a house and garden drawn to a scale of 1 : 2500. Find, by scaling, the dimensions of the plot and the plan area of the house. Hence calculate, in square metres, the area of the garden.

Fig. 30.6

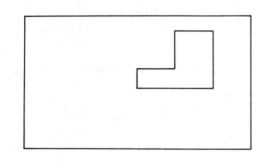

7. Fig. 30.7 represents the lounge and kitchen of a house.

    (a) The lounge floor is to be covered with a fitted carpet. Calculate, in square metres, the area of carpet required.

Fig. 30.7

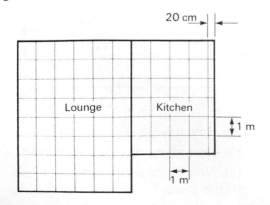

(b) The kitchen floor is to be tiled with tiles 15 cm square. How many tiles are required?

Areas on Maps and Drawings

Consider a field which is shown on a map. The field itself is simply an **enlargement** of the representation of the field on the map. The scale factor of the enlargement is the scale, expressed as a ratio, to which the map has been drawn.

It will be remembered (from Chapter 26) that the ratio of the areas of similar shapes is equal to the square of the enlargement factor. Hence to find the actual area of the field we multiply the area on the map by the square of the scale.

Example 5

A drawing is made to a scale of 1 : 1000. The area of a plot of ground is shown as 40 cm^2 on the drawing. What is the actual area of the plot?

$$\text{Actual area of plot} = 40 \times 1000^2 \text{ cm}^2$$
$$= \frac{40 \times 1000^2}{100^2} \text{ m}^2$$
$$= 4000 \text{ m}^2$$

Example 6

The scale of a map is 1 cm to 5 km. What area does 8 cm^2 on the map represent on the ground?

$$1 \text{ cm} = 5 \text{ km}$$
$$= 5000 \text{ m}$$
$$= 500\,000 \text{ cm}$$
$$\text{Hence the map scale} = 1 : 500\,000$$
$$\text{Area on the ground} = 8 \times 500\,000^2 \text{ cm}^2$$
$$= \frac{8 \times 500\,000^2}{100^2 \times 1000^2} \text{ km}^2$$
$$= 200 \text{ km}^2$$

Exercise 30.3

1. The scale of a map is $1:10\,000$. What area, in square metres, does $4\,\mathrm{cm}^2$ on the map represent?

2. The scale of a drawing is $2\,\mathrm{cm}$ to $1\,\mathrm{m}$. The lounge of a house has an area of $48\,\mathrm{cm}^2$ on the drawing. What is the actual area of the lounge?

3. The scale of a map is $1\,\mathrm{cm}$ to $2\,\mathrm{km}$. How many square kilometres are represented by $5\,\mathrm{cm}^2$ on the map?

4. The scale of a drawing is $5\,\mathrm{cm}$ to $1\,\mathrm{km}$. A field is represented on the map by a trapezium (Fig. 30.8). What is the actual area of the field?

Fig. 30.8

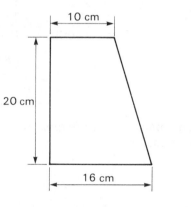

5. A, B and C are three points on a map whose scale is $1:10\,000$.
   $AB = 7\,\mathrm{cm}$, $AC = 5\,\mathrm{cm}$ and $\angle ABC = \angle BAC = 50°$. Calculate

   (a) the area of triangle ABC

   (b) the actual area represented by triangle ABC

6. A triangular plot of land when measured on a map has sides of $8\,\mathrm{cm}$, $9\,\mathrm{cm}$ and $11\,\mathrm{cm}$. If the scale of the map is $1:10\,000$, what is the actual area of the plot in hectares?
   (1 hectare $= 10\,000\,\mathrm{m}^2$.)

Volumes and Scales

Engineers and architects often make models of houses and buildings before starting work on them. The object and its model are similar solids. Therefore to find the volume of the object we must multiply the volume of the model by the cube of the scale to which the model is made.

Example 7

A model of a school is made to a scale of $1:50$. The volume of the model is $6\,\mathrm{m}^3$. What is the volume of the school?

$$\text{Volume of school} = 6 \times 50^3\,\mathrm{m}^3$$
$$= 750\,000\,\mathrm{m}^3$$

Exercise 30.4

1. A model of a library is made to a scale of $1:20$. Its volume is $3\,\mathrm{m}^3$. Find the volume of the library.

2. The model of a train is made to a scale of $1:40$. Its volume is $24\,\mathrm{cm}^3$. Find, in cubic metres, the actual volume of the train.

3. The fuselage of an aeroplane has a volume of $600\,\mathrm{m}^3$. A model of the aeroplane is made to a scale of $1\,\mathrm{cm} = 10\,\mathrm{m}$. Find the volume of the fuselage on the model.

4. On the model of a church, the spire has a volume of $0.5\,\mathrm{m}^3$. If the model is made to a scale of $1:25$, find the actual volume of the spire.

5. The model of a house consists of a cuboid $7\,\mathrm{cm}$ by $5\,\mathrm{cm}$ by $3\,\mathrm{cm}$. If the scale of the model is $1\,\mathrm{cm} = 2\,\mathrm{m}$, find the actual volume of the house.

Gradient

When used in connection with roads and railways, the gradient is used to denote the ratio of the vertical distance to the corresponding distance measured along the slope. Thus in Fig. 30.9,

$$\text{Gradient} = \frac{BC}{AC}$$

Fig. 30.9

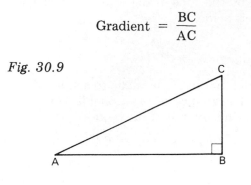

Example 8

(a) Two points P and Q are linked by a straight road with a uniform gradient. If Q is 50 m higher than P and the distance PQ measured along the road is 800 m, what is the gradient of the road?

Fig. 30.10

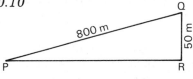

Referring to Fig. 30.10 we see that

$$\text{Gradient} = \frac{RQ}{PQ}$$

$$= \frac{50}{800}$$

$$= \frac{1}{16}$$

The gradient of the road is 1 in 16.

Nowadays gradients are often expressed as percentages, in which case

$$\text{Gradient} = \frac{1}{16} \times \frac{100}{1}$$

$$= 6\tfrac{1}{4}\%$$

(b) A man walks up a road whose gradient is 1 in 15. How much higher is he than when he started if he walks a distance of 3000 m?

Fig. 30.11

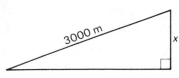

From Fig. 30.11,

$$\text{Gradient of road} = \frac{x}{3000}$$

$$\frac{x}{3000} = \frac{1}{15}$$

$$x = \frac{3000}{15}$$

$$= 200 \text{ m}$$

The man is 200 m higher than when he started.

Exercise 30.5

1. Two points A and B are connected by a straight road with a uniform gradient. If A is 40 m higher than B and the distance AB (measured along the road) is 1 km, what is the gradient of the road?

2. Two points P and Q whose heights differ by 50 m are connected by a straight road 2 km long. Find the gradient of the road.

3. A road has a gradient of 1 in 15. A man walks 1.5 km up the road. How much higher is he than when he first started?

4. A man walks up a road whose gradient is 1 in 30. He walks a distance of 900 m. How much higher is he than when he first started?

5. A road has a gradient of $12\frac{1}{2}\%$. What angle does the road make with the horizontal?

6. A mountain track is inclined at $12°$ to the horizontal. Express the gradient of the track as

    (a) $1:n$ (b) a percentage

7. A woman walks 100 m up a slope whose gradient is 5% and then a further 80 m up a slope whose gradient is $12\frac{1}{2}\%$. How much higher is the woman than when she first started?

Contours

The contours on a map join places which have the same height. Fig. 30.12 shows a small hill. The bottom view, the plan, is the one which would be shown on a map.

Fig. 30.12

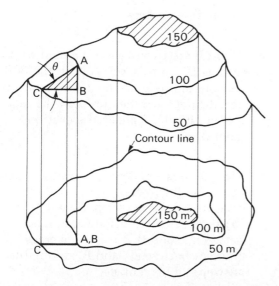

Suppose that A and C are two positions on a path up the hill, the path being taken as a straight line. It will be seen that C is on the 50 m contour and that A is on the 100 m contour. B is directly below A (within the hill) and on the same contour as C. A and B appear as the same point on the map but the height AB is 50 m. The distance AC measured on the map corresponds to the length BC on the elevation. The angle of elevation is θ and $\tan\theta = AB/BC$. The gradient is AB/AC which is $\sin\theta$.

Example 9

(a) Two points P and Q whose heights differ by 50 m are shown 0.60 cm apart on a map whose scale is $1:25\,000$. Calculate the angle of elevation of the line PQ.

$$0.60\,\text{cm} = 0.60 \times 25\,000\,\text{cm}$$

$$= \frac{0.6 \times 25\,000}{100}\,\text{m}$$

$$= 150\,\text{m on the ground}$$

From Fig. 30.13

$$\tan\theta = \frac{RQ}{PR}$$

$$= \frac{50}{150}$$

$$= 0.3333$$

$$\theta = 18°26'$$

Therefore the angle of elevation of PQ is $18°26'$.

Fig. 30.13

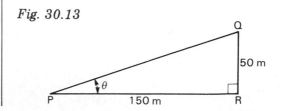

(b) Two points A and B are linked by a straight road with a uniform gradient. A is on the 100 m contour and B is on the 125 m contour. If the scale of the map is 10 cm to 1 km and the distance AB on the map is 4 cm, find the gradient of the road stating the answer **(i)** in the form $n:1$, **(ii)** as a percentage. The map is shown in Fig. 30.14.

Fig. 30.14

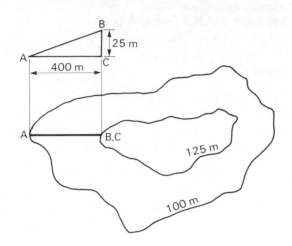

Since 10 cm = 1 km on the ground

1 cm = 0.1 km

= 100 m on the ground

and 4 cm = 4 × 100 m

= 400 m on the ground

In triangle ABC,

$$\tan \theta = \frac{BC}{AC}$$

$$= \frac{25}{400}$$

$$= 0.0625$$

$$\theta = 3°35'$$

$$\text{Gradient} = \sin \theta$$

$$= \sin 3°35'$$

$$= 0.0625$$

(i) As a ratio,

$$\text{Gradient} = 1:0.0625$$

$$= \frac{1}{0.0625}:1$$

$$= 16:1$$

(ii) As a percentage,

$$\text{Gradient} = 0.0625 \times 100\%$$

$$= 6.25\%$$

$$= 6\tfrac{1}{4}\%$$

Exercise 30.6

1. Two points P and Q whose heights differ by 150 m are shown 0.8 cm apart on a map whose scale is $1:10\,000$. Calculate the angle of elevation of the line PQ.

2. Two points P and Q whose heights differ by 150 m are shown 0.8 cm apart on a map whose scale is 1 cm = 500 m. Calculate the gradient of the line PQ, stating the answer

   (a) in the form $1:n$

   (b) as a percentage

3. A and B are two places on a map whose scale is 10 cm to 1 km. A is on the 500 m contour and B is on the 250 m contour. The distance AB measured on the map is 8 cm. Find the angle of elevation from B to A.

4. Fig. 30.15 is a scale diagram of a contour map. Use the map to find **(a)** the difference in height between A and B, **(b)** the horizontal distance, in metres, between A and B, **(c)** the average gradient between A and B, stating the answer as a percentage.

Fig. 30.15

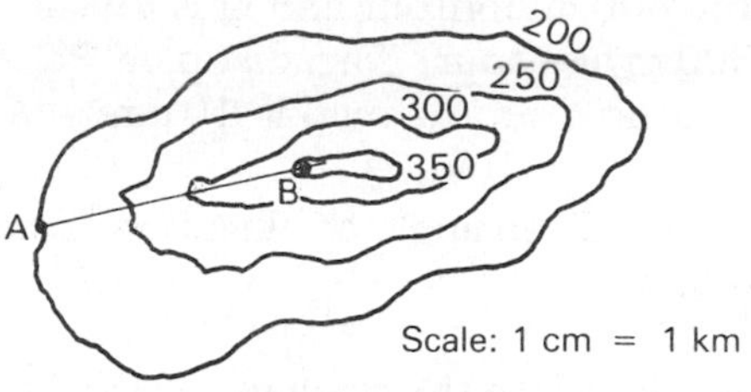

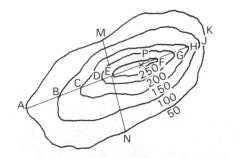

Scale: 1 cm = 1 km

The contours are measured in metres.

5. Fig. 30.16 is a scale drawing (scale: 2 cm to 2 km) of part of a contour map of a given area.

   (a) What is the difference in level between B and G?

   (b) Find the difference in height between the highest and lowest points on the drawing if the summit is at a height of 286 m.

   (c) What is the horizontal distance between A and C and between D and J?

   (d) Find the average angle of slope of the roadway ME.

Fig. 30.16

Miscellaneous Exercise 30

1. A map is drawn to a scale of $1:25\,000$. What length of road, in metres, is represented by 2 cm on the map?

2. A rectangular kitchen is $3\frac{1}{2}$ m wide, $4\frac{1}{2}$ m long and $2\frac{1}{2}$ m high. A model of the house is made to a scale of 1 cm to 2 m. Work out the dimensions of the kitchen on the model.

3. The building plans of a house are drawn to a scale of $1:50$.

   (a) Find the length of the house if it is represented by a length of 20 cm on the drawing.

   (b) The actual height of the house is to be 7 m. Find the distance, in centimetres, which represents this dimension on the plan.

   (c) On the plan, the area of the kitchen floor is $80\ \text{cm}^2$. Calculate the actual area of the kitchen floor, stating the answer in square metres.

4. A model of a public building is made to a scale of $1:20$.

   (a) The model has a length of 4 m. What is the actual length of the building?

   (b) The actual floor area of one of the rooms is $2000\ \text{m}^2$. What is the area on the model?

   (c) The volume of the model is $36\ \text{m}^3$. What is the actual volume, in cubic metres, of the building?

5. The plan of a house and garden is drawn to a scale of $1:60$.

   (a) Find the distance on the plan which represents an actual distance of 30 m.

   (b) On the plan, the garden measures 60 cm by 30 cm. What is the actual area, in square metres, of the garden?

   (c) On the plan, the volume of the house is shown to be $90\ \text{cm}^3$. What is the actual volume, in cubic metres, of the house?

6. Fig. 30.17 shows the plan of a bungalow. Copy and complete the following table:

Name of room	Length (m)	Breadth (m)	Area (m^2)
Lounge			
Dining room			
Bedroom 1			
Bedroom 2			
Hall			
Bathroom			
WC			
Kitchen			

How many windows and doors does the bungalow possess?

Fig. 30.17

7. Two points P and Q are two places on a map whose scale is $1:10\,000$. P is on the 800 m contour and Q is on the 600 m contour. The distance PQ measured on the map is 10 cm.

(a) Find the angle of elevation from Q to P.

(b) Calculate the gradient of the line PQ in the form $1:n$.

(c) State the gradient as a percentage.

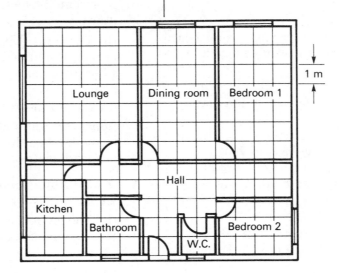

Charts, Tables and Diagrams

Tables

Conversion tables of various types are used in business, science and engineering.

Ready reckoners provide a quick way of multiplying sums of money. The calculator has largely superseded the ready reckoner but many small businesses and shops still use it. Table 31.1 shows an extract from a ready reckoner giving the price of N articles at 26p each.

TABLE 31.1

N	£	N	£	N	£	N	£
22	5.72	64	16.64	106	27.56	525	136.50
23	5.98	65	16.90	107	27.82	550	143.00
24	6.24	66	17.16	108	28.08	600	156.00
25	6.50	67	17.42	109	28.34	625	162.50
26	6.76	68	17.68	110	28.60	650	169.00
27	7.02	69	17.94	111	28.86	700	182.00
28	7.28	70	18.20	112	29.12	750	195.00
29	7.54	71	18.46	113	29.38	800	208.00
30	7.80	72	18.72	114	29.64	900	234.00

Example 1

Use Table 31.1 to find the cost of
(a) 28 articles at 26p each,
(b) 579 articles at 26p each,
(c) $7\frac{1}{2}$ metres of cloth at 26p per metre,
(d) 72.9 kg of foodstuff at 26p per kilogram.

    (a) Directly from the table:

      Cost of 28 articles at 26p each $= £7.28$

(b) $579 = 550 + 29$

Cost of 550 articles at 26p each
$$= £143.00$$

Cost of 29 articles at 26p each
$$= \quad £7.54\ +$$

Cost of 579 articles at 26p each
$$= \underline{£150.54}$$

(c) $\qquad 7\frac{1}{2} = 7.5$

$$= \frac{750}{100}$$

From the table:

Cost of 750 metres of
cloth at 26p per metre $= £195.00$

Cost of $7\frac{1}{2}$ metres of
cloth at 26p per metre $= £195.00 \div 100$

$$= £1.95$$

(d) $\qquad 72.9 = \frac{729}{10}$

$$= \frac{700 + 29}{10}$$

Cost of 700 kg of foodstuff
at 26p per kilogram $= £182.00$

Cost of 29 kg of foodstuff
at 26p per kilogram $= \quad £7.54\ +$

Cost of 729 kg of foodstuff
at 26p per kilogram $= \underline{£189.54}$

Cost of 72.9 kg of foodstuff
at 26p per kilogram $= £189.54 \div 10$

$$= £18.95$$

Conversion tables are frequently used by engineers and scientists. Table 31.2 shows part of a temperature conversion table.

TABLE 31.2

TEMPERATURE	Degrees Fahrenheit to degrees Celsius								
°F	0	1	2	3	4	5	6	7	8	9
0	−17.8	−17.2	−16.7	−16.1	−15.6	−15.0	−14.4	−13.9	−13.3	−12.8
10	−12.2	−11.7	−11.1	−10.6	−10.0	−9.4	−8.9	−8.3	−7.8	−7.2
20	−6.7	−6.1	−5.6	−5.0	−4.4	−3.9	−3.3	−2.8	−2.2	−1.7
30	−1.1	−0.6	0	0.6	1.1	1.7	2.2	2.8	3.3	3.9
40	4.4	5.0	5.6	6.1	6.7	7.2	7.8	8.3	8.9	9.4
50	10.0	10.6	11.1	11.7	12.2	12.8	13.3	13.9	14.4	15.0
60	15.6	16.1	16.7	17.2	17.8	18.3	18.9	19.4	20.0	20.6
70	21.1	21.7	22.2	22.8	23.3	23.9	24.4	25.0	25.6	26.1
80	26.7	27.2	27.8	28.3	28.9	29.4	30.0	30.6	31.1	31.7
90	32.2	32.8	33.3	33.9	34.4	35.0	35.6	36.1	36.7	37.2
100	37.8	38.3	38.9	39.4	40.0	40.6	41.1	41.7	42.2	42.8
110	43.3	43.9	44.4	45.0	45.6	46.1	46.7	47.2	47.8	48.3
120	48.9	49.4	50.0	50.6	51.1	51.7	52.2	52.8	53.8	53.9
130	54.4	55.0	55.6	56.1	56.7	57.2	57.8	58.3	58.9	59.4
140	60.0	60.6	61.1	61.7	62.2	62.8	63.3	63.9	64.4	65.0
150	65.6	66.1	66.7	67.2	67.8	68.3	68.9	69.4	70.0	70.6
160	71.1	71.7	72.2	72.8	73.3	73.9	74.4	75.0	75.6	76.1
170	76.7	77.2	77.8	78.3	78.9	79.4	80.0	80.6	81.1	81.7
180	82.2	82.8	83.3	83.9	84.4	85.0	85.6	86.1	86.7	87.2
190	87.8	88.3	88.9	89.4	90.0	90.6	91.1	91.7	92.2	92.8
Interpolation: deg F:	0.1	0.2	0.3	0.4	0.5	0.6	0.7	0.8	0.9
deg C:	0.1	0.1	0.2	0.2	0.3	0.3	0.4	0.4	0.5

Example 2

Convert **(a)** 54°F to degrees Celsius,
(b) 122.7°F to degrees Celsius,
(c) 35°C to degrees Fahrenheit,
(d) 62.7°C to degrees Fahrenheit.

(a) We first find 50 in the first column and move along this row until we find the column headed 4. We find the number 12.2 and hence 54°F = 12.2°C.

(b) To convert 122.7°F into degrees Celsius we make use of the figures given under the heading 'interpolation'. From the table we find 122°F = 50°C. Looking at the figures given at the foot of the table we can find 0.7°F and this is equivalent to 0.4°C which is shown immediately below. Thus

$$122.7°F = 50°C + 0.4°C$$

$$= 50.4°C$$

(c) To convert degrees Celsius into degrees Fahrenheit we use the table in reverse. To convert 35°C into degrees Fahrenheit we search in the body of the table until we find the figure 35. This occurs in the column headed 5 in the row starting with 90. Hence 35°C is equivalent to 95°F.

(d) To convert 62.7°C into degrees Fahrenheit we look in the body of the table for a number as close to 62.7 as possible, but less than 62.7. This number is 62.2 corresponding to 144°F. Now 62.2°C is 0.5°C less than 62.7°C. Using the interpolation figures we see that 0.5°C corresponds to 0.9°F. Therefore

$$62.7°C = 62.2°C + 0.5°C$$

$$= 144°F + 0.9°F$$

$$= 144.9°F$$

Mileage charts are often set out like Table 31.3.

TABLE 31.3

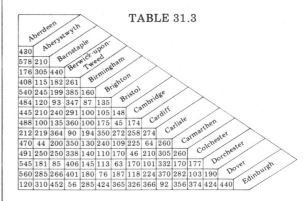

Example 3

Use Table 31.3 to find the distance between
(a) Aberdeen and Carlisle,
(b) Barnstaple and Cambridge.

(a) To find the distance from Aberdeen to Carlisle, find Aberdeen and Carlisle in the sloping list. Move vertically down the column for Aberdeen and horizontally along the row for Carlisle until the two movements coincide. The figure given in the table at this point is 212 and thus the distance between Aberdeen and Carlisle is 212 miles.

(b) Find Barnstaple in the sloping list and move down the third column. The row for Cambridge is the seventh row from the top. In the third column of the seventh row we find the figure 240 and hence the distance between Barnstaple and Cambridge is 240 miles.

Parallel Scale Conversion Charts

A system of parallel scales may be used when we wish to convert from one set of units to another related set. Fig. 31.1 is a chart relating degrees Fahrenheit and degrees Celsius.

Fig. 31.1

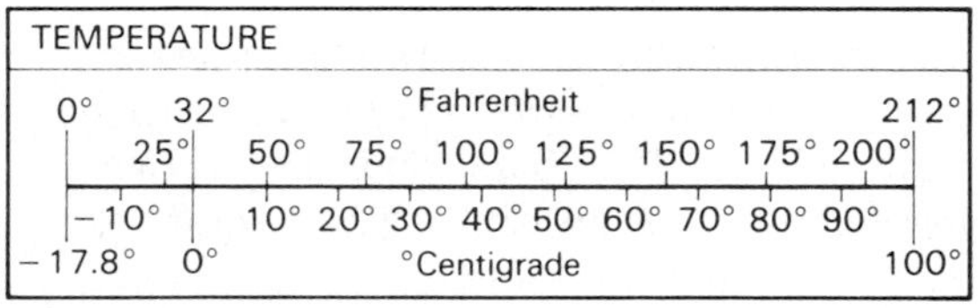

TEMPERATURE

Example 4

Using the chart in Fig. 31.1 convert (a) 50°F to degrees Celsius, (b) 75°F to degrees Celsius, (c) 20°C to degrees Fahrenheit.

(a) From the chart we see that 50°F corresponds with 10°C and therefore 50°F is equivalent to 10°C.

(b) From the chart we see that 75°F corresponds roughly with 24°C. Note carefully that when using the chart we cannot expect accuracy greater than 1°.

(c) From the chart we see that 20°C is about 68°F.

Exercise 31.1

1. Use the ready reckoner on page 293 to find the cost of
   (a) 24 articles costing 26p each
   (b) 102 articles costing 26p each
   (c) 827 articles costing 26p each
   (d) 672 articles costing 26p each
   (e) $6\frac{1}{2}$ metres of materal costing 26p per metre
   (f) 9.3 lbs of fruit costing 26p per pound
   (g) 82.2 litres of fluid costing 26p per litre

2. Use the temperature conversion table on page 294 to convert
   (a) 83°F to degrees Celsius
   (b) 114.6°F to degrees Celsius
   (c) 55°C to degrees Fahrenheit
   (d) 68.7°C to degrees Fahrenheit

3. Use the mileage chart on page 294 to find the distances between
   (a) Aberdeen and Cardiff
   (b) Birmingham and Carmarthen
   (c) Bristol and Dorchester

4. Table 31.4 shows a comparison between gradients expressed as a ratio and gradients expressed as a percentage. Use the table to convert
   (a) a gradient of 1 : 7 to a percentage
   (b) a gradient of 7.1% to a ratio

TABLE 31.4

Gradients	
Ratio	%
1 : 3	33.3
1 : 4	25
1 : 5	20
1 : 6	16.7
1 : 7	14.3
1 : 8	12.5
1 : 9	11.1
1 : 10	10
1 : 11	9.1
1 : 12	8.3
1 : 13	7.7
1 : 14	7.1
1 : 15	6.7
1 : 16	6.3
1 : 17	5.9
1 : 18	5.6
1 : 19	5.3
1 : 20	5

5. Table 31.5 allows conversion from miles to kilometres (and also miles per hour to kilometres per hour) to be made. Use the table to convert

(a) 15 miles to kilometres

(b) 48.27 km/h to miles per hour

(c) 58 miles to kilometres

(d) 568 miles per hour to kilometres per hour

TABLE 31.5

Distance and Speed					
Miles to kilometres					
Miles per hour to kilometres per hour					
1	1.60	20	32.18	75	120.7
2	3.21	25	40.23	80	128.7
3	4.82	30	48.27	85	136.8
4	6.43	35	56.32	90	144.8
5	8.04	40	64.37	95	152.9
6	9.65	45	72.41	100	160.9
7	11.26	50	80.46	200	321.9
8	12.87	55	88.51	300	482.8
9	14.48	60	96.55	400	643.7
10	16.09	65	104.60	500	804.7
15	24.13	70	112.70	1000	1609.3
1 mile = 1.606344 km					
1 kilometre = 0.621371 miles					

6. Use the chart (Fig. 31.1) relating temperatures in degrees Fahrenheit to degrees Celsius to convert

(a) $100\,^{\circ}$F to degrees Celsius

(b) $80\,^{\circ}$C to degrees Fahrenheit

(c) $64\,^{\circ}$F to degrees Celsius

(d) $48\,^{\circ}$C to degrees Fahrenheit

7. Fig. 31.2 is a chart relating pounds avoirdupois to kilograms. Use the chart to convert

(a) 3.4 lb to kilograms

(b) 568 lb to kilograms

(c) 1.8 kg to pounds

Fig. 31.2

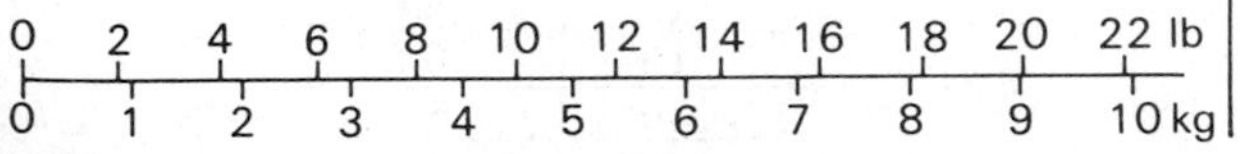

8. Table 31.6 shows a comparison of tyre pressures. Convert (a) 22 lb/in² into kilograms per square centimetre, (b) $2.10\,\text{kg/cm}^2$ into pounds per square inch.   (c) Estimate the tyre pressure in pounds per square inch corresponding to $3\,\text{kg/cm}^2$.

TABLE 31.6

Tyre Pressures					
Pounds per sq in to kg per sq cm					
16	1.12	26	1.83	40	2.80
18	1.26	28	1.96	50	3.50
20	1.40	30	2.10	55	3.85
22	1.54	32	2.24	60	4.20
24	1.68	36	2.52	65	4.55

Simple Bar Charts

The information in a simple bar chart is represented by a series of bars all of the same width. The bars may be drawn vertically or horizontally. The height (or length) of the bars represents the magnitude of the figures given. A simple bar chart shows clearly the size of each item of information but it is not easy to obtain the total of all the items from the diagram.

Example 5

A family spends its weekly income of £300 as follows:

Food	£100
Clothes	£50
Fuel	£40
Mortgage	£90
Other expenses	£20
Total	£300

Draw (a) a vertical bar chart, (b) a horizontal bar chart to represent this information.

Fig. 31.3 shows the information in the form of a vertical bar chart whilst Fig. 31.4 shows the information in the form of a horizontal bar chart.

Fig. 31.3

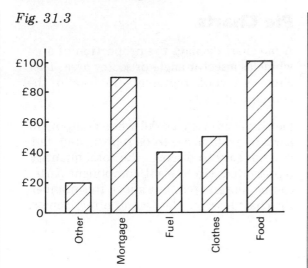

Fig. 31.4

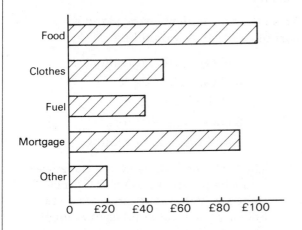

Year	Population (millions)
1750	728
1800	906
1850	1171
1900	1608
1950	2504

When drawing a chronological bar chart, time is always marked off along the horizontal axis, as shown in Fig. 31.5. The chart clearly shows how the population of the world has increased over this period of 200 years.

Fig. 31.5

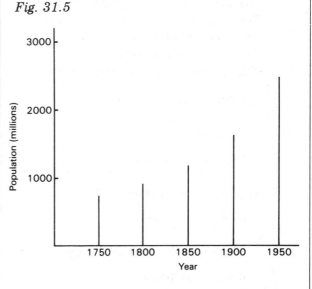

Chronological Bar Charts

A chronological bar chart compares quantities over periods of time. It consists of a number of vertical lines (see also page 305).

Example 6

The figures below give the world population in millions of people from 1750 to 1950. Draw a chronological bar chart to represent this information.

Proportionate Bar Charts

The proportionate bar chart relies on heights (if drawn vertically) or lengths (if drawn horizontally) to convey proportions of the whole. It should be the same width throughout its height or length.

The proportionate bar chart shows clearly the total of all the items, but it is rather difficult to obtain the proportion of each item accurately.

Example 7

The table below shows the number of people employed on various kinds of work in a factory. Represent this information in the form of a proportionate bar chart.

Type of personnel	Number employed
Unskilled workers	45
Craftsmen	27
Draughtsmen	5
Clerical workers	8
Total	85

Suppose that the total height of the proportionate bar chart is to be 6 cm. The heights of the component parts must first be calculated and then drawn accurately using a rule.

45 unskilled workers are represented by

$$\frac{45}{85} \times 6\,\text{cm} = 3.18\,\text{cm}$$

27 craftsmen are represented by

$$\frac{27}{85} \times 6\,\text{cm} = 1.91\,\text{cm}$$

5 draughtsmen are represented by

$$\frac{5}{85} \times 6\,\text{cm} = 0.35\,\text{cm}$$

8 clerical workers are represented by

$$\frac{8}{85} \times 6\,\text{cm} = 0.56\,\text{cm}$$

The proportionate bar chart is shown below.

Fig. 31.6

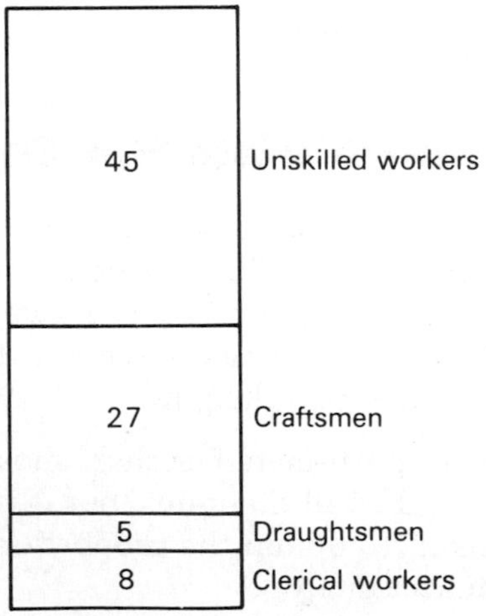

45	Unskilled workers
27	Craftsmen
5	Draughtsmen
8	Clerical workers

Pie Charts

A pie chart displays the proportion of a whole as a sector angle or sector area. The circle as a whole represents the total of the component parts.

Pie charts are very useful when component parts of a whole are to be represented, but it is not easy to discover the total quantity represented. Up to eight component parts can be represented, but above this number the chart loses its clarity and effectiveness.

Example 8

25 sixth form students were asked what sort of career they would like.

         8 chose work with animals.

         3 chose office work.

         2 chose teaching.

         5 chose outdoor work.

         7 said they did not know.

The first step is to calculate the angles at the centre of the pie chart. Since there are $360°$ in the circle we divide $360°$ by the total number of students quizzed. This gives the angle for one pupil. To find the angle for each section, we multiply the angle for one pupil by the number of pupils in the section and correct the product to the nearest degree. The work may be set out as follows:

$$360° \div 25 = 14.4°$$

Career	Number of pupils	Sector angle
Work with animals	8	$8 \times 14.4° = 115°$
Office work	3	$3 \times 14.4° = 43°$
Teaching	2	$2 \times 14.4° = 29°$
Outdoor work	5	$5 \times 14.4° = 72°$
Don't know	7	$7 \times 14.4° = 101°$

The pie chart is shown in Fig. 31.7.

Fig. 31.7

Line Graphs

Line graphs are sometimes used instead of chronological bar charts.

Example 9

Represent the information given in Example 6 in the form of a line graph.

> The line graph is shown in Fig. 31.8. As with chronological bar charts the year is always taken along the horizontal axis. The points are joined by straight lines because no information is given about the world population between the years given in the table.

Fig. 31.8

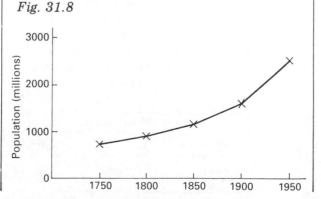

Pictograms

Pictograms are diagrams in the form of pictures which are used to present information to those who are unskilled in dealing with figures or to those who have only a limited interest in the topic depicted.

Example 10

The table below shows the output of bicycles for the years 1980 to 1984.

Year	1980	1981	1982	1983	1984
Output	2000	4000	7000	8500	9000

Represent this information in the form of a pictogram.

> The pictogram is shown in Fig. 31.9. It will be seen that each bicycle in the diagram represents an output of 2000 bicycles. Part of a symbol is shown in 1982, 1983 and 1984 to represent a fraction of 2000 but clearly this is not a precise way of representing the output.

Fig. 31.9

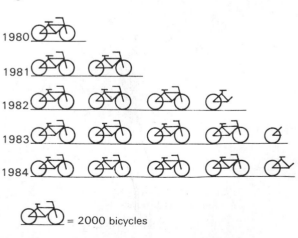

A method not recommended is shown in Fig. 31.10. Comparison is difficult because the reader is not sure whether to compare heights, areas or volumes.

Fig. 31.10

Sales of milk in 1960 and 1980 (millions of litres)

Exercise 31.2

1. Fig. 31.11 is a pictogram showing the method by which first year boys come to school. How many
   (a) come by bus
   (b) come by car

Fig. 31.11

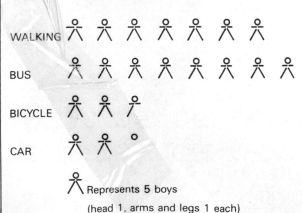

Represents 5 boys

(head 1, arms and legs 1 each)

2. Draw a proportionate bar chart 5 cm long for the figures shown below which represent the way in which people travel to work in Gloucester.

Type of transport	Numbers using
Bus	780
Private motoring	420
Other (foot, bicycle, motor cycle, etc.)	160
Total	1360

3. Draw a proportionate bar chart 10 cm long to represent the following information, which relates to the way in which a family spends its income.

Item	Amount spent (£)
Food and drink	95
Housing	42
Transport	29
Clothing	33
Other	51
Total	250

4. The table below shows the results of a survey of the colours of doors on a housing estate. Draw a vertical bar chart to represent this information.

Colour of door	Number of houses
White	85
Red	17
Green	43
Brown	70
Blue	15

5. The figures below show the results of a survey to find the favourite sports of a group of boys. Draw a pie chart to represent this information.

Sport	Number of boys
Athletics	20
Cricket	58
Football	32
Hockey	10

6. The table below shows the output of bicycles from a certain factory for the years 1980–1984. Represent this information in the form of

(a) a chronological bar chart

(b) a line graph

Year	1980	1981	1982	1983	1984
Number of bicycles	2000	4000	7000	8500	9000

7. The table below shows the number of houses completed in a certain town for the years 1978–1983. Represent this information by means of

(a) a chronological bar chart

(b) a line graph

Year	1978	1979	1980	1981	1982	1983
Number of houses	81	69	73	84	80	120

8. The pie chart (Fig. 31.12) shows a local election result. There were three candidates, White, Green and Brown. The angle representing the votes cast for White is $144°$ and that for Green is $90°$.

(a) Work out the angle representing the votes cast for Brown.

(b) Calculate the votes cast for each candidate if the total number voting was 20 000.

Fig. 31.12

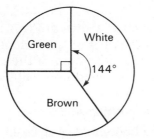

9. A pie chart was drawn to show the number of tons of various crops grown by a group of farmers. The sector angles corresponding to each crop was as follows:

Crop	Sector angle (degrees)
Vegetables	137
Grain	93
Potatoes	80
Fruit	50

If the total amount grown was 500 tons, calculate the amount of each crop grown.

10. The information shown below gives the output of motor tyres made by the Treadwell Tyre Company for the first six months of 1985. Draw

(a) a chronological bar chart

(b) a line graph, to represent this information

Month	Jan	Feb	Mar	Apr	May	June
Output (thousands)	40	43	39	38	37	45

11. Fig. 31.13 shows how Mrs. Lever spends her housekeeping money. If she receives £50 per week, how much does she spend on each item shown in the diagram?

Fig. 31.13

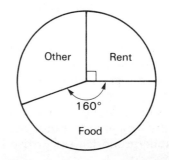

Mental Test 31

1. Use the mileage chart shown in Table 31.7 to find the distance from

   (a) Exeter to Gloucester

   (b) Exeter to Lincoln

   (c) Hereford to Liverpool

   (d) Kendal to Manchester

TABLE 31.7

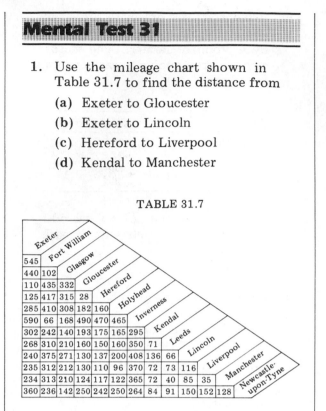

2. Fig. 31.14 shows a diagram which can be used to change degrees Celsius to degrees Fahrenheit and vice-versa. Use the diagram to change

   (a) $-40\,^{\circ}$C to degrees Fahrenheit

   (b) $20\,^{\circ}$C to degrees Fahrenheit

   (c) $100\,^{\circ}$F to degrees Celsius

Fig. 31.14

3. Fig. 31.15 is a simple bar chart which shows the way that commuters in the south-east region travel to work. Use the chart to find the number of commuters in the sample who travelled to work by (a) private motoring, (b) bus and underground, (c) British Rail.

Fig. 31.15

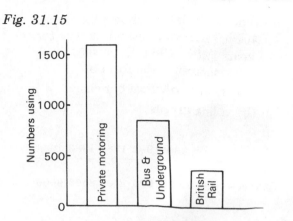

4. The chronological bar chart in Fig. 31.16 shows the number of colour television sets sold in southern England during the period 1970-5. Use the diagram to estimate the numbers of television sets sold in the years 1972 and 1975.

Fig. 31.16

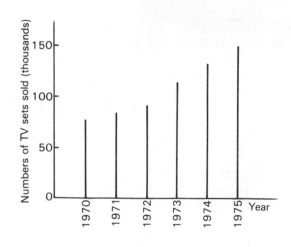

5. The pie chart shown in Fig. 31.17 depicts the sales of the various departments of a large store. If the sales for the week depicted were £600 000, find the value of

   (a) the clothing sold

   (b) the household goods sold

Fig. 31.17

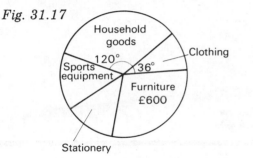

Fig. 31.18

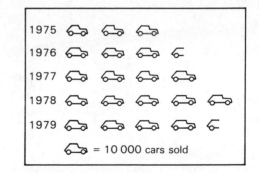

6. The pictogram (Fig. 31.18) shows the sales of cars in the years 1975–1979.

(a) In which year were the most cars sold?

(b) How many cars were sold in 1978?

(c) How many cars were sold in 1976?

(d) Estimate the total number of cars sold in the years represented on the pictogram.

7. Mrs. Wood has £100 per week for housekeeping. A pie chart is drawn which illustrates how she spends the money. If she spends £20 on food, work out the angle of the sector representing this amount.

32

Graphs

In newspapers, business reports and govern-ment publications great use is made of pictorial illustrations to help the reader understand the report. The most common form of illustration is the graph.

Axes of Reference

The first step in drawing a graph is to draw two lines at right angles to each other as shown in Fig. 32.1. These lines are called the **axes of reference**. The vertical axis is usually called the *y*-axis and the horizontal axis is called the *x*-axis.

Fig. 32.1

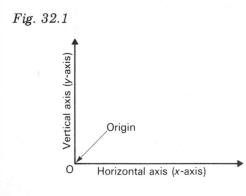

Scales

The number of units represented by a unit length along an axis is called the **scale**. For instance 1 cm could represent 2 units and we say that the scale is 1 cm = 2 units. The scales need not be the same along both axes.

The most useful scales are 1 cm to 1, 2, 5 and 10 units. Some multiples of these are also satisfactory, for instance 1 cm to 20, 50 and 100 units.

No matter what scale is chosen it is very important that the scale is easy to read. When graph paper is used the scale will depend upon the type of graph paper available and its size.

Cartesian Coordinates

Cartesian coordinates are used to mark the points on a graph. In Fig. 32.2, values of *x* are to be plotted against values of *y*. The point P has been plotted so that $x = 4$ and $y = 6$. The values of 4 and 6 are called the **rectangular coordinates** of P. For brevity the point P is said to be the point (4, 6). Note carefully that the value of *x* is always given first.

Fig. 32.2

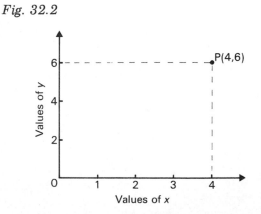

Drawing a Graph

Every graph shows a relationship between two sets of numbers.

Example 1

Tom and Martha made a count of the number of motor cars which passed their school between 10.00 a.m. and 11.00 a.m. and the number of passengers, including the driver, who travelled in each car. The results were as follows:

Number of passengers	1	2	3	4
Number of cars	46	40	12	5

Draw a graph to represent this information.

One way of representing this information on a graph is to draw a bar chart (Fig. 32.3). It consists of a number of vertical lines whose heights represent the number of cars carrying the number of passengers given on the horizontal axis.

Fig. 32.3

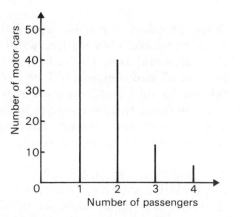

Quantities like time, age, etc. which increase at a steady rate are usually taken on the horizontal axis.

Exercise 32.1

1. The children in a class at school are given a test periodically. One child obtained the following marks out of 50.

Number of test	1	2	3	4	5	6
Score	18	23	32	36	42	47

Draw a bar chart to represent this information. Plot the scores on the vertical axis.

2. The table below gives the temperature at 12.00 noon on seven successive days. Plot a bar chart to represent this information. Take the days on the horizontal axis.

Day	1	2	3	4	5	6	7
Temperature ($^\circ$C)	15	21	26	24	17	25	28

3. The information given below shows the number of television sets sold by a wholesaler during the years 1980 to 1985.

Year	1980	1981	1982	1983	1984	1985
Number sold (thousands)	35	43	58	67	77	92

Draw a bar chart with the year taken on the horizontal axis.

4. The table below shows the amount of fruit grown by a farmer in seven successive years.

Year	1	2	3	4	5	6	7
Fruit grown (tonnes)	23	45	34	63	32	54	70

Draw a bar chart with the year taken on the horizontal axis.

5. In a primary school the ages of the children were as follows:

Age (years)	5	6	7	8	9	10
Number of children	21	19	36	42	28	32

Draw a bar chart taking age on the horizontal axis.

Example 2

The table below shows corresponding values of x and y.

x	0	2	4	6	8	10
y	4	10	16	22	28	34

Plot a graph of this information and use it to find the value of y when $x = 7$ and the value of x when $y = 19$.

It is usualy to plot x along the horizontal axis and y along the vertical axis. When the points are plotted it is seen that a straight line passes through all of them, as shown in Fig. 32.4.

Fig. 32.4

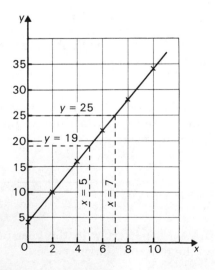

When a graph is a straight line or a smooth curve it can be used to find corresponding values of x and y not given in the table of values. Thus to find the value of y when $x = 7$, draw the vertical and horizontal lines shown in the diagram and it is found that when $x = 7$, $y = 25$. Also it is found that when $y = 19$, $x = 5$.

Using a graph in this way to find values not given in the original table is called **interpolation**.

Example 3

The table below shows corresponding values of P and Q.

P	5	6	7	8	9
Q	57	79	105	135	169

Plot P horizontally and draw the graph which shows the relationship between P and Q. From the graph find the value of Q when $P = 7.5$.

Since the values of P range from 5 to 9, we can make 5 our starting point on the horizontal axis. The values of Q start at 57 and hence it will be convenient to start at 40 on the vertical axis. By doing this we can have larger scales on both axes, thereby allowing us to plot the graph more accurately.

The graph is shown in Fig. 32.5 and it is a smooth curve. We can therefore use the graph to estimate values intermediate to those given in the original table.

From the graph we see that when $P = 7.5$, $Q = 120$.

Fig. 32.5

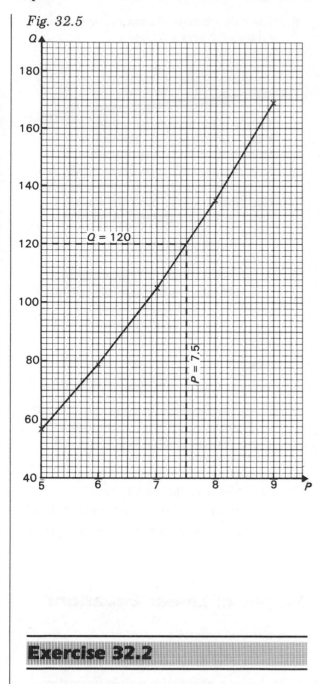

Fig. 32.6

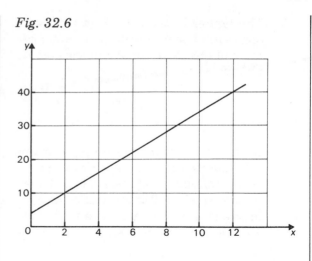

2. Fig. 32.7 shows a graph of distance (in metres) plotted against time (in seconds). From the graph find

(a) the distances corresponding to times of 2 s, 3 s and 5 s

(b) the times corresponding to distances of 4 m, 28 m and 54 m

Fig. 32.7

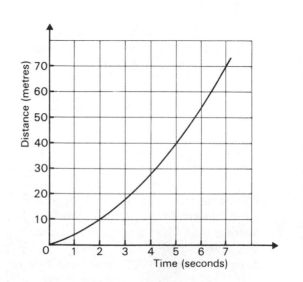

Exercise 32.2

1. Fig. 32.6 shows a graph with values of y plotted against values of x. Use the graph to find

(a) the values of y corresponding to $x = 0$, $x = 4$, $x = 9$ and $x = 6.5$

(b) the values of x corresponding to $y = 16$, $y = 22$, $y = 37$ and $y = 20.5$

3. The figures in the table below show corresponding values of the number of units of electricity used and the cost in pounds.

Units used	0	300	600	900	1200
Cost (£)	2	8	14	20	26

Draw a graph of this information taking the number of units used along the horizontal axis. Use a scale of 1 large square equals 200 units and 1 large square on the vertical axis to represent £5. Use your graph to find

(a) the cost of 800 units

(b) the number of units used when the cost is £12

4. The following table gives corresponding values of P and Q:

P	0	2	4	6	8	10
Q	2	12	22	32	42	52

Take P along the horizontal axis using a scale of 1 large square to represent 2 units. On the vertical axis use 1 large square to represent 10 units. Draw a graph to show the relationship between P and Q and use it to find the value of Q when $P = 5$ and to find the value of P when $Q = 37$.

5. The speed of a body (v metres per second) measured at various times (t seconds) was as follows:

t	2	4	6	8	10	12
v	6.4	7.7	9.0	10.3	11.7	13.0

Plot a graph of this information taking t on the horizontal axis. Use scales of 1 large square to represent 2 seconds on the horizontal axis and 1 large square to represent 2 metres per second on the vertical axis. Use your graph to estimate the speed when $t = 7$ seconds and the time at which $v = 9.5$ metres per second.

6. The table below shows corresponding values of x and y:

x	0	1	2	3	4	5
y	8	11	20	35	56	83

Taking x along the horizontal axis plot a graph showing the relationship between x and y. Suitable scales are: 1 large square to 1 unit horizontally and 1 large square to 10 units vertically. Use your graph to find the value of y when $x = 3.5$ and the value of x when $y = 27$.

7. A quantity of gas is contained in a cylinder and subjected to various pressures. The volume occupied by the gas under different pressures is as follows:

Pressure	1	2	3	4	5	6
Volume	2.4	1.2	0.8	0.6	0.48	0.4

Plot a graph of volume against pressure with pressure on the horizontal axis. Use a scale of 1 large square to 1 unit on the horizontal axis and 1 large square to 0.5 units on the vertical axis. Use your graph to find

(a) the volume when the pressure is 3.5

(b) the pressure when the volume is 0.5

Graphs of Linear Equations

Consider the equation

$$y = 2x + 5$$

We can give x any value we please and so calculate a corresponding value for y.

When $x = 0$ $y = 2 \times 0 + 5 = 5$

When $x = 1$ $y = 2 \times 1 + 5 = 7$

When $x = 2$ $y = 2 \times 2 + 5 = 9$

and so on.

The value of y depends upon the value allocated to x. We therefore call y the **dependent variable** and x the **independent variable.**

When drawing a graph of an equation it is usual to take the values of the independent variable along the horizontal axis, and hence this is often called the x-axis. The values of the dependent variable are then taken along the vertical axis, usually called the y-axis.

In plotting graphs of equations we may have to include coordinates which are positive and negative. To represent these on the graph we make use of the number lines used for directed numbers as shown in Fig. 32.8.

Fig. 32.8

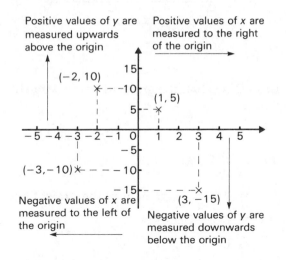

Example 4

Draw the graph of $y = 2x - 5$ for values of x between -3 and 4.

Having decided upon some values for x we calculate the corresponding values of y by substituting the chosen values of x in the given equation.
Thus when $x = -3$,

$$y = 2 \times (-3) - 5$$
$$= -6 - 5$$
$$= -11$$

For convenience the calculations are tabulated as shown below.

x	-3	-2	-1	0	1	2	3	4
$2x$	-6	-4	-2	0	2	4	6	8
-5	-5	-5	-5	-5	-5	-5	-5	-5
y	-11	-9	-7	-5	-3	-1	1	3

The graph is plotted in Fig. 32.9 and it is seen to be a straight line.

Fig. 32.9

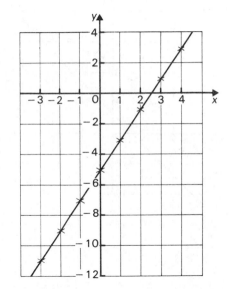

Equations of the type $y = 2x - 5$ where the powers of the variables x and y are 1, always give a graph which is a straight line. For this reason they are often called **linear equations.**

In order to draw the graph of a linear equation it is only necessary to plot two points. It is safer, however, to plot three points, the third point acting as a check on the other two.

Example 5

By means of a graph show the relationship between x and y in the equation $y = 5x + 3$. Plot the graph for values of x between -3 and 3.

Since $y = 5x + 3$ is a linear equation we need only take three points:

x	-3	0	3
y	-12	3	18

The graph is shown in Fig. 32.10.

Fig. 32.10

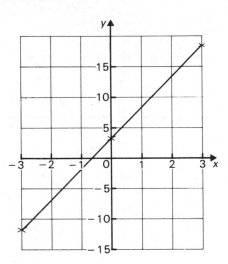

The Equation of a Straight Line

Every linear equation may be written in the standard form:

$$y = mx + c$$

Hence $y = 3x - 7$ is the standard form with $m = 3$ and $c = -7$.

The equation $y = 4 - 5x$ may be put in the standard form by rearranging it to give $y = -5x + 4$.

We then see that $m = -5$ and $c = 4$.

The Meaning of m and c in the Equation of a Straight Line

The point B is any point on the straight line shown in Fig. 32.11. It has the coordinates x and y. Point A is where the straight line cuts the x-axis and it has the coordinates $x = 0$ and $y = c$.

Fig. 32.11

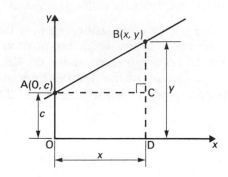

BC/AC is called the gradient of the straight line.

Now

$$BC = \frac{BC}{AC} \times AC$$

$$= AC \times \text{the gradient of the straight line}$$

$$y = BC + CD$$

$$= BC + AO$$

$$= AC \times \text{the gradient of the straight line} + AO$$

$$= AC \times \text{the gradient of the straight line} + c$$

But

$$y = mx + c$$

Hence

$$m = \text{the gradient of the straight line}$$

$$c = \text{the intercept on the } y\text{-axis}$$

Fig. 32.12 shows the difference between a positive and a negative gradient.

Fig. 32.12

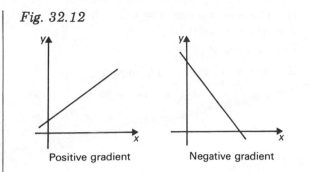

Positive gradient   Negative gradient

Example 6

The table below shows corresponding values of x and y. Plot a graph and hence find the equation connecting x and y.

x	3	5	8	12
y	13	17	23	31

The graph is shown in Fig. 32.13. It is seen to be a straight line which slopes upwards from left to right. Since the true origin is at the intersection of the two axes, c is the intercept on the y-axis, and from the graph we see that $c = 7$.

We now have to find the value of m. Since m is the gradient of the straight line we draw the right-angled triangle PQR, making the sides reasonably long because a small triangle will give less accurate results.

Using the scales of the x and y-axes, PQ = 16 units and RQ = 8 units. Hence

$$m = \frac{\text{Vertical height}}{\text{Base}}$$

$$= \frac{PQ}{RQ}$$

$$= \frac{16}{8}$$

$$= 2$$

Since the straight line slopes upwards from left to right m is positive. Therefore the equation connecting x and y is

$$y = 2x + 7$$

Fig. 32.13

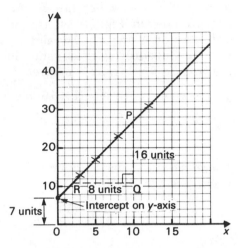

Example 7

The table below shows corresponding values of x and y.

x	10	15	18	20
y	30	20	14	10

Plot the graph and hence find the equation connecting x and y.

The graph is shown in Fig. 32.14. It will be noticed that the starting points on the x and y-axes is 10. Since the true origin is not shown on the graph we cannot find the value of c by direct measurement. We can, however, find the value of m by drawing the right-angled triangle PQR.

$$m = \frac{PQ}{QR}$$

$$= \frac{10}{5}$$

$$= 2$$

Since the straight line slopes downwards from left to right, the gradient is negative and $m = -2$. The equation of the line is

$$y = -2x + c$$

To find the value of c we choose a point, such as A, which lies on the straight line. In Fig. 32.14, the co-ordinates of A are $x = 16$ and $y = 18$. Substituting these values in the equation $y = -2x + c$ gives

$$18 = -2 \times 16 + c$$

$$18 = -32 + c$$

$$c = 50$$

Therefore the equation connecting x and y is

$$y = -2x + 50$$

Fig. 32.14

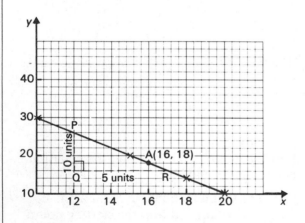

Example 8

(a) The straight line $y = 3x + c$ passes through the point $(2, 8)$.
Find the value of c.

When $x = 2$, $y = 8$, hence

$$8 = 3 \times 2 + c$$

$$8 = 6 + c$$

$$c = 2$$

(b) The straight line $y = mx - 3$ passes through the point $(4, -11)$.
Find the value of m.

When $x = 4$, $y = -11$, hence

$$-11 = 4m - 3$$

$$-8 = 4m$$

$$m = -2$$

Exercise 32.3

1. Draw the graph of $y = 3x + 5$ for values of x between 0 and 6.

2. Draw the graph of $y = 3 - 2x$ for values of x between 3 and 7.

3. Draw the graph of $y = 3x - 4$ for values of x between -4 and 3.

4. Draw the graph of $y = 3 - x$ for values of x between -2 and 6.

The following equations represent straight lines. State in each case the gradient of the line and the intercept on the y-axis.

5. $y = x + 3$
6. $y = -3x + 4$

7. $y = 2 - 5x$
8. $y = 4x - 3$

9. $y = -5x - 2$

10. Find the values of m and c if the straight line $y = mx + c$ passes through the point $(-2, 5)$ and has a gradient of 4.

11. Find the values of m and c if the straight line $y = mx + c$ passes through the point $(3, 4)$ and the intercept on the y-axis is -2.

12. The following table gives values of x and y which are connected by an equation of the type $y = mx + c$. Plot the graph and from it find the values of m and c.

x	4	6	10	12
y	16	22	34	40

Use scales of 1 large square to 2 units along the x-axis and 1 large square to 5 units on the y-axis.

13. Using the table below, plot a graph showing the relationship between x and y. Hence find the equation connecting x and y. Use scales of 1 large square to 2 units horizontally and 1 large square to 10 units vertically.

x	10	15	18	22
y	52	72	84	100

14. Using a scale of 1 large square to 1 unit on the horizontal axis and 1 large square to 2 units on the vertical axis draw a pair of axes on graph paper for values of x between -5 and 5 and for values of y between -10 and 10.

(a) Plot the points P(3, 10) and Q(-3, 4) and then join P and Q.

(b) Use your graph to find an equation of the line PQ in the form $y = ax + b$.

15. The table below shows corresponding values of P and Q. Plot a graph showing the relationship between P and Q. Suitable scales are 1 large square to 5 units on both axes. Take P along the horizontal axis and hence find the equation connecting P and Q.

P	15	18	20	28
Q	45	30	20	-20

Experimental Data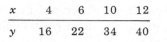

One of the most important applications of the straight-line equation is in the determination of an equation connecting two quantities when values have been obtained in an experiment.

Example 9

In an experiment carried out with a lifting machine the effort E and the load W were found to have the values given in the table below.

W	15	25	40	50	60
E	2.75	3.80	5.75	7.00	8.20

Plot these figures and obtain the equation connecting E and W, which is thought to be of the type $E = aW + b$.

Note that if E and W are connected by an equation of the type $E = aW + b$, then the graph must be a straight line (here a and b have been used instead of m and c in the straight-line equation, which is not unusual). Note also that E is the dependent variable and W the independent variable, so that W is taken on the horizontal axis and E on the vertical axis.

On plotting the points (Fig. 32.15) it will be seen that they deviate slightly from a straight line; since the data are experimental we must expect errors in measurement and observation, and slight deviations from a straight line are to be expected. Although the straight line will not pass through some of the points, an attempt must be made to ensure an even spread of points above and below the line.

Fig. 32.15

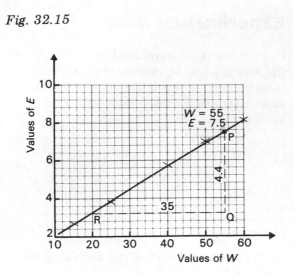

To find the gradient of the line (i.e. the value of *a*) we draw the right-angled triangle PQR.

$$a = \frac{PQ}{RQ}$$

$$= \frac{4.4}{35}$$

$$= 0.13$$

To find the value of *b* we find the coordinates of the point P, which lies on the line. These are $W = 55$ and $E = 7.5$. Substituting these coordinates in $E = aW + b$ gives

$$7.5 = 55 \times 0.13 + b$$

$$7.5 = 7.15 + b$$

$$b = 0.35$$

Therefore the equation connecting *E* and *W* is

$$E = 0.13W + 0.35$$

Exercise 32.4

1. The following observed values of *P* and *Q* are supposed to be related by the linear equation $P = aQ + b$, but there are experimental errors. Find by plotting the graph the most probable values of *a* and *b*.

Q	2.5	3.5	4.4	5.8	7.5	9.6	12.0	15.1
P	13.6	17.6	22.2	28.0	35.5	47.4	56.1	74.6

2. In an experiment carried out with a machine the effort *E* and the load *W* were found to have the values given in the table below. The equation connecting *E* and *W* is thought to be of the type $E = aW + b$. By drawing a graph check if this is so and from the graph find the most probable values of *a* and *b*.

W	10	30	50	60	80	100
E	8.9	19.1	29	33	45	54

3. A test on a metal-filament lamp gave the following values of resistance *R* (ohms) at various voltages *V* (volts).

V	62	75	89	100	120
R	100	117	135	149	175

These results are expected to agree with an equation of the type $R = mV + c$. Test this assumption by drawing a graph, and from it find suitable values for *m* and *c*.

4. During an experiment to verify Ohm's law the following results were obtained:

E (volts)	0	1.0	2.0	2.5	3.7	4.1
I (amps)	0	0.24	0.50	0.63	0.92	1.05

Plot these values with *I* on the horizontal axis and find an equation connecting *E* and *I*.

5. The following figures show how the resistance of a conductor R (ohms) varies as the temperature t ($^\circ$C) increases.

t	25	50	75	100	150
R	20.7	21.5	22.3	23.0	24.5

Plot these results and show that $R = at + b$.

Hence estimate values for a and b.

Miscellaneous Exercise 32

1. The graph shown in Fig. 32.16 shows the relation between the number of units used and the cost of a telephone bill. Use the graph to find

(a) the telephone bill when 360 units are used

(b) the number of units used when the bill is £14.75

(c) The bill is made up of a rental charge plus so much per unit used. What is the rental charge?

Fig. 32.16

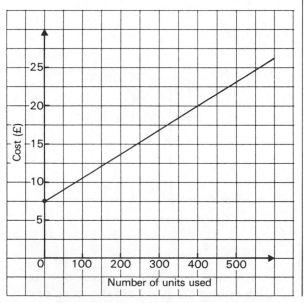

Fig. 32.17

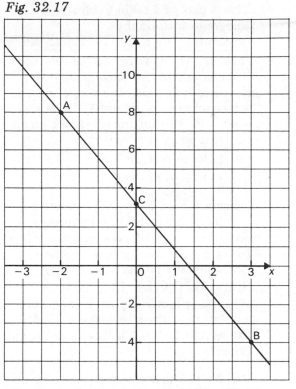

2. Copy the graph shown in Fig. 32.17.

(a) On your graph write down the coordinates of the points A, B and C.

(b) Plot the points D$(-3, -2)$ and E$(3, 7)$.

3. A car was tested for fuel consumption at various speeds.

Speed in mile/h	10	30	50	70	80
Fuel consumption in mile/gal	20	35	40	30	23

(a) Plot the points and draw a smooth curve through them. Suitable scales are one large square to represent 5 mile/gal along the horizontal axis and one large square to represent 20 mile/h on the vertical axis.

(b) Use your graph to answer the following questions:

 (i) What is the fuel consumption when the speed is 40 miles/h?

(ii) At what speeds is the consumption of fuel 25 miles/gal?

(iii) What is the most economical speed?

4. This is part of a table to change kilograms into pounds.

kilograms	5	10	15	20	25	30
pounds	11	22	33	44	55	66

Using scales of one large square to represent 5 kg along the x-axis and one large square to represent 10 lb on the y-axis plot this information on a graph. Use your graph to convert

(a) 12 kg into pounds

(b) 35 lb into kilograms

5. Using a scale of one large square to 2 units on both the x and y-axes, construct a pair of axes on graph paper for values of x and y from -8 to $+8$. Then plot the points A$(-7, -4)$, B$(6, -4)$, C$(6, 5)$, D$(-4, 5)$. Join the points in alphabetical order with straight lines to make a quadrilateral ABCD. What is this quadrilateral called?

6. Fig. 32.18 shows the relation between the number of units of electricity used and the cost.

(a) Find the cost when 360 units are used.

(b) Find the number of units used when the cost is £17.

(c) What is the standing charge?

(d) What is the cost per unit used?

Fig. 32.18

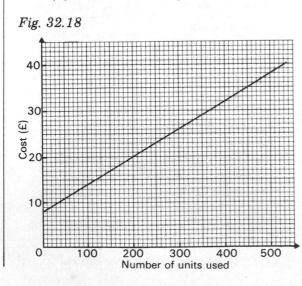

7. Using a scale of 1 large square to 1 unit on the x-axis and 1 large square to 2 units on the y-axis construct a pair of axes on graph paper for values of x between -4 and 5 and values of y between -10 and 10.

(a) On your graph plot the points A$(4, 8)$ and B$(-3, 1)$.

(b) Use your graph to find the coordinates of the point where the line AB cuts the y-axis.

(c) Find the gradient of the line AB.

(d) Write down the equation of the line AB in the form $y = ax + b$.

8. The equation of the straight line PQ is expressed in the form $y = mx + c$. Given that P is the point $(0, 3)$ and the gradient of PQ is 4:

(a) Write down the equation of the line PQ.

(b) The point R has its x-coordinate equal to -2. What is its y-coordinate?

9. In Fig. 32.19, RP is the line $y = 2x + 3$. If ON is 3 units, find the length of PN. What is the area of the trapezium AONP?

Fig. 32.19

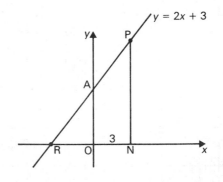

10. The line $y = 2x + c$ passes through the point $(3, 10)$. Find the value of c.

11. Using the graph of Fig. 32.20

   (a) find the coordinates of the point A

   (b) find the gradient of the line AB

   (c) write down the equation of the line AB

Fig. 32.20

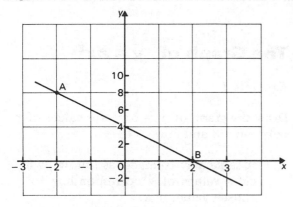

12. Fig. 32.21 shows a straight line passing through the points A and B. The equation of the line AB is $y = \frac{1}{2}x + 6$.

   (a) Write down the coordinates of the point A.

   (b) Find the coordinates of the point B.

   (c) Calculate the area of the triangle AOB.

Fig. 32.21

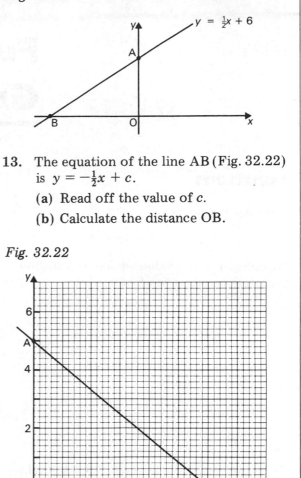

13. The equation of the line AB (Fig. 32.22) is $y = -\frac{1}{2}x + c$.

   (a) Read off the value of c.

   (b) Calculate the distance OB.

Fig. 32.22

33 Further Graphical Work

Functions

Consider the equation

$$y = 3x^2 - 4x + 5$$

We can give x any value we like and so find the corresponding value of y. Thus

if $x = 2$,

$$y = 3 \times 2^2 - 4 \times 2 + 5$$

$$= 9$$

if $x = 5$,

$$y = 3 \times 5^2 - 4 \times 5 + 5$$

$$= 60$$

Since the value of y depends upon the value allocated to x it is called the **dependent variable**. Because we can give x any value we like it is called the **independent variable**.

Because the value of y depends upon the value allocated to x we say that y is a function of x. Instead of writing

$$y = 3x^2 - 4x + 5$$

we can write

$$f(x) = 3x^2 - 4x + 5$$

meaning that the function of x is $3x^2 - 4x + 5$.

Note carefully that $f(x)$ does not mean f multiplied by x. It is simply shorthand for the statement 'is a function of x'.

The Graph of $y = ax^2$

Example 1

Draw the graph of $y = 5x^2$, for values of x between -3 and 3.

A table may be made, as follows, which gives values of y corresponding to chosen values of x.

x	-3	-2	-1	0	1	2	3
y	45	20	5	0	5	20	45

The graph is shown in Fig. 33.1 and it is seen to just touch the x-axis at the origin. It is also symmetrical about the y-axis.

Fig. 33.1

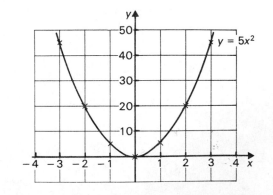

Graphs of Quadratic Functions

The function $f(x) = ax^2 + bx + c$ is called a quadratic function of x. The constants a, b and c can take any numerical value. For instance

$f(x) = x^2 - 36$
in which $a = 1$, $b = 0$ and $c = -36$

$f(x) = 5x^2 + 7x + 8$
in which $a = 5$, $b = 7$ and $c = 8$

$f(x) = 2.5x^2 - 3.1x - 2.9$
in which $a = 2.5$, $b = -3.1$ and $c = -2.9$

A quadratic function may contain only the square of the unknown quantity, as in the first of the above functions, or it may contain both the square and the first power of the unknown quantity, as in the second and third of the above functions.

When a quadratic function is plotted it always gives a smooth curve called a **parabola**.

Example 2

Plot the graph of $f(x) = 5x^2 + 13x - 6$ for values of x between -5 and 2.

A table may be made, as follows, which gives values of $f(x)$ corresponding to chosen values of x

x	-5	-4	-3	-2	-1	0	1	2
$5x^2$	125	80	45	20	5	0	5	20
$13x$	-65	-52	-39	-26	-13	0	13	26
-6	-6	-6	-6	-6	-6	-6	-6	-6
$f(x)$	54	22	0	-12	-14	-6	12	40

The graph of the function is shown in Fig. 33.2.

Fig. 33.2

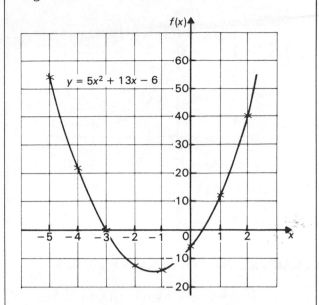

The Reciprocal Function

The function $f(x) = a/x$ is called the **reciprocal function**, a being a constant.

Example 3

Plot the graph of $f(x) = \dfrac{5}{x}$ for values of x between -5 and 5.

x	-5	-4	-3	-2	-1	0	1	2	3	4	5
$f(x)$	-1	-1.25	-1.67	-2.5	-5		5	2.5	1.67	1.25	1

Note that the reciprocal of a negative number is a negative number, e.g. $\dfrac{1}{-5} = -0.2$, whilst the reciprocal of a positive number is a positive number, for instance $\dfrac{1}{5} = 0.2$.

It is impossible to find the reciprocal of 0 because we cannot divide by 0. Hence the graph consists of two branches as shown in Fig. 33.3.

Fig. 33.3

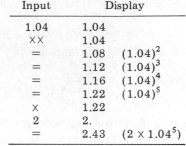

Curves of Natural Growth

Curves showing population growth, multiplication of bacteria, growth of invested money, etc. are called **curves of natural growth.** Expressed mathematically the exponential growth function is $f(x) = an^x$ where x is called the **exponent** and n the **base,** a being a constant.

Example 4

The population of a certain country is increasing at a rate of 4% per annum. If the population in 1975 was 2 000 000, draw a graph to represent the population growth up to the year 2000.

Since the increase is 4%,

$$n = 1 + \frac{4}{100} = 1.04.$$

Hence the graph may be constructed using the equation

$$y = 2 \times 1.04^x$$

where x is the number of years after 1975 and y is the corresponding population in millions.

To calculate the value of y which corresponds to a chosen value of x we proceed as follows:

When the year is 1980, $x = 5$ (i.e. 5 years after 1975) and

$$y = 2 \times 1.04^5$$
$$= 2.43$$

It is necessary to use a calculator, as follows:

Input	Display	
1.04	1.04	
× ×	1.04	
=	1.08	$(1.04)^2$
=	1.12	$(1.04)^3$
=	1.16	$(1.04)^4$
=	1.22	$(1.04)^5$
×	1.22	
2	2.	
=	2.43	(2×1.04^5)

The calculation of y for other values of x is obtained in a similar way.

Year	1975	1980	1985	1990	1995	2000
x	0	5	10	15	20	25
y	2	2.43	2.96	3.60	4.38	5.33

The graph is shown in Fig. 33.4, where it will be seen that the curve rises more steeply as the number of years increases. This is typical of graphs depicting the law of natural growth.

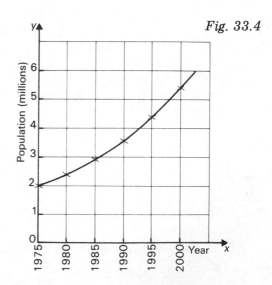

Fig. 33.4

The Gradient of a Curve

In mathematics and science we often need to know the rate of change of one variable with respect to another. For instance speed is the rate of change of distance with respect to time.

Consider the graph of $y = x^2$, part of which is shown in Fig. 33.5. As the values of x increase so do the values of y, but they do not increase at the same rate. A glance at Fig. 33.5 shows that the values of y increase faster when x is large because the gradient of the curve is increasing.

To find the rate of change of y with respect to x at a particular point we need to find the gradient of the curve at that point. If we draw a tangent to the curve at the point, the gradient of the tangent will be the same as the gradient of the curve.

Fig. 33.5

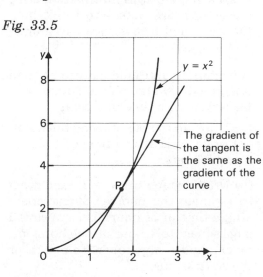

Example 5

Draw the graph of $y = x^2 - 3x + 7$ for values of x between -4 and $+4$. Hence find the gradient of the curve at the points where $x = -3$ and $x = 2$.

To plot the curve we draw up the following table:

x	-4	-3	-2	-1	0	1	2	3	4
y	35	25	17	11	7	5	5	7	11

The point where $x = -3$ is the point $(-3, 25)$. We draw a tangent to the curve at this point as shown in Fig. 33.6. The gradient is found by constructing a right-angled triangle (which should be as large as possible for accuracy), as shown, and measuring the length of its height and base. The gradient is negative because the tangent slopes downwards from left to right.

$$\text{Gradient at the point } (-3, 25) = \frac{\text{Height}}{\text{Base}}$$

$$= -\frac{27}{3}$$

$$= -9$$

The negative gradient indicates that the values of y are decreasing as the values of x increase.

At the point where $x = 2$ the value of y is 5. Hence by drawing a tangent and a right-angled triangle at the point $(2, 5)$,

$$\text{Gradient at the point } (2, 5) = \frac{3}{3} = 1$$

The gradient is positive because the tangent slopes upwards from left to right. A positive gradient indicates that the values of y are increasing as the values of x increase.

Fig. 33.6

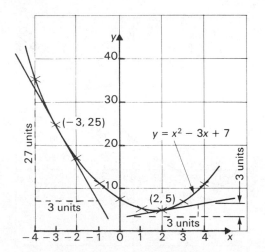

Example 6

Draw the graph of $f(x) = x^2 + 3x - 2$ for values of x between -1 and 4. Hence find the value of $f(x)$ where the gradient of the curve is 7.

To plot the graph the following table is made:

x	-1	0	1	2	3	4
y	-4	-2	2	8	16	26

The graph is shown in Fig. 33.7.

To obtain the line whose gradient is 7 we draw the right-angled triangle ABC making (for convenience) $AB = 2$ units to the scale on the x-axis and $BC = 14$ units to the scale on the y-axis.

$$\text{Gradient of AC} = \frac{14}{2}$$

$$= 7$$

Using set squares draw a tangent to the curve so that it is parallel to the line AC. From the diagram it will be seen that the tangent just touches the curve at the point $(2, 8)$. Hence when the gradient of the curve is 7, the value of $f(x)$ is 8.

Fig. 33.7

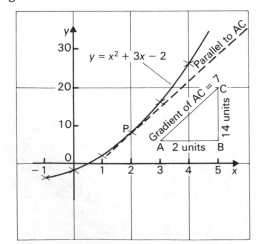

Exercise 33.1

1. Draw the graph of $f(x) = 3x^2$ for values of x between -3 and $+3$.

2. Plot the graph of $y = 2x^2 - 7x - 5$ for values of x between -4 and 12.

3. Plot the graph of $f(x) = x^2 - 4x + 4$ for values of x between -3 and $+3$.

4. Draw a graph of $f(x) = 4/x$ for values of x between -4 and $+4$.

5. The figures in the table below give corresponding values of x and y for a certain reciprocal function.

x	-4	-3	-2	-1	0	1	2	3	4
y	-2	-2.67	-4	-8		8	4	2.67	2

Draw a graph of this information using a scale of 1 large square to 1 unit on the x-axis and 1 large square to 2 units on the y-axis.

6. A bacterium multiplies according to the law $N = 2^T$ where N is the number of bacteria after a time of T hours.

   (a) Draw up a table showing the number of bacteria after a time of 1, 2, 3, 4 and 5 hours.

   (b) Draw a graph of $N = 2^T$ and use it to estimate the number of bacteria after a time of $3\frac{1}{2}$ hours. Use scales of 1 large square to 1 hour on the horizontal axis and 1 large square to 5 units on the vertical axis.

7. The following table shows how a sum of £8000 appreciates over a period of 7 years.

Year	1	2	3	4	5	6	7
Amount (£)	8800	9680	10 650	11 710	12 880	14 170	15 590

   (a) Draw a graph to depict this information using scales of 1 large square to 1 year on the horizontal axis and 1 large square to £1000 on the vertical axis.

(b) What is the amount after $4\frac{1}{2}$ years?

(c) Estimate the number of years for the amount to reach £13 500.

8. The population of a territory is increasing at the rate of 5% per annum. In 1960 the population was 2 million. Estimate the population in the year 2000, assuming that the rate of growth of the population remains the same.

9. Draw the graph of $y = 3x^2 + 7x + 3$ for values of x between -5 and $+5$. Hence find the gradient of the curve at the points where $x = -2$ and $x = 2$.

10. Draw the graph of $f(x) = 2x^2 - 5$ for values of x between -2 and $+3$. Hence find the gradient of the curve at the points where $x = -1$ and $x = 2$.

11. If $y = (1 + x)(5 - 2x)$, copy and complete the table below.

x	-2	$-1\frac{1}{2}$	-1	0	$\frac{1}{2}$	1	$1\frac{1}{2}$	2	3
y	-9			0	5		6	5	3	-4

Hence draw the graph of $y = (1 + x)(5 - 2x)$. Find the value of x at which the gradient of the curve is -2.

12. If $f(x) = x^2 - 5x + 4$, find, by plotting the curve for values of x between 4 and 12, the value of x at which the gradient of the curve is 11.

Distance–Time Graphs

In Chapter 12 it was shown that

$$\text{Average speed} = \frac{\text{Total distance travelled}}{\text{Total time taken}}$$

If the distance is measured in metres and the time in seconds, then the speed is measured in metres per second (m/s). If the distance is measured in kilometres and the time in hours, then the speed is measured in kilometres per hour (km/h).

Example 7

(a) A car travels a distance of 100 km in 2 hours. Calculate its average speed.

$$\text{Average speed} = \frac{100}{2} \text{ km/h}$$
$$= 50 \text{ km/h}$$

(b) A body travels a distance of 80 m in 4 s. Find its average speed.

$$\text{Average speed} = \frac{80}{4} \text{ m/s}$$
$$= 20 \text{ m/s}$$

Since

$$\text{Distance} = \text{Speed} \times \text{Time}$$

when the speed is constant the distance travelled is proportional to the time taken. The graph of distance against time will therefore be a straight line passing through the origin. The gradient of this graph will represent the speed.

Example 8

Fig. 33.8 shows a distance–time graph for a car. Find the average speed of the car.

Fig. 33.8

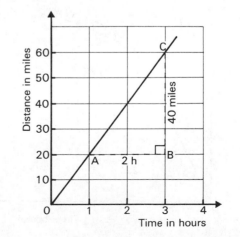

The average speed of the car is given by the gradient of the graph. Drawing the right-angled triangle ABC we see that when AB = 2 hours, BC = 40 miles. Hence

$$\text{Gradient of line} = \frac{BC}{AB}$$

$$= \frac{40}{2}$$

$$= 20$$

Hence the average speed of the car is 20 miles per hour.

Example 9

A man travels a distance of 120 km in 2 hours by car. He then cycles 20 km in $1\frac{1}{2}$ hours and finally walks a distance of 8 km in 1 hour, all at constant speed. Draw a graph to illustrate this journey and from it find the average speed for the entire journey.

Fig. 33.9

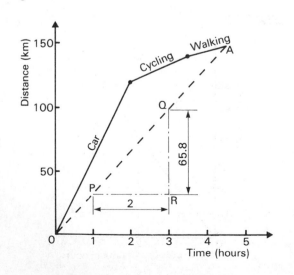

The graph is drawn in Fig. 33.9 and it consists of three straight lines. The average speed for the entire journey is found by drawing the straight line OA and finding its gradient.

To find the gradient of OA, the right-angled triangle PQR is drawn, which for accuracy should be as large as possible.

$$\text{Gradient of OA} = \frac{QR}{PR}$$

$$= \frac{65.8}{2}$$

$$= 32.9$$

Hence the average speed for the entire journey is 32.9 km/h.

Exercise 33.2

1. A vehicle travels 300 km in 5 hours. Calculate its average speed.

2. A car travels 400 km at an average speed of 80 km/h. How long did the journey take?

3. A train travels for 5 hours at an average speed of 60 km/h. How far did it travel?

4. A vehicle travels a distance of 250 km in a time of 5 hours. Draw a distance-time graph to depict the journey and from it find the average speed of the vehicle.

5. A car travels a distance of 120 km in 3 hours. It then changes speed and travels a further distance of 80 km in $2\frac{1}{2}$ hours. Assuming that the two speeds are constant, draw a distance–time graph and from it determine the average speed for the complete journey.

6. A girl cycles a distance of 20 km in 100 minutes. She rests for 20 minutes and then cycles a further 10 km, which

takes her 60 minutes. Draw a distance–time graph to represent the journey and from it find the average speed for the entire journey.

7. A man travels a distance of 90 km by car, which takes him $1\frac{1}{2}$ hours. He then cycles 18 km in a time of $1\frac{1}{2}$ hours. He rests for 15 minutes before continuing his journey on foot, during which he walks 8 km in 2 hours. By drawing a distance–time graph find his average speed for the entire journey, assuming that he travels at constant speed for each of the three parts of his journey.

Example 10

Two towns A and B are 120 km apart. A motorist starts out from A at 9.00 a.m. and drives towards B at a constant speed of 60 km/h. At 9.30 a.m. a second motorist sets out from B and drives at a constant speed of 90 km/h towards A. How far from A will the two cars meet?

The distance–time graphs for the two motorists are drawn on the same axes, as shown in Fig. 33.10. Since the speed of the first motorist is constant the graph is a straight line. If 9.00 a.m. is taken as zero time, then this line will pass through the origin.

Half an hour later the second motorist sets out from B, which is 120 km from A. As time progresses he gets nearer to A. Hence the distance–time graph for the second motorist is also a straight line, but it slopes downwards from left to right.

The vertical coordinate of the point X, where the two graphs intersect, gives the distance from A where the two motorists meet. From the graph, this distance is found to be 66 km.

Fig. 33.10

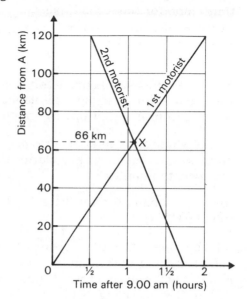

Exercise 33.3

1. A cyclist starts from a town A at 9.00 a.m. and cycles towards town B at an average speed of 15 km/h. At 10.00 a.m. a motorist starts the same journey but travels at 60 km/h. When and where will the motorist overtake the cyclist?

2. A train left London for Newcastle, a distance of 420 km, at 1500 hours. It stopped at Grantham (180 km from London) for 12 minutes. It stopped again at York (315 km from London) for 12 minutes. Except for the halts, it travelled at a steady 70 km/h. A second train left Newcastle for London at 1530 hours and it travelled non-stop at 90 km/h.

(a) Draw a distance–time graph for the first train.

(b) At what time did the first train reach Newcastle?

(c) On the same axes draw a distance-time graph for the second train.

(d) When and where did the two trains cross?

3. Fig. 33.11 shows an incomplete distance-time graph for a squad of soldiers walking from A to B. Copy the graph.

(a) At what speed, in kilometres per hour, was the squad walking during the first twenty minutes?

(b) Up to 2 p.m., for how long did the squad rest?

(c) The squad started off again at 2 p.m. and at the end of 3 hours and 40 minutes had walked a total distance of 22 km. On your copy of Fig. 33.11 complete the graph for the last 1 hour and 40 minutes, assuming that the squad walked at a uniform speed during this period.

(d) By drawing a suitable straight line on the graph find the average speed of the squad for the entire journey, including resting time.

Fig. 33.11

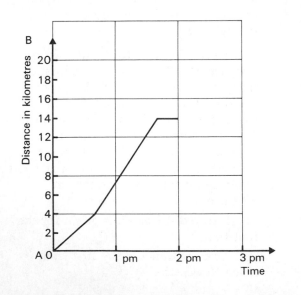

4. A woman leaves home at 9 a.m. and returns at 4.30 p.m. The graph (Fig. 33.12) shows her distance from home, in kilometres, at various times during the day.

Fig. 33.12

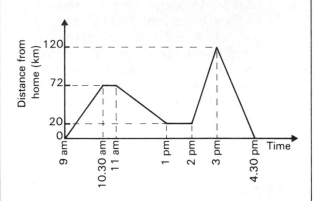

(a) Calculate her average speed between 9 a.m. and 10.30 a.m.

(b) Find the total number of kilometres she travels between 9 a.m. and 4.30 p.m.

(c) How far is she from home at 2.45 p.m.?

(d) For how many hours during the day is she moving towards home?

5. A young man sets out at noon to cycle from P to Q, a distance of 200 km. He cycles at a steady 20 km/h for $4\frac{1}{2}$ hours then stops and rests for 1 hour. He then cycles on and reaches Q at 2200 hours. A motorist using the same route sets out from Q at 1800 hours and travels towards P. He maintains a steady speed of 60 km/h.

On the same axes draw graphs of these two journeys using scales of 1 cm = 1 hour horizontally and 1 cm = 40 km vertically.

(a) At what time do the motorist and the cyclist pass each other?

(b) How far is the cyclist from Q when the motorist passes him?

(c) At what speed does the cyclist travel during the second part of his journey?

(d) At what times were the motorist and cyclist 20 km apart?

Speed–Time Graphs

If a speed–time graph is drawn (Fig. 33.13), the area under the graph gives the distance travelled. The gradient of the curve gives the acceleration, because acceleration is rate of change of speed with respect to time. If the speed is measured in metres per second then the acceleration will be measured in metres per second per second (m/s^2).

Fig. 33.13

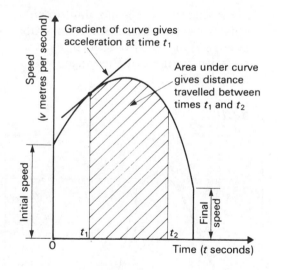

If the acceleration is uniform the speed–time graph will be a straight line (Fig. 33.14). When the speed is increasing the graph has a positive gradient (i.e. it slopes upwards to the right). If the speed is decreasing (i.e. there is deceleration, sometimes called retardation) the graph has a negative gradient and it slopes downwards to the right.

Fig. 33.14

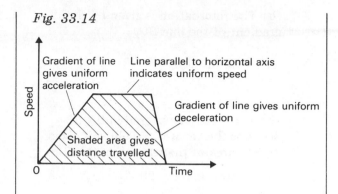

Example 11

A car starting from rest attains a speed of 20 m/s after 5 seconds. It continues at this speed for 15 seconds and then slows down and comes to rest in a further 8 seconds. If the acceleration and retardation are constant, draw a speed–time graph and from it find

(a) the acceleration of the car

(b) the retardation of the car

(c) the distance travelled in the total time of 28 seconds

The speed–time graph is shown in Fig. 33.15.

Fig. 33.15

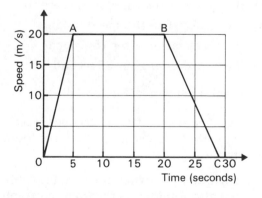

(a) The acceleration is given by the gradient of the line OA

$$\text{Acceleration} = \frac{20}{5}$$

$$= 4 \, m/s^2$$

(b) The retardation is given by the gradient of the line BC

$$\text{Retardation} = \frac{20}{8}$$

$$= 2.5 \text{ m/s}^2$$

(c) The distance covered in the 28 seconds the car was travelling is given by the area of the trapezium OABC

$$\text{Distance travelled} = 20 \times \tfrac{1}{2} \times (15 + 28)$$

$$= 20 \times \tfrac{1}{2} \times 43$$

$$= 430 \text{ metres}$$

Example 12

The table below gives the speed of a car v metres per second after a time t seconds.

t	0	5	10	15	20	25
v	0	2.4	5.0	7.5	9.5	10.2

t	30	35	40	45	50	55
v	9.2	5.2	2.7	2.3	2.7	3.5

Draw a smooth curve to show how v varies with t. Use the graph to find

(a) the speed of the car after 47 seconds

(b) the times when the speed is 4 m/s

(c) the acceleration after 15 seconds

(d) the retardation after 35 seconds

The graph is shown in Fig. 33.16.

(a) The speed after 47 seconds is read directly from the graph and it is found to be 2.4 m/s.

(b) The times when the speed is 4 m/s are also found directly from the graph. They are 8 seconds and 37 seconds.

(c) To find the acceleration at a time of 15 seconds draw a tangent to the curve at the point A and find its gradient by constructing the right-angled triangle ABC.

$$\text{Gradient} = \frac{AB}{BC}$$

$$= \frac{4}{10}$$

$$= 0.4$$

Hence the acceleration is 0.4 m/s².

(d) To find the retardation at a time of 35 seconds draw a tangent to the curve at the point D. Then by constructing the right-angled triangle EFG

$$\text{Gradient} = \frac{EF}{FG}$$

$$= \frac{5.6}{10}$$

$$= 0.56$$

Hence the retardation is 0.56 m/s².

Note that the car is slowing down at this time and therefore retardation occurs. This is also shown by the negative gradient of the tangent at the point D.

Fig. 33.16

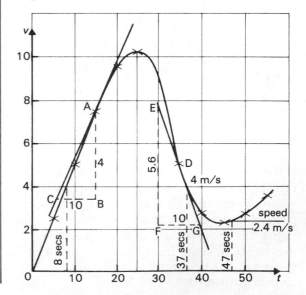

Exercise 33.4

1. Fig. 33.17 shows a number of speed–time graphs. In each case state the distance travelled.

Fig. 33.17

(a)

(b)

(c)

(d)

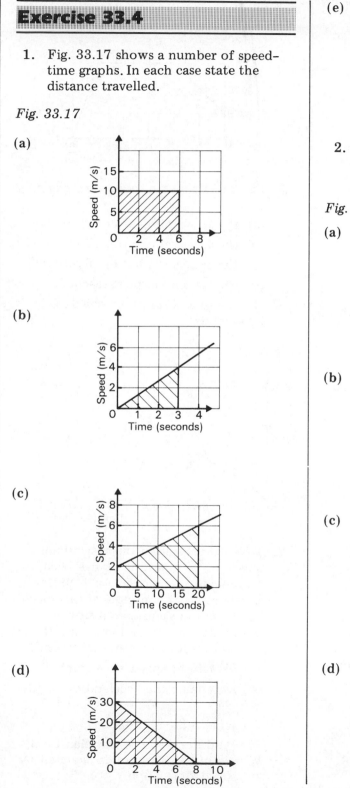

(e)

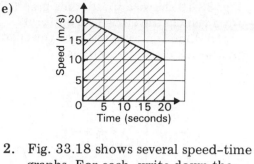

2. Fig. 33.18 shows several speed–time graphs. For each, write down the acceleration or retardation.

Fig. 33.18

(a)

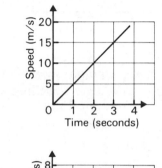

(b)

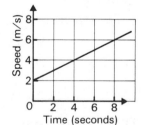

(c)

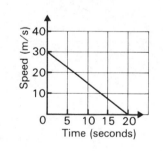

(d)

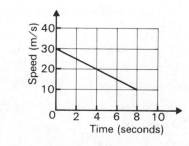

3. Fig. 33.19 shows a speed–time graph for a vehicle travelling at a uniform speed. Find the total distance travelled by the vehicle in 15 seconds.

Fig. 33.19

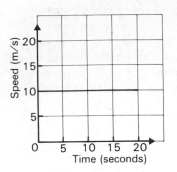

4. The speed–time graph shown in Fig. 33.20 shows a car travelling with constant acceleration. Find
   (a) the acceleration
   (b) the initial speed
   (c) the distance travelled in 30 seconds

Fig. 33.20

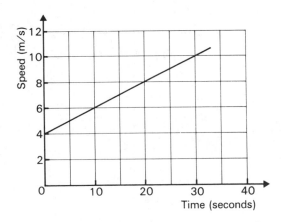

5. From the speed–time graph shown in Fig. 33.21, find
   (a) the acceleration
   (b) the retardation
   (c) the initial speed
   (d) the distance travelled in the first 10 seconds

Fig. 33.21

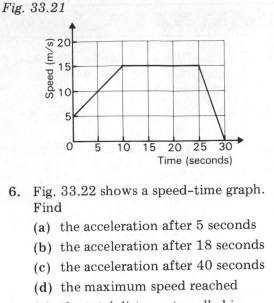

6. Fig. 33.22 shows a speed–time graph. Find
   (a) the acceleration after 5 seconds
   (b) the acceleration after 18 seconds
   (c) the acceleration after 40 seconds
   (d) the maximum speed reached
   (e) the total distance travelled in 50 seconds

Fig. 33.22

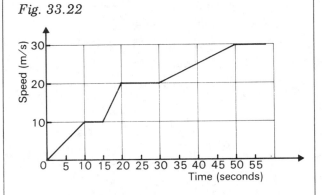

7. A vehicle starting from rest attains a speed of 16 m/s after it has been travelling for 8 seconds with uniform acceleration. It continues at this speed for 15 seconds and then it slows down with uniform retardation until it finally comes to rest in a further 10 seconds.
   (a) Sketch the speed–time graph.
   (b) Determine the acceleration of the vehicle.
   (c) What is the retardation?
   (d) Find the distance travelled by the vehicle in the 33 seconds depicted by the speed–time graph.

8. A car has an initial speed of 5 m/s. It then accelerates uniformly for 6 seconds at $0.5\,\text{m/s}^2$. It then proceeds at this speed for a further 25 seconds. Draw the speed–time graph and use it to find the total distance travelled in the time of 31 seconds.

9. The speed of a body, v metres per second, at various times t seconds is shown in the table below.

t	0	1	2	3	4	5	6	7	8
v	0	1	2	6	12	20	30	42	56

(a) Draw the graph showing how v varies with t. Horizontally, take 1 cm to represent 1 second and vertically, take 1 cm to represent 10 m/s.

(b) From the graph find the acceleration after times of 2 seconds and 6 seconds.

10. A body moves a distance s metres in a time t seconds so that $s = t^3 - 3t^2 + 8$. Plot a distance–time graph for this equation taking values of t between 0 and 5 seconds in half-second intervals. Hence use the graph to find the speed of the body after 3 seconds.

Location of a Point in Space

The location of a point in space requires coordinates referred to the origin O and three mutually perpendicular axes, Ox, Oy and Oz (Fig. 33.23).

Fig. 33.23

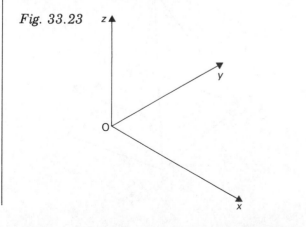

Consider the point P (Fig. 34.24). Its co-ordinates are $x = 5$, $y = 3$ and $z = 2$. These coordinates are written $(5, 3, 2)$.

Fig. 33.24

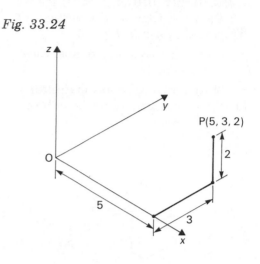

Example 13

Write down the coordinates of the points A, B, C, D, E, F, G and H which are the vertices of the cuboid shown in Fig. 33.25.

Fig. 33.25

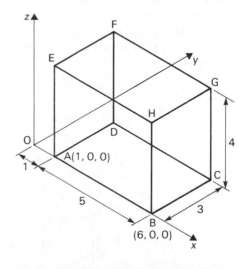

The coordinates are: A$(1, 0, 0)$, B$(6, 0, 0)$, C$(6, 3, 0)$, D$(1, 3, 0)$, E$(1, 0, 4)$, F$(1, 3, 4)$, G$(6, 3, 4)$ and H$(6, 0, 4)$.

Exercise 33.5

1. A(2, 0, 0), B(7, 0, 0), C(7, 4, 0) and E(2, 0, 3) are four of the vertices of the cuboid shown in Fig. 33.26.

   (a) Copy the diagram and label these four vertices on it.

   (b) Write down the coordinates for the other four vertices of the cuboid.

Fig. 33.26

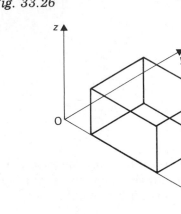

2. Fig. 33.27 shows a cuboid. The co-ordinates of three of the vertices are A(4, 0, 0), B(4, 0, 3) and C(2, 2, 0). Write down the coordinates of the vertex D.

Fig. 33.27

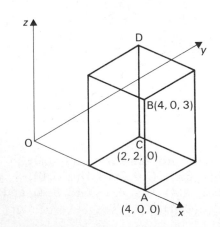

3. Fig. 33.28 shows a cuboid having a length of 6 units, a width of 5 units and a height of 7 units. The coordinates of the vertices A and B are (3, 0, 0) and (9, 0, 0) respectively. Write down the coordinates of the remaining 6 vertices.

Fig. 33.28

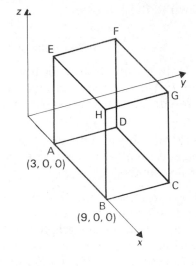

4. Fig. 33.29 shows a pyramid with a square base. Find the coordinates of the centre of the base O and the apex E.

Fig. 33.29

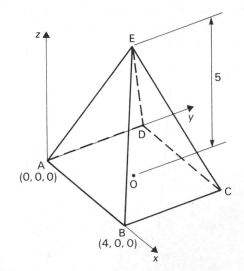

5. For the cube shown in Fig. 33.30, write down the coordinates for each of the eight vertices.

Fig. 33.30

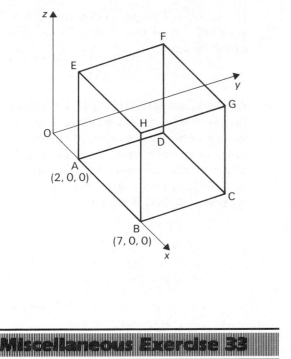

Miscellaneous Exercise 33

1. Part of the timetable for morning trains between four towns, Altown, Beeford, Colville and Denton is given below.

Altown	dep 0820	dep 0830	dep 0945
Beeford		arr 0857	arr 1015
		dep 0900	dep 1017
Colville			arr 1035
			dep 1037
Denton	arr 0910	arr 0930	arr 1054

(a) How long does the 0820 train from Altown take to make the journey to Denton?

(b) The 0945 train from Altown averages 66 km/h on its journey to Beeford. How far is it from Altown to Beeford?

(c) Denton is 60 km from Altown. What is the average speed, in kilometres per hour, for the journey of the 0820 train from Altown?

(d) The distance–time graph for one of the trains is sketched in Fig. 33.31. To which of the trains does it apply?

Fig. 33.31

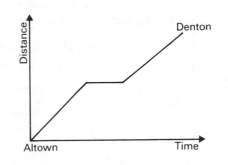

2. Copy and complete the table below, given that $y = x^2 - 6x + 8$.

x	−1	0	1	2	2.5	3	3.5	4	5
x^2	1				6.25		12.25		25
$-6x$	6				−15		−21		−30
8	8				8		8		8
y	15				−0.75		−0.75		3

Use the completed table to plot the graph of $y = x^2 - 6x + 8$, taking 2 cm to represent 1 unit on the x-axis and 1 cm to represent 1 unit on the y-axis. Hence find the gradient of the curve at the points where $x = 1$ and $x = 4$.

3. The table below gives the population (in millions) of a certain country during the years stated.

Year	1950	1955	1960	1965	1970	1975	1980
Population	3	3.48	4.03	4.67	5.42	6.28	7.28

(a) Draw a graph of this information taking the year on the horizontal axis. Suitable scales are: 1 large square to 5 years horizontally and 1 large square to 1 million of population vertically.

(b) The equation connecting year and population is $p = 3 \times 1.03^x$ where p is the population at the end of x years. Use this equation to estimate the population in the year 2000 assuming that the rate of increase of the population remains the same.

4. During the first stage of the vertical ascent of a rocket, its speed v metres per second t seconds after the start is given by the formula $v = 20t + 8t^2$.

Copy and complete the following table.

t	0	5	10	15	20	25	30
$20t$		100				500	
$8t^2$		200				5000	
v		300				5500	

(a) Use the completed table to draw a graph showing how v varies with t. Suitable scales are 1 large square to 5 seconds horizontally and 1 large square to 1000 metres per second vertically.

(b) Use your graph to determine the acceleration of the rocket 20 seconds after the start.

5. A person invests £10 000. If the interest rate is 10% per annum, draw a graph showing the appreciation of the investment over a period of 10 years. Use 1 large square to represent 2 years horizontally and 1 large square to represent £5000 vertically.

Transformational Geometry

Introduction

When we are given a geometrical shape there are many things which we can do with it.

(1) We can move the shape from one position to another by sliding it in one direction. Thus we can slide a parallelogram from position A to position A′ (Fig. 34.1). We say that the parallelogram A has been **translated** to the new position A′.

Fig. 34.1

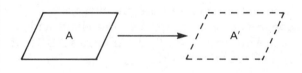

(2) A mirror reflects an image. In Fig. 34.2, the triangle B′ is a **reflection** of the triangle B in the mirror line m.

Fig. 34.2

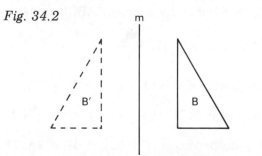

(3) We can rotate the shape from one position to another. In Fig. 34.3, the triangle C has been given a **rotation** about P so that it takes up the position C′.

Fig. 34.3

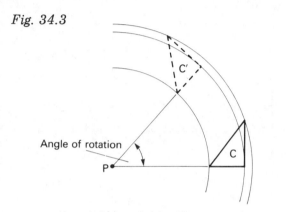

(4) A slide projector enlarges pictures. Working in a similar way we can enlarge geometric shapes. Thus in Fig. 34.4, rectangle D′ is an enlargement of the rectangle D.

Fig. 34.4

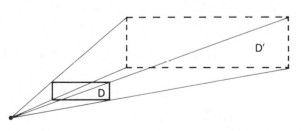

In each case we say that the figure has been **transformed**. The operation applied is called a **transformation**. The original figure is called the **pre-image** and the transformed figure is called the **image**. For example, in Fig. 34.1, A′ is the image of A.

A transformation is often referred to as a **mapping**. A transformation therefore maps a pre-image on to an image.

Translation

If every point in a line or a shape moves the same distance in the same direction, the transformation is called a **translation**. Thus in Fig. 34.5, every point in the line AB has moved 2 units to the right and 3 units upwards. That is $A(2, 1)$ translates to $A'(4, 4)$ and $B(6, 3)$ translates to $B'(8, 6)$.

Fig. 34.5

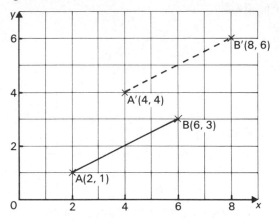

Similarly in Fig. 34.6, the triangle ABC has been translated to A'B'C'. Every point in ABC has been displaced by 4 units to the left and 2 units upwards. The translation maps the point $A(1, 1)$ on to $A'(-3, 3)$ and $B(3, 1)$ on to $B'(-1, 3)$ and $C(1, 4)$ on to $C'(-3, 6)$.

Fig. 34.6

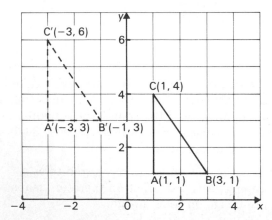

Generally this transformation maps (x, y) on to $(x - 4, y + 2)$. If this translation is denoted by T_1 we may write

$$T_1 : (x, y) \rightarrow (x - 4, y + 2)$$

The translation is defined by the constants -4 and $+2$ which are added to the coordinates of the original point. We may therefore define the translation by the **column vector** $\begin{pmatrix} -4 \\ 2 \end{pmatrix}$.

Properties of Translations

If two figures are related by a translation, the corresponding segments of these figures are equal in length and they are also parallel. Thus in Fig. 34.6, $AB = A'B'$ and $AB \parallel A'B'$; $AC = A'C'$ and $AC \parallel A'C'$; $BC = B'C'$ and $BC \parallel B'C'$. When a figure undergoes a translation, the image and the pre-image are **congruent** i.e. they are the same shape and size. Thus in Fig. 34.6, the triangles ABC and A'B'C' are congruent.

Example 1

The rectangle ABCD is formed by joining four points, $A(1, 5)$, $B(5, 5)$, $C(5, 3)$ and $D(1, 3)$. What is the image of ABCD under the translation $\begin{pmatrix} 3 \\ 4 \end{pmatrix}$?

The image of A is $(1 + 3, 5 + 4) = (4, 9)$

The image of B is $(5 + 3, 5 + 4) = (8, 9)$

The image of C is $(5 + 3, 3 + 4) = (8, 7)$

The image of D is $(1 + 3, 3 + 4) = (4, 7)$

The rectangle ABCD and its image under the translation are shown in Fig. 34.7.

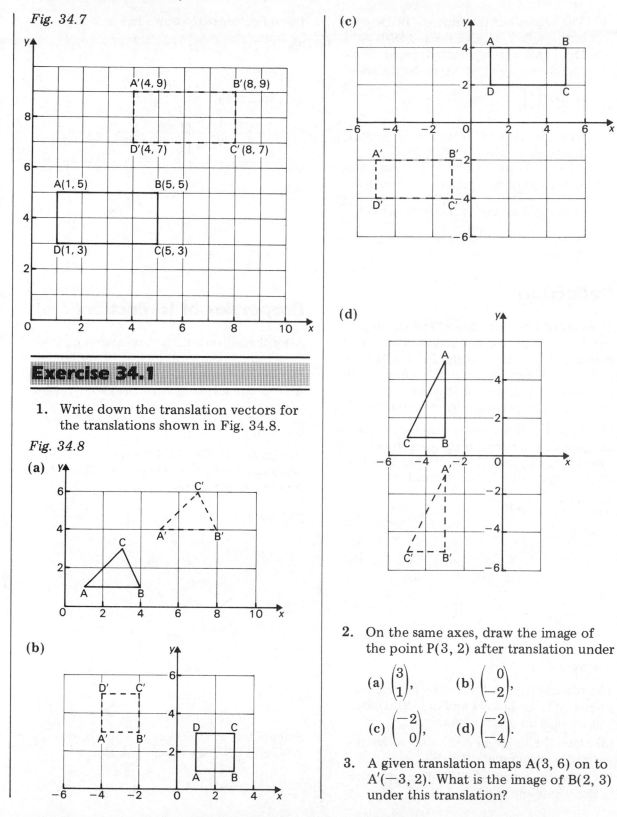

Fig. 34.7

Exercise 34.1

1. Write down the translation vectors for the translations shown in Fig. 34.8.

Fig. 34.8

(a)

(b)

(c)

(d)

2. On the same axes, draw the image of the point P(3, 2) after translation under

(a) $\begin{pmatrix} 3 \\ 1 \end{pmatrix}$,   (b) $\begin{pmatrix} 0 \\ -2 \end{pmatrix}$,

(c) $\begin{pmatrix} -2 \\ 0 \end{pmatrix}$,   (d) $\begin{pmatrix} -2 \\ -4 \end{pmatrix}$.

3. A given translation maps A(3, 6) on to A'(−3, 2). What is the image of B(2, 3) under this translation?

4. The vertices of the square ABCD are A(0, 1), B(2, 1), C(2, 3) and D(0, 3). Draw this square on graph paper. Draw also the image of ABCD under the translation $\binom{5}{4}$.

5. The triangle ABC is formed by joining the three points A(−2, −2), B(−1, −4) and C(−1, −1). Draw this triangle on graph paper and show its image A′B′C′ under the translation $\binom{2}{-1}$.

State the coordinates of the points A′, B′ and C′.

Reflection

The idea of a reflection is a familiar one. If you look in a mirror you see an image of yourself. A reflection is another transformation and it is very useful in the study of the properties of geometric figures and in forming wallpaper and carpet patterns.

If a point P is reflected in a mirror so that its image is P′, the mirror is the perpendicular bisector of PP′ (Fig. 34.9). We say that P′ is the **reflection** of P in the mirror line m.

Fig. 34.9

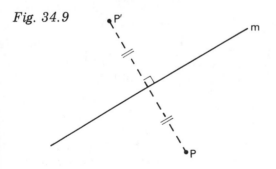

Example 2

The triangle ABC is formed by joining the points A(1, 1), B(2, 1) and C(1, 2). Draw this triangle on graph paper.

(a) Draw the image of ABC after reflection in the *x*-axis and mark it A′B′C′.

(b) Draw the image of ABC after reflection in the *y*-axis and mark it A″B″C″.

The reflections are shown in Fig. 34.10

Fig. 34.10

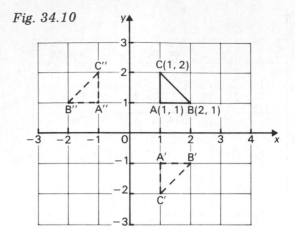

Properties of Reflection

After the reflection, the pre-image and the image are congruent figures. Thus in Fig. 34.10, triangles ABC, A′B′C′ and A″B″C″ are all congruent triangles.

Example 3

In Fig. 34.11, XY is a reflection of AB, m being the mirror line. Given that BC = 4 cm, find the length of CX.

Fig. 34.11

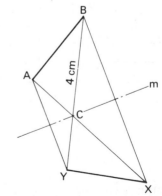

Since XY is a reflection of AB, triangles ABC and CXY are congruent. Hence

$$CX = BC$$
$$= 4\ cm$$

Look at the kite drawn in Fig. 34.12. We see that it is symmetrical about the line AC. This means that △ABC is a reflection of △ADC, AC being the mirror line (△ is shorthand notation for 'triangle'). It follows that

(i)   AC is the perpendicular bisector of BD, since the point B is a reflection of the point D,
(ii) AC bisects the angles BAD and BCD,
(iii) Triangles ABC and ADC are congruent.

Fig. 34.12

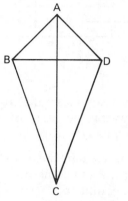

If a figure contains its own reflection in a line m, then the figure is said to be symmetrical about m. The line m is the mirror line or the axis of symmetry.

Example 4

Complete the figure in Fig. 34.13 so that it has two axes of symmetry.

Fig. 34.13

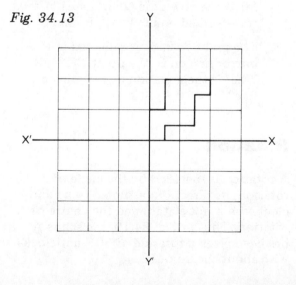

The figure is completed by reflecting it in the line YY' and then reflecting the double figure in the line XX'. The finished figure then has two axes of symmetry, XX' and YY', as shown in Fig. 34.14.

Fig. 34.14

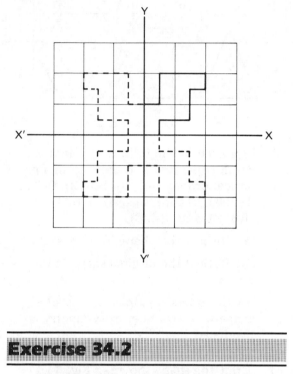

Exercise 34.2

1. Reflect the shape in Fig. 34.15 in
   (a) the *x*-axis,   (b) the *y*-axis.

Fig. 34.15

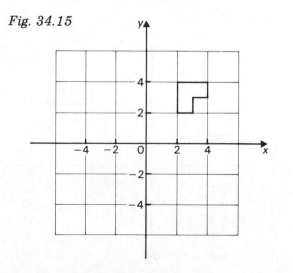

2. Copy the shapes shown in Fig. 34.16 and show all the lines of symmetry for each shape.

Fig. 34.16

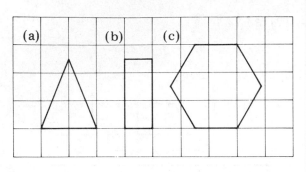

3. Plot each of the following points on graph paper and join them up in alphabetical order: A(0, 0), B(0, 2), C(2, 4), D(2, 6), E(4, 6), F(4, 4), G(6, 4), H(6, 2) and I(4, 0).

    (a) Reflect this shape in the x-axis.

    (b) Reflect the original shape in the y-axis.

    (c) Complete the drawing so that a shape with two lines of symmetry is produced.

4. Draw the shape shown in Fig. 34.17 and then reflect it in

    (a) the x-axis,   (b) the y-axis.

Fig. 34.17

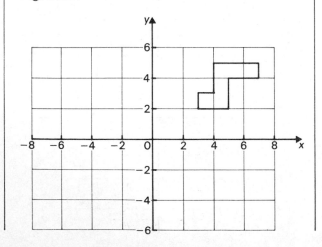

5. Complete the shape shown in Fig. 34.18 so that it has two axes of symmetry.

Fig. 34.18

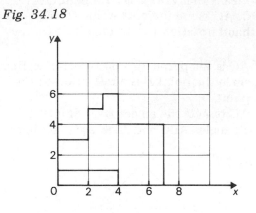

6. Draw an isosceles right-angled triangle. By drawing the axis of symmetry, obtain an angle of $45°$.

7. PQRS is a square such that P is the point (2, 2), Q is the point (6, 2) and R is the point (6, 6).

    (a) Write down the coordinates of the point S.

    (b) Draw the square on graph paper using a scale of 1 large square to 1 unit.

    (c) Draw the straight line $y = x$ on the graph paper using the same axes.

    (d) Reflect PQRS in the line $y = x$.

    (e) Write down the coordinates of the transformed points P$'$, Q$'$, R$'$ and S$'$.

8. The point A(2, 6) is reflected in a mirror line on to the point B(2, −2). What is the equation of the mirror line?

Rotation

A rotation is specified by the angle of rotation, its direction (assumed to be anticlockwise if not stated) and the centre of rotation. Thus in Fig. 34.19, the shape A has been given a rotation of $60°$, anticlockwise about the point P.

Fig. 34.19

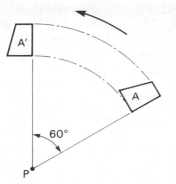

After a rotation the pre-image and the image are congruent.

Example 5

Triangle ABC has vertices A(3, 4), B(7, 4) and C(7, 6). Plot these points on graph paper and join them to form the triangle ABC. Show the image of ABC after a clockwise rotation of $90°$ about the origin and mark it A'B'C'. Write down the coordinates of the points A', B' and C'.

Triangle ABC is drawn in Fig. 34.20. The image A'B'C' has vertices A'(4, −3), B'(4, −7) and C'(6, −7).

Fig. 34.20

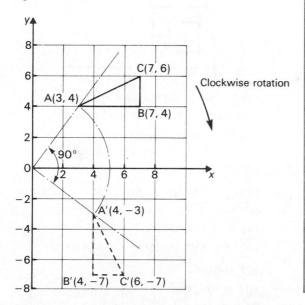

Rotational Symmetry

A figure has rotational symmetry if it coincides with its image by rotation about a point.

In Fig. 34.21, when the rectangle ABCD is rotated through $180°$ about O it will take up the position shown in Fig. 34.22. If the lettering is removed, there is no change in the two rectangles.

Fig. 34.21

Fig. 34.22

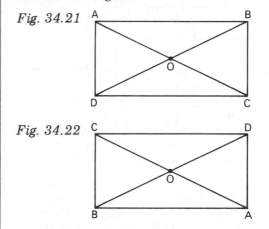

The rectangle ABCD has only two positions which are the same under rotation. We say that the rectangle has rotational symmetry of order 2.

Example 6

Draw the diagram of Fig. 34.23, then rotate it successively through $90°$, $180°$ and $270°$ about O.

(a) What is the order of rotational symmetry for the original shape?

(b) What is the order of rotational symmetry for the completed figure?

Fig. 34.23

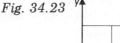

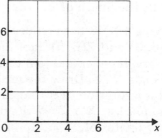

The shape after rotations of 90°, 180° and 270° is drawn in Fig. 34.24 where the images are shown by dotted lines.

Fig. 34.24

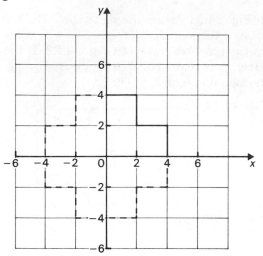

(a) The original shape has rotational symmetry of order 1 since it only gives the same shape after a rotation of 360°.

(b) The completed shape has rotational symmetry of order 4 because the shape appears unchanged after rotations of 90°, 180°, 270° and 360°.

Exercise 34.3

1. Using the origin as the centre of rotation, draw the pre-image of the point P(4, 3) and the image of P after a rotation of 30° clockwise. What are the coordinates of P', the image of P?

2. Triangle XYZ has vertices X(3, 3), Y(6, 3) and Z(6, 5). Plot these points on graph paper and join them to form triangle XYZ. Draw the image of XYZ after a rotation of 90° about the origin and mark it X'Y'Z'. Write down the coordinates of X', Y' and Z'.

3. Using O as the centre of rotation, draw the image of the rectangle ABCD (Fig. 34.25) after it has been given a rotation of 45° clockwise. Write down the coordinates of the transformed points A', B', C' and D'.

Fig. 34.25

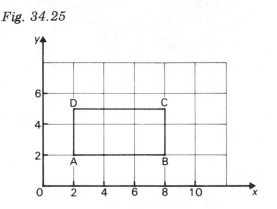

4. Draw the shape shown in Fig. 34.26. Rotate it through 90°, then 180° and finally through 270° anticlockwise.

Fig. 34.26

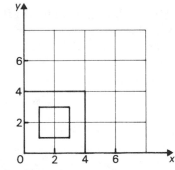

(a) Write down the order of rotational symmetry of the original shape.

(b) What is the order of rotational symmetry of the figure obtained after the rotations?

5. State the order of rotational symmetry of

(a) a rhombus

(b) a parallelogram

(c) a kite

(d) a square

Similarities

Similarities are transformations which multiply all lengths by a scale factor. If the scale factor is greater than 1 the transformation is called an **enlargement**. If the scale factor is less than 1, the transformation is called a **reduction**.

With both enlargements and reductions, angles are preserved and hence the image and the pre-image are similar figures. Thus in Fig. 34.27, the enlargement factor is $\frac{6}{2} = 3$ and ABC and A′B′C′ are similar triangles. The lengths of the sides of A′B′C′ are 3 times the lengths of the corresponding sides of ABC.

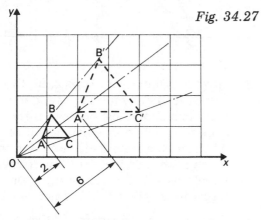

Fig. 34.27

To specify an enlargement completely we must state the scale factor and also the centre of the enlargement.

Example 7

Draw on graph paper the rectangle ABCD with vertices A(6, 4), B(16, 4), C(16, 7), D(6, 7). Draw the image of ABCD if it is enlarged by a scale factor of 2, the centre of the enlargement being the origin.

> The rectangle ABCD is shown in Fig. 34.28. To draw the image, each of the points (x, y) is mapped on to $(x′, y′)$ by the mapping $(x, y) \rightarrow (2x, 2y)$.
> Thus A(6, 4) → A′(12, 8),
> B(16, 4) → B′(32, 8), etc.
> The image of ABCD is A′B′C′D′, and the transformation enlarges ABCD by a scale factor of 2.

Fig. 34.28

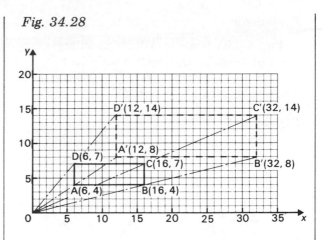

Example 8

Draw on graph paper the rectangle with vertices A(14, 13), B(22, 13), C(22, 25), D(14, 25). Draw A′B′C′D′ the image of ABCD if it is reduced by a scale factor of $\frac{1}{4}$, the centre of the reduction being P($-2, -3$).

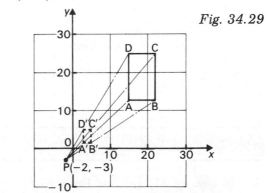

Fig. 34.29

The first step is to plot the points A, B, C, D and P (Fig. 34.29). Next, the radial lines PA, PB, PC and PD are drawn. Since the scale factor is $\frac{1}{4}$, PA′ is $\frac{1}{4}$ of PA, PB′ is $\frac{1}{4}$ of PB and so on. The points A′, B′, C′ and D′ can now be marked as shown in Fig. 34.29. The image of ABCD is A′B′C′D′ with A′(2, 1), B′(4, 1), C′(4, 4) and D′(2, 4). Note carefully that A′B′ is parallel to AB, that B′C′ is parallel to BC, that C′D′ is parallel to CD and that A′D′ is parallel to AD.

Example 9

In Fig. 34.30, triangle ABC is an enlargement of triangle ADE with a scale factor
of 4. If BC = 12 cm and AE = 3.5 cm,
find the lengths of DE and AC.

Fig. 34.30

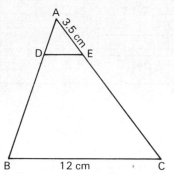

Because the scale factor is 4, the corresponding sides of triangle ABC are
4 times as long as those of triangle ADE.

$$DE = \tfrac{1}{4} \text{ of } 12 \text{ cm}$$
$$= 3 \text{ cm}$$
$$AC = 4 \times 3.5 \text{ cm}$$
$$= 14 \text{ cm}$$

Exercise 34.4

1. In Fig. 34.31, **(a)** find the scale factor
   of the enlargement, **(b)** state the coordinates of the centre of the enlargement.

Fig. 34.31

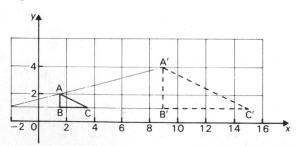

2. Copy the diagram shown in Fig. 34.32.
   Draw the lines joining A to A', B to B'
   and C to C'. Hence find

   (a) the scale of the enlargement

   (b) the coordinates of the centre of
   the enlargement

Fig. 34.32

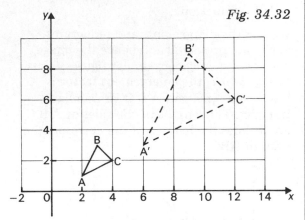

3. Plot the points A(2, 0), B(2, 2), C(4, 3)
   and D(4, 0) on graph paper and join
   them in alphabetical order to form a
   trapezium. Draw an enlargement of
   ABCD with a scale factor of 2 and the
   origin as the centre of the enlargement.
   Label this A'B'C'D'. How many times
   bigger in area is A'B'C'D' than ABCD?

4. For each of the diagrams of Fig. 34.33,
   write down the coordinates of the centre
   of the enlargement or reduction and
   state the scale factor. In each case, A
   has been mapped on to A'.

(a) *Fig. 34.33*

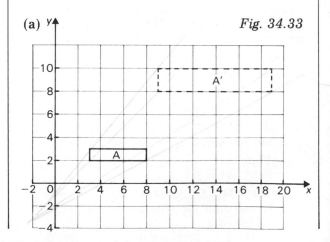

(b)

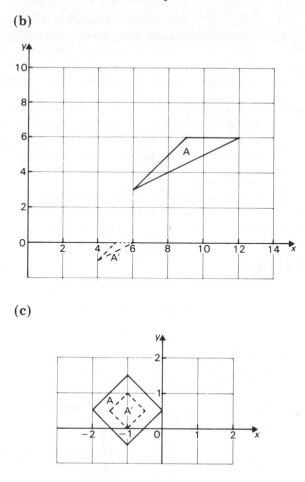

(c)

5. Draw on graph paper the quadrilateral ABCD which is formed by joining the points A(1, 1), B(1, 4), C(7, 7) and D(4, 1). ABCD is reduced by a scale factor of $\frac{1}{2}$ with P(−6, 4) as the centre of the reduction. Draw A′B′C′D′, the image of ABCD, stating the coordinates of A′, B′, C′ and D′.

6. Draw on graph paper the x and y-axes for values on both axes between −5 and 5, using a scale of 1 cm to 1 unit.

(a) Plot the point A(2, 1) and plot points B, C and D as follows:

B is the reflection of A in the x-axis;
C is the image of A after a rotation of 180° about the origin;
D is the reflection of A in the line y = x.

(b) Join the points A, B, C and D to form the quadrilateral ABCD.

(c) ABCD is enlarged by a scale factor of 2 with the origin as the centre of the enlargement. A → A′, B → B′, C → C′ and D → D′. Draw the quadrilateral A′B′C′D′ and state the coordinates of the points A′, B′, C′ and D′.

7. In Fig. 34.34, CD is mapped on to AB by a reduction. Find the magnitude of the reduction and then find the length of AC.

Fig. 34.34

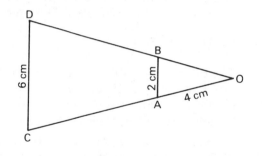

8. Rectangle ABCD (Fig. 34.35) is an enlargement of rectangle EBGF with a scale factor of 4. Find the lengths of EF and CD.

Fig. 34.35

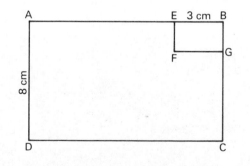

Miscellaneous Exercise 34

1. The point A(4, −2) is reflected in the line $y = x$. Find the coordinates of A′, the image of A.

2. P is the point (4, 3). Plot this point on graph paper.

   (a) The image of P when reflected in the x-axis is the point A. Mark A on your graph paper and state its coordinates.

   (b) The image of P when reflected in the y-axis is the point B. Mark B on your graph paper and state its coordinates.

3. Write down the coordinates (p, q) of the image of the point R(3, 1) under the translation $\begin{pmatrix} 3 \\ 2 \end{pmatrix}$.

4. The point P(3, −2) is given a rotation of 60° anticlockwise about the point (2, 1). Write down the coordinates of P′, the image of P, after the rotation.

5. A given translation maps P(3, 5) on to P′(7, 8). What is the image of the point Q(−2, 4) under this translation.

6. Draw on graph paper the triangle ABC formed by joining A(0, 0), B(2, 0) and C(2, 1). Draw its image under an enlargement whose centre is O (the origin) and whose scale factor is 2.

7. A trapezium is formed by joining, in alphabetical order, the points A(2, 1), B(7, 1), C(6, 3) and D(3, 3). Draw the trapezium on graph paper and indicate on it all the lines of symmetry.

8. Describe fully the single transformation which maps the shaded triangle in Fig. 34.36 on to   (a) R,   (b) T.

Fig. 34.36

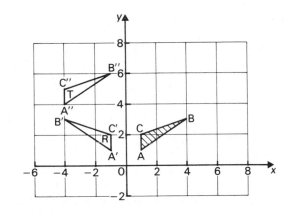

Introduction to Statistics

Statistics is the name given to the science of collecting and analysing facts. Originally, statistics used only information about the state, hence the name statistics. Nowadays, however, in almost all business reports, newspapers and government publications use is made of statistical methods.

Statistical methods range from the use of tables and diagrams to the use of statistical averages, the object being to assist the reader to understand what conclusions can be made from the quantities under discussion.

Raw Data

Raw data is collected information which is not arranged in any kind of order.

Consider the marks obtained by 50 students in a test:

```
4 3 3 5 5 6 5 8 7 6 7 8 9 5 4
1 8 7 5 6 6 7 5 2 5 2 6 9 5 7
6 5 6 2 8 6 7 3 3 8 7 6 5 5 6
4 3 4 5 7
```

This is an example of raw data and we see that the information is not organized in any order.

Frequency Distributions

Frequency distributions are used to organize raw data. A tally chart is usually used to do this.

Example 1

Using a tally chart, form a frequency distribution for the data given above.

On examining the data we see that the lowest mark is 1 whilst the highest mark is 9. The marks 1 to 9 inclusive are written in column 1 of the tally chart. We now take each figure in the raw data, just as it comes and we place a tally mark in column 2 of the tally chart opposite the appropriate mark.

The fifth tally mark for each number is usually made in an oblique direction, thereby tying the tally marks into bundles of five. This makes counting easier.

When each test mark has been entered, the tally marks are counted and the numerical value found is recorded in the column headed frequency. Hence the frequency is the number of times each mark occurs in the raw data.

From the tally chart it will be seen that the mark 1 occurs once (a frequency of 1), mark 2 occurs three times (a frequency of 3), mark 6 occurs ten times (a frequency of 10) and so on.

TALLY CHART 1

Mark	Tally	Frequency
1	I	1
2	I I I	3
3	⊥⊦⊦⊤	5
4	I I I I	4
5	⊥⊦⊦⊤ ⊥⊦⊦⊤ I I	12
6	⊥⊦⊦⊤ ⊥⊦⊦⊤	10
7	⊥⊦⊦⊤ I I I	8
8	⊥⊦⊦⊤	5
9	I I	2

Grouped Distributions

Grouped distributions are used when a large amount of numerical data has to be organised. The data is then grouped into classes or categories.

Example 2

The following is a record of the heights of 100 people in a random sample. The heights are given in centimetres.

```
146 194 136 157 117 151 164 131
187 166 158 140 145 176 126 146
175 194 185 126 178 129 156 152
113 187 157 136 152 158 157 149
180 117 129 180 169 127 168 182
193 164 169 153 152 177 198 137
185 122 147 173 165 108 142 148
163 164 169 159 155 147 153 142
146 130 153 166 165 151 148 153
154 151 155 158 137 151 141 153
150 145 152 154 154 140 151 148
150 157 151 125 154 140 184 148
130 160 131 134
```

Draw up a tally chart for the classes 100–109 cm, 110–119 cm, ... , 190–199 cm, and hence form a grouped frequency distribution.

TALLY CHART 2

Height (cm)	Tally	Frequency
100–109	I	1
110–119	I I I	3
120–129	⊞⊞ I I	7
130–139	⊞⊞ I I I I	9
140–149	⊞⊞ ⊞⊞ ⊞⊞ I I I	18
150–159	⊞⊞ ⊞⊞ ⊞⊞	
	⊞⊞ ⊞⊞ ⊞⊞ I I	32
160–169	⊞⊞ ⊞⊞ I I I	13
170–179	⊞⊞	5
180–189	⊞⊞ I I I	8
190–199	I I I I	4

The class interval for the first class in Tally Chart 2 is 100–109 cm. The end numbers 100 and 109 are called the **class limits** for the first class.

The first number 100 is called the **lower class limit** whilst the second number 109 is called the **upper class limit.** Thus for the sixth class the lower class limit is 150 and the upper class limit is 159.

In Tally Chart 2, the heights have been recorded to the nearest centimetre. The class interval 150–159 cm theoretically includes all the heights from 149.5 cm to 159.5 cm. These numbers are called the **lower** and **upper class boundaries** respectively.

The width of a class interval is the difference between lower and upper class boundaries, i.e.

Width of class interval
$$= \text{Upper class boundary}$$
$$- \text{Lower class boundary}$$

Thus for the sixth class of Tally Chart 2

$$\text{Width of class interval} = 159.5 - 149.5$$
$$= 10 \text{ cm}$$

Example 3

The figures in the table below show part of a grouped frequency distribution. State the upper and lower class boundaries for the second class and hence find the width of the class interval.

Mass (kg)	Frequency
350–379	27
380–409	34
410–439	43

For the second class the upper and lower class boundaries are 409.5 kg and 379.5 kg respectively.

$$\text{Width of class interval} = 409.5 \text{ kg} - 379.5 \text{ kg}$$
$$= 30.0 \text{ kg}$$

Variables

A **continuous variable** is a variable which can take any value between two given values. For instance the height of an individual, which can be 159 cm, 165.2 cm or 176.831 cm, depending upon the accuracy of the measurement, is a continuous variable.

A **discrete variable** is a variable which can have only certain values. An example is the number of children in a family, which can take the value 0, 1, 2, 3, ... but not $2\frac{1}{2}$, $3\frac{1}{4}$, ... A second example is the number of pound coins in a bank, which can only be a whole number. Note that a discrete variable need not necessarily be a whole number. For instance the size of shoes can be 5, $5\frac{1}{2}$, 6, $6\frac{1}{2}$, 7, ... but not $5\frac{1}{3}$, $6\frac{1}{4}$, ... Hence the size of a shoe is a discrete variable.

Histograms

A **histogram** is a diagram which is used to represent a frequency distribution. It consist of a set of rectangles whose **areas** represent the frequencies of the various classes. If all the classes have the same width (which is usually the case) then the rectangles will all have the same width. The frequencies are then represented by the heights of the rectangles. Fig. 35.1 shows the histogram for the frequency distribution in Example 1.

Fig. 35.1

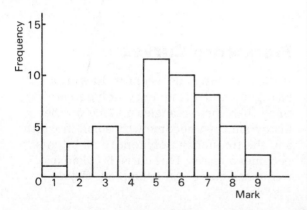

A histogram for a grouped frequency distribution may be drawn by using the mid-points of the class intervals as the centre of the rectangles. The histogram for the frequency distribution in Example 2 is shown in Fig. 35.2. Note that the extremes of each rectangle represent the lower and upper class boundaries.

Fig. 35.2

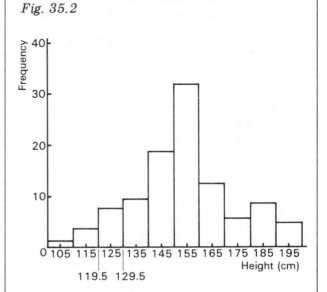

Example 4

Six coins were tossed 80 times and after each toss the number of tails was recorded as shown in the table below. Represent this information in the form of a suitable histogram.

Number of tails	Number of tosses (frequency)
0	3
1	9
2	15
3	25
4	17
5	9
6	2
Total	80

Fig. 35.3 shows a histogram for a discrete distribution. It has been drawn to represent the information given in Example 4 and takes the form of a vertical bar chart in which the bars have zero width.

Discrete data, however, are often represented by a histogram like that shown in Fig. 35.4, despite the fact that we are treating the data as though they were continuous.

Fig. 35.3

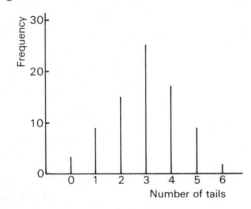

Fig. 35.4

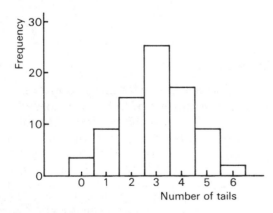

Frequency Polygons

A **frequency polygon** provides a second way of representing a frequency distribution. It is drawn by joining, with straight lines, the mid-points of the rectangles making up the histogram of the distribution.

Example 5

Draw a frequency polygon for the information given in the following table:

Age of employee	15-19	20-24	25-29	30-34	35-39
Frequency	5	23	58	104	141

Age of employee	40-44	45-49	50-54	55-59
Frequency	98	43	19	6

The frequency polygon is drawn in Fig. 35.5. It is customary to add the extensions PQ and RS to the next lower and next higher mid-points as shown in the diagram. If this is done then the area of the polygon is equal to the area of the histogram.

Fig. 35.5

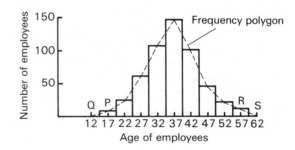

Frequency Curves

If the class intervals used for drawing a histogram and a frequency polygon are very small then the rectangles making up the histogram then become very small in width and the frequency polygon, to all intents, becomes a curve. This curve is called a **frequency curve**. An example is shown in Fig. 35.6.

Fig. 35.6

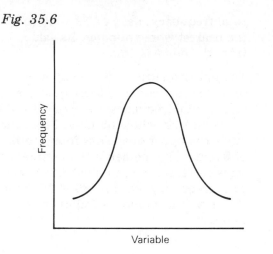

Cumulative Frequency Distribution

A cumulative frequency curve is an alternative way of representing a frequency distribution. The way in which a cumulative frequency is obtained is shown in Example 6.

Example 6

The diameters of 200 ball bearings were measured with the results shown in the table below.

Diameter (mm)	5.94–5.96	5.97–5.599	6.00–6.02
Number	8	37	90

Diameter (mm)	6.03–6.05	6.06–6.08
Number	52	13

Draw up a cumulative frequency distribution and from it draw a cumulative frequency curve. Then from it find the number of ball bearings with a diameter:

(a) less than 5.98 mm

(b) between 6.00 mm and 6.05 mm

(c) greater than 6.07 mm.

The cumulative frequencies are given in the table below:

Diameter (mm)	Cumulative frequency
Less than 5.965	8
Less than 5.995	8 + 37 = 45
Less than 6.025	45 + 90 = 135
Less than 6.055	135 + 52 = 187
Less than 6.085	187 + 13 = 200

The cumulative frequency curve (sometimes called an ogive) is shown in Fig. 35.7.

Fig. 35.7

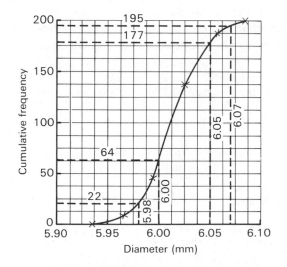

(a) The cumulative frequency corresponding to a diameter of 5.98 mm is 22. So 22 ball bearings have a diameter less than 5.98 mm.

(b) The cumulative frequencies corresponding to 6.00 mm and 6.05 mm are 64 and 177 respectively. Therefore the number of ball bearings with a diameter between 6.00 mm and 6.05 mm is 177 − 64 = 113.

(c) The cumulative frequency corresponding to a diameter of 6.07 mm is 195. So the number of ball bearings with a diameter greater than 6.07 mm is 200 − 195 = 5.

Quartiles

The median (page 357) divides a set of values into two equal parts. Using a similar method we can divide the set of values into four equal parts. The values which so divide the set are called the **quartiles**. They are often denoted by the symbols Q_1, Q_2, Q_3 and Q_4, Q_1 being the first or lower quartile and Q_3 being the third or upper quartile. Q_2 is the second quartile and is the median of the distribution.

Example 7

From the information given in the table below find the upper and lower quartiles and the median of the distribution which relates to the number of seats in various theatres.

Number of seats	Up to 250	251–500	501–750
Number of theatres	8	14	25

Number of seats	751–1000	1001–1250	1251–1500
Number of theatres	16	11	5

Number of seats	More than 1500
Number of theatres	1

The first step is to construct a cumulative frequency table as follows:

Number of seats	Cumulative frequency
Less than 250	8
Less than 500	8 + 14 = 22
Less than 750	22 + 25 = 47
Less than 1000	47 + 16 = 63
Less than 1250	63 + 11 = 74
Less than 1500	74 + 5 = 79

The cumulative frequency curve can now be drawn (Fig. 35.8). The lower quartile is the value of the variable corresponding to one-quarter of the total frequency, i.e. $\frac{1}{4}$ of 79 = 20 to the nearest whole number. Its value, from the curve is 460.

The upper quartile is the number of seats corresponding to three-quarters of the total frequency, i.e. $\frac{3}{4}$ of 79 = 60 to the nearest whole number. From the curve the upper quartile is found to be 950 seats. The median is the number of seats corresponding to one-half of the total frequency, i.e. $\frac{1}{2}$ of 80 = 40. From the curve the median is found to be 530 seats.

Fig. 35.8

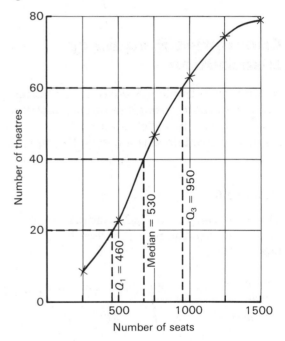

Measures of Dispersion

A statistical average gives some idea about the position of a distribution (Fig. 35.9) and hence statistical averages are often called measures of location. We now need a measure which will define the spread or dispersion of the data. There are several of these but we will confine ourselves to discussing the range, the interquartile range and the semi-interquartile range.

Fig. 35.9

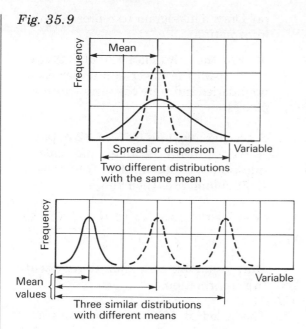

Two different distributions
with the same mean

Three similar distributions
with different means

The **range** is the difference between the largest and smallest observations in a distribution. That is,

$$\text{Range} = \text{Largest observation} - \text{Smallest observation}$$

Example 8

(a) The salaries paid in a small office are £285, £310, £420, £342 and £410 per week. Find the range of these salaries.

$$\text{Lowest salary} = £285$$

$$\text{Largest salary} = £420$$

$$\text{Range of salaries} = £420 - £285$$
$$= £135$$

(b) The table below gives the maximum loads to be liften by a crane. Find the range of these loads.

Maximum load (tonnes)	8.3–8.5	8.6–8.8	8.9–9.1	9.2–9.4
Frequency	2	8	6	1

The lower boundary of the first class is 8.25 tonnes and the upper boundary of the last class is 9.45 tonnes. So

$$\text{Range of loads} = (9.45 - 8.25) \text{ tonnes}$$
$$= 1.2 \text{ tonnes}$$

The range gives some idea of the spread of the data but it depends upon the extreme values of the data. It gives no information about the way in which the data is dispersed and hence it is seldom used as a measure of dispersion for a frequency distribution. However, when dealing with small samples as in Example 8 it is a very effective measure of dispersion.

The **quartile deviation** is the difference between the upper quartile and the lower quartile, that is

$$\text{Quartile deviation} = Q_3 - Q_1$$

The semi-interquartile range is one-half of the interquartile range, that is

$$\text{Semi-interquartile range} = \tfrac{1}{2}(Q_3 - Q_1)$$

Example 9

An examination of the wages paid by a certain firm showed that the upper quartile was £410 per week whilst the lower quartile was £340 per week. Find the interquartile range and the semi-interquartile range.

We are given that $Q_3 = £410$ and $Q_1 = £340$

$$\text{Interquartile range} = £410 - £340$$
$$= £70 \text{ per week}$$

$$\text{Semi-interquartile range} = \tfrac{1}{2}(£410 - £340)$$
$$= \tfrac{1}{2} \times £70$$
$$= £35 \text{ per week}$$

The quartile ranges have the advantage that extreme values in distribution are ignored. However, these measures of dispersion only cover that half of the distribution which is centred on the median. Therefore they do not show the dispersion of the distribution as a whole. Small values of either of these measures show that the data has only a small amount of spread between the quartiles. Both these measures of dispersion are widely used in commercial and educational statistics.

Note that for a frequency distribution a cumulative frequency curve is needed to find the upper and lower quartiles and hence the interquartile range and the semi-interquartile range.

Exercise 35.1

1. The following marks were obtained by 50 students during a test:

Mark	1	2	3	4	5	6	7	8
Frequency	2	4	8	14	9	6	5	2

Draw a suitable histogram to represent this information.

2. The marks obtained in a recent mathematics examination were as follows:

15 37 42 19 27 42 37 27   7 33
30 12   6 31 24 25 50 29 38 35

Use the above information to obtain a grouped frequency distribution with classes 1-10, 11-20, 21-30, 31-40 and 41-50. Draw a histogram to represent this frequency distribution.

3. The distances in metres achieved by 80 competitors in a discus-throwing competition were grouped in class intervals of 10 cm. The results were as follows:

Class interval (m)	Frequency
20.16-20.25	1
20.26-20.35	4
20.36-20.45	8
20.46-20.55	12
20.56-20.65	14
20.66-20.75	15
20.76-20.85	11
20.86-20.95	7
20.96-21.05	5
21.06-21.15	3

(a) Draw a histogram to represent this information.

(b) For the third class (20.36-20.45) write down the lower and upper class boundaries and state the class width in metres.

4. A teacher asked his class of 32 pupils to write down how long a particular question took them. The following table summarises their times:

Time (min)	10	11	12	13	14	15	16	17	18	19	20
Frequency	1	1	2	7	4	5	3	3	2	2	2

Draw a suitable histogram to represent this information.

5. The speed of 120 vehicles passing a police checkpoint was measured to the nearest kilometre per hour. The results are summarised below:

Speed (km/h)	Number of vehicles
30-34	1
35-39	4
40-44	9
45-49	17
50-54	23
55-59	31
60-64	18
65-69	9
70-74	6
75-79	2

(a) For the third class (40-44) write down the lower and upper class boundaries. Hence calculate the class width.

(b) Draw a histogram to represent this information.

(c) State whether the speed is a continuous or a discrete variable.

6. Read the following passage and record the number of letters in each word to form a grouped frequency distribution with classes as follows: 1 or 2 letters, 3 or 4 letters, 5 or 6 letters, 7 or 8 letters, 9 or 10 letters, 11 or 12 letters. Hence

draw a histogram to represent the frequency distribution.

'The morning was cool and cloudless. Barbara rose early and with a growing feeling of excitement hurried through her breakfast. The forecast had been good and as she slipped out of the house she gave no thought to the possibility of a change in the weather. She arrived at the quay to find an old boatman waiting for her. The sea was calm, there was no hint of the storm to come.'

7. Classify each of the following as a continuous or a discrete variable:

   (a) the diameter of motor car pistons

   (b) the number of dresses sold by a shop in one day

   (c) the weight of fertiliser in a sack

   (d) the number of five-pound notes in a wallet

   (e) the daily temperature at noon

   (f) the lifetime of electric light bulbs

   (g) the temperature of a furnace

   (h) the number of telephone calls received by the police in one day

8. Fig. 35.10 shows the number of children in a sample classified by weight, in groups of 2 kg.

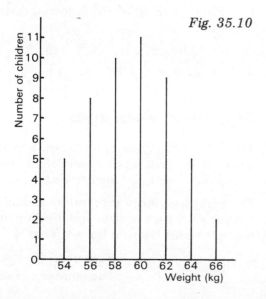

Fig. 35.10

(a) Find the number of children in the survey.

(b) Is the weight a continuous or a discrete variable?

(c) Represent the information in a conventional histogram.

9. The figures below are measurements of the noise level, in dBA units, at 36 discos.

93	90	98	88	103	92	89	82
86	87	91	89	85	95	86	86
94	85	103	92	88	102	99	85
98	100	95	98	105	86	100	92
96	91	87	88				

(a) Display this information in the form of a grouped frequency table using intervals of 81–85, 86–90, etc.

(b) Construct a histogram and hence draw a frequency polygon.

(c) Make a table showing the cumulative frequencies.

(d) Draw a cumulative frequency curve and from it estimate the number of discos at which the sound level exceeds 94 dBA.

10. A school decided that all 900 of its pupils should take an intelligence test. The following table of results was obtained:

Range of mark	Frequency
0–9	5
10–19	33
20–29	82
30–39	148
40–49	270
50–59	215
60–69	87
70–79	44
80–89	12
90–99	4

(a) Construct a frequency polygon.

(b) Draw up a cumulative frequency table.

(c) Draw a cumulative frequency curve on graph paper. Use the longer side of the paper for the horizontal axis. Recommended scales are 2 cm to 10 marks on the horizontal axis and 2 cm to 100 pupils vertically.

(d) Estimate how many pupils scored
(i) less than 33 marks
(ii) more than 65 marks.

11. An examination of the wages paid by a certain company showed that the upper quartile was £336 whilst the lower quartile was £272 per week. What was the semi-interquartile range?

12. The following table shows the weights (in kilograms) of 250 boys, each weight being recorded to the nearest 100 grams.

Weight (kg)	44.0–47.9	48.0–51.9	52.0–55.9
Frequency	3	17	50

Weight (kg)	56.0–57.9	58.0–59.9	60.0–63.9
Frequency	45	46	57

Weight (kg)	64.0–67.9	68.0–71.9
Frequency	23	9

(a) Draw up a cumulative frequency table.

(b) Construct a cumulative frequency curve and from it estimate the semi-interquartile range.

13. 32 children in a class were asked to estimate the length of a metal rod to the nearest centimetre. The table below shows the results obtained.

Estimated length (cm)	Frequency
35	1
36	3
37	4
38	8
39	6
40	5
41	3
42	2

(a) Draw up a cumulative frequency table and plot a cumulative frequency curve. Take 2 cm to represent 1 cm on the horizontal axis and 2 cm to represent 5 units on the vertical axis.

(b) From your curve find the median and the semi-interquartile range.

14. 100 students took a test in mathematics. The table below shows the distribution of the scores obtained by the students.

Score	36–40	41–45	46–50	51–55	56–60
Frequency	1	2	5	10	16
Score	61–65	66–70	71–75	76–80	81–85
Frequency	28	19	11	5	2
Score	86–90				
Frequency	1				

Construct an ogive and from it obtain the median score and the interquartile range.

15. The salaries of five office workers are £297.50, £278.00, £432.00, £324.25 and £373.75. Determine the range of these salaries.

16. The largest of 50 measurements is 29.88 cm. If the range of the measurements is 0.12, what is the smallest measurement?

17. Find the range of the following distribution:

Length (mm)	167	168	169	170	171
Frequency	2	8	15	6	3

Statistical Averages

There are three types of statistical average. They are the arithmetic mean (often just called the mean), the median and the mode.

The **arithmetic mean** is found by adding all the values in a set of data and dividing by the number of items in the set. That is

$$\text{Arithmetic mean} = \frac{\text{Sum of all the values}}{\text{The number of values}}$$

Example 10

Find the arithmetic mean of 3, 5, 8, 9 and 10.

$$\text{Arithmetic mean} = \frac{3 + 5 + 8 + 9 + 10}{5}$$

$$= \frac{35}{5}$$

$$= 7$$

Example 11

The values 3, 4, 6, 8, 9 and x have a mean of 7. What is the value of x?

$$\frac{3 + 4 + 6 + 8 + 9 + x}{6} = 7$$

$$\frac{30 + x}{6} = 7$$

$$30 + x = 42$$

$$x = 12$$

When a set of values is arranged in ascending (or descending) order, the **median** is the value which lies half-way along the sequence.

Example 12

Find the median of the numbers 5, 4, 2, 8, 7, 2, 9, 7 and 3.

Arranging the numbers in ascending order we have

$$2, 2, 3, 4, 5, 7, 7, 8, 9$$

The median is 5 because there are four numbers below this value and four numbers above it.

When there is an even number of values in a set, the median is found by taking the mean of the two middle values.

Example 13

Find the median of 8, 5, 4, 7, 9, 10, 2 and 5.

Arranging the numbers in descending order we have

$$10, 9, 8, 7, 5, 5, 4, 2$$

The middle two values are 7 and 5 and therefore

$$\text{Median} = \frac{7 + 5}{2}$$

$$= 6$$

The **mode** of a set of values is that value which occurs most frequently.

Example 14

Find the mode of 11, 12, 12, 12, 13, 13, 13, 13, 13, 14, 14, 14, 15 and 15.

Looking at the set of numbers we see that 13 is the number which occurs most frequently (i.e. most times). Hence the mode is 13.

Example 15

The marks obtained in a test taken by a class of 30 children were as follows:

Mark	1	2	3	4	5
Frequency	3	7	12	6	2

Find **(a)** the mean mark, **(b)** the median mark, **(c)** the modal mark.

(a) Mean mark

$$= \frac{(1 \times 3) + (2 \times 7) + (3 \times 12) + (4 \times 6) + (5 \times 2)}{3 + 7 + 12 + 6 + 2}$$

$$= \frac{3 + 14 + 36 + 24 + 10}{30}$$

$$= \frac{87}{30}$$

$$= 2.9$$

The mean of a distribution may be calculated by using the memory keys on a calculator.

Input	Display
1	1.
×	1.
3	3.
M+	3.
2	2.
×	2.
7	7.
M+	14.
3	3.
×	3.
12	12.
M+	36.
4	4.
×	4.
6	6.
M+	24.
5	5.
×	5.
2	2.
M+	10.
MR	87.
÷	87.
30	30.
=	2.9

So, as before, mean mark = 2.9

(b) Since there are 30 children the median mark must lie between the 15th and 16th items in the distribution. We now look at the 15th and 16th items and we find that these are both 3. Hence the median mark is 3.

(c) The mode is the most frequently occurring mark. Looking at the table we see that 3 marks occurs twelve times. Hence the modal mark is 3.

Comparison of Statistical Averages

The mean: Advantages

(1) The mean is the best known average.

(2) It can be calculated exactly.

(3) It makes use of all the data.

(4) It can be used in further statistical work.

Disadvantages

(1) It is greatly affected by extreme values.

(2) When the data are discrete it can give an impossible value (e.g. 15.362 runs per innings).

(3) It cannot be obtained graphically.

The median: Advantages

(1) The median is not affected by extreme values.

(2) It can be obtained even if some of the values in the distribution are unknown.

(3) If the number of items in the distribution is odd it can represent an actual value in the distribution.

Disadvantages

(1) When only a few values are available the median may not be characteristic of the group.

(2) It cannot be used for further statistical calculations.

The mode: Advantages

(1) The mode is not affected by extreme values.

(2) It is easy to obtain from a histogram or a frequency distribution.

(3) To determine its value only values near to the modal value are needed.

Disadvantages

(1) There may be no mode or there may be more than one mode.

(2) It cannot be used in further statistical calculations.

Which Average To Use

The **mean** is the most familiar kind of average and it is extensively used in business, engineering, etc. It is easy to understand but in some cases it may be definitely misleading. For instance if the hourly rates of pay for five office workers are £4.05, £4.32, £4.81, £9.40 and £3.52, the mean rate of pay is

£5.22, However, this value is greatly affected by the extreme value of £9.40. Therefore the mean gives a false impression of the wages paid in the office; it gives the impression that the wages are higher than they really are.

The **median** is not affected by extreme values and it will give a better indication of the wages paid in the office discussed above. The median wage is £4.32 which gives a better idea of the wages actually paid.

The **mode** is used when the commonest value in a distribution is required. For instance a clothing manufacturer will not be particularly interested in the mean waist size of women because almost certainly this will not be a stock size for his skirts. What he needs is a frequency distribution for his stock sizes, size 12, 14, 16, etc. From this he can obtain the modal size and organise his production accordingly. However, the average used will depend upon the particular set of circumstances.

Exercise 35.2

1. The heights of five men are 177.8 cm, 175.3 cm, 174.8 cm, 179.1 cm and 176.5 cm. What is the mean height of these five men?

2. Calculate the mean of the numbers 5, 8, 9 and 10.

3. The numbers 3, 5, 8, 10 and x have a mean of 8. Determine the value of x.

4. Find the median of the numbers 5, 3, 8, 5, 4, 2 and 8.

5. Find the median of the numbers 2, 4, 6, 5, 9, 6, 4 and 7.

6. Find the mode of 3, 5, 4, 8, 3, 6, 5, 9, 5, 4 and 7.

7. Thirteen people were asked to guess the weight of a cake to the nearest half-kilogram. Their estimates were as follows: $3\frac{1}{2}$, $2\frac{1}{2}$, 2, 1, $3\frac{1}{2}$, 2, $3\frac{1}{2}$, 3, 3, 1, $1\frac{1}{2}$, $2\frac{1}{2}$, $3\frac{1}{2}$ kg. What was
    (a) the estimated mean weight
    (b) the estimated modal weight
    (c) the estimated median weight

8. What additional number must be included in the following list to make the median of the six numbers equal to $3\frac{1}{2}$?

    3, 5, 2, 2, 8

9. Six people occupy a lift. Their masses to the nearest kilogram are 84, 67, 73, 76, 80 and 82 kg. Find
    (a) their median mass
    (b) their mean mass

10. During a census a check was made to find out how many people were living in each house in a certain street. The results are given in the table:

Number of people per house	1	2	3	4	5	6	7
Number of houses	3	9	12	9	4	2	1

    (a) How many houses were checked?
    (b) State the modal number of people per house.
    (c) Calculate the mean number of people per house.
    (d) Determine the median number of people per house.

11. The marks of 100 students were as follows:

Mark	1	2	3	4	5	6	7	8	9
Frequency	2	8	20	32	18	9	6	3	2

    (a) Calculate the mean mark.
    (b) What is the modal mark?
    (c) Find the median mark.

12. The following table shows the number of runs scored by a sample of cricketers in a limited-over competition.

Number of runs scored	18	19	20	21	22	23	24	25	26	27
Frequency	5	10	12	28	36	41	30	25	10	3

(a) How many cricketers are in the sample?

(b) What is the modal score?

(c) What is the median score?

(d) What is the mean score correct to one decimal place?

Equiprobable Events

When an unbiased die is rolled, each of the six faces has an equal chance of landing uppermost. The throwing of 1, 2, 3, 4, 5 or 6 are called **equiprobable events**.

Probability

Suppose that we want to deal an ace from a pack of 52 playing cards. Each time we deal an ace we say that there has been a **favourable outcome** or a success. Since there are four aces in a pack, the number of equiprobable events which produces a favourable outcome (or the number of successes) is 4. The probability of obtaining a favourable outcome is calculated from:

$$\text{Probability} = \frac{\substack{\text{Number of equiprobable}\\ \text{events which produces a}\\ \text{favourable outcome}}}{\substack{\text{Total number of}\\ \text{equiprobable events}}}$$

$$= \frac{\text{Number of successes}}{\text{Total number of outcomes}}$$

All probabilities have a value between 0 and 1. A probability of 0 occurs when an event can never happen. A probability of 1 occurs when an event is certain to happen (e.g. picking the winner of a one-horse race).

The **total probability** covering all possible events is 1. That is

Probability of success
+ Probability of failure = 1

Example 16

One letter is chosen from the word 'horizon'. What is the probability that it will be

(a) the letter r,   (b) the letter o,

(c) a vowel,   (d) a consonant?

(a) Since there are seven letters in the given word, there is a total of 7 equiprobable events. Only one of these is favourable because there is only one r in the word. Hence

$$\text{Probability of choosing the letter r} = \frac{1}{7}$$

(b) In this case there are two favourable outcomes because the letter o occurs twice in the given word. Hence

$$\text{Probability of choosing the letter o} = \frac{2}{7}$$

(c) The vowels are the letters a, e, i, o and u. There are three vowels in the given word, hence the number of favourable outcomes is 3.

$$\text{Probability of choosing a vowel} = \frac{3}{7}$$

(d) If a letter is not a vowel it is a consonant. Hence

Probability of choosing a vowel
+ Probability of not choosing a vowel = 1

Probability of not choosing a vowel (i.e. of choosing a consonant) $= 1 - \dfrac{3}{7} = \dfrac{4}{7}$

Alternatively we see that the letters h, r, z and n are consonants and therefore the number of favourable outcomes is 4. Hence, as before

$$\text{Probability of choosing a consonant} = \frac{4}{7}$$

Exercise 35.3

1. A die is rolled. Calculate the probability that the result will be
   (a) a 3
   (b) a score less than 4
   (c) an even number

2. A letter is chosen from the word 'terrific'. Determine the probability that it will be
   (a) an f   (b) an r
   (c) a vowel   (d) a consonant

3. A bag contains 3 red balls, 5 blue balls and 2 green balls. A ball is chosen at random from the bag. Find the probability that it will be
   (a) green   (b) blue
   (c) red   (d) not red

4. There are seven tomatoes in a bag. Four of them are red and the remainder are green. A tomato is picked at random from the bag. What is the probability that it is red?

5. 20 discs numbered 1 to 20 are placed in a box and one disc is chosen at random. Determine the probability that the number on the disc will be
   (a) even   (b) a multiple of 3
   (c) more than 8 (d) less than 8

6. A card is drawn from a deck of 52 playing cards. Find the probability that it will be
   (a) the king of hearts
   (b) a queen
   (c) an ace, king, queen or jack
   (d) the jack of spades or the ace of diamonds

7. Two dice are thrown together and their scores added. Determine the probability that the sum will be
   (a) 7
   (b) less than 5
   (c) more than 9

8. From a shuffled pack of 52 playing cards two cards are dealt. What is the probability of
   (a) the first card being an ace
   (b) the second card also being an ace

9. The number of players in the first-team squad of a local football team is 23. Only eleven players can play in a game. What is the probability that John will *not* be chosen to play?

10. It states on the box of a certain brand of matches that the average contents is 36 matches. The contents of 9 different boxes were counted with the following results: 32, 37, 29, 35, 37, 32, 36, 40, 36. What is the probability of selecting a box from this sample which contains more than 36 matches?

Relative Frequency and Probability

Although it is possible to calculate many probabilities theoretically as previously shown, in a great many cases we have to rely on probabilities obtained by experiment. The method of relative frequencies is usually used in such cases to find the probabilities.

$$\text{Relative frequency} = \frac{\text{Class frequency}}{\text{Total frequency}}$$

Example 17

During a certain period of the year, the number of blooms on each of a number of rose bushes was counted with the following results:

Number of blooms	5	6	7	8	9	10	11	12
Frequency			15	20	25	27	22	17	14	10

Draw a table showing the relative frequencies and from it find the probability of finding a rose bush with

(a) less than 7 blooms

(b) exactly 9 blooms

(c) 10 or more blooms.

Total frequency = 15 + 20 + 25 + 27
                        + 22 + 17 + 14 + 10

                  = 150

Number of blooms	5	6	7	8
Relative frequency	0.10	0.13	0.17	0.18

Number of blooms	9	10	11	12
Relative frequency	0.15	0.11	0.09	0.07

(a) Probability of finding a rose bush with less than 7 blooms
    = 0.10 + 0.13
    = 0.23

(b) Probability of finding a rose bush with exactly 9 blooms
    = 0.15

(c) Probability of finding a rose bush with 10 or more blooms
    = 0.11 + 0.09 + 0.07
    = 0.27

Note that the total of the relative frequencies is equal to 1 because this is the probability of covering all possible events.

Exercise 35.4

In this exercise calculate the relative frequencies and hence find the required probabilities.

1. Pupils in a school made a survey of the colours of front doors on a housing estate. The table below summarises the information they obtained:

Colour of door	Red	Yellow	Green
Frequency	26	9	4

Colour of door	Blue	White	Other
Frequency	15	35	11

Determine the probability of

(a) finding a house with a white door

(b) finding a house with a door that is not white

2. During a survey to find out how many people were living in each house in a certain street the following results were obtained:

Number of people per house	1	2	3	4	5	6	7
Number of houses	2	8	14	10	4	2	0

Calculate the probability that the number of people in a house selected at random will

(a) exceed 4

(b) be less than 3

(c) be exactly 6

3. The table below shows the scores obtained at ten-pin bowling:

Score	0	1	2	3	4	5	6	7	8	9	10
Frequency	1	10	12	22	25	22	16	14	12	10	6

Determine the probability that a player, chosen at random, will score

(a) more than 4

(b) less than 7

(c) exactly 9

4. The number of goals scored on a Saturday afternoon during the football season were as follows:

Number of goals	0	1	2	3	4	5
Number of teams	5	12	14	8	7	4

    Calculate the probability that one team, chosen at random, scored

    (a) 2 goals

    (b) more than 2 goals

    (c) less than 3 goals on this particular Saturday

5. The pocket money, in pence, received by 25 young children during one week of term is given below:

60	120	40	35	100	100	45
50	70	125	55	150	50	75
80	65	80	100	30	50	75
50	40	110	75			

    Calculate the probability that one child, chosen at random, received

    (a) exactly 50 pence

    (b) more than 60 pence

    (c) less than 40 pence

The Addition Law of Probability

If two events could not happen at the same time the events are said to be **mutually exclusive.** For instance in a single roll of a fair die a 3 or 4 can occur but not both. Hence the events of scoring a 3 and a 4 in a single roll of a die are mutually exclusive.

Similarly it is impossible to cut a jack and a king in a single cut of a pack of cards. Hence these two events are mutually exclusive.

If $p_1, p_2, p_3, \ldots$ are the separate probabilities of a set of mutually exclusive events, then the probability of *one* of these events occurring is

$$P = p_1 + p_2 + p_3 + \ldots$$

Example 18

A die with faces numbered 1 to 6 is rolled once. What is the probability of scoring either a 3 or a 4?

$$\text{Probability of scoring a } 3 = p_1$$
$$= \frac{1}{6}$$

$$\text{Probability of scoring a } 4 = p_2$$
$$= \frac{1}{6}$$

$$\text{Probability of scoring a 3 or a 4} = p_1 + p_2$$
$$= \frac{1}{6} + \frac{1}{6}$$
$$= \frac{1}{3}$$

Sometimes the addition law is called the OR law because we seek the probability that one event OR another event will happen.

The Multiplication Law of Probability

An **independent** event is one which has no effect on subsequent events. For instance if a die is rolled twice, what happens on the first roll does not affect what happens on the second roll. Hence the two rolls of the die are independent events.

The events of cutting an ace from a pack of cards and tossing a head in a single toss of a fair coin are also independent events because what happens with the cut has no effect on the way in which the coin lands. If $p_1, p_2, p_3, \ldots$ are the separate probabilities of a set of independent events then the probability that *all* events will occur is

$$P = p_1 \times p_2 \times p_3 \times \ldots$$

Example 19

(a) If a coin is tossed and a card is dealt from a pack of 52 playing cards, calculate the probability that the result will be a head and an ace.

$$\text{Probability of dealing an ace} = p_1$$

$$= \frac{4}{52}$$

$$= \frac{1}{13}$$

$$\text{Probability of tossing a head} = p_2$$

$$= \frac{1}{2}$$

Probability of dealing an ace and tossing a head

$$= p_1 \times p_2$$

$$= \frac{1}{13} \times \frac{1}{2}$$

$$= \frac{1}{26}$$

(b) A coin is tossed four times. What is the probability that each toss will result in a head?

The four tosses of the coin are independent events. Therefore

Probability of tossing four heads

$$= \frac{1}{2} \times \frac{1}{2} \times \frac{1}{2} \times \frac{1}{2}$$

$$= \frac{1}{16}$$

The multiplication law is often called the AND law because we seek the probability that one event AND another event will occur.

The Probability Tree

Example 20

A coin is tossed three times. Find the probability that (a) a head will occur on all three tosses, (b) only one head will appear on three tosses.

We can work out these probabilities by using a probability tree (Fig. 35.11). Possible heads are shown in a full line and possible tails in a dotted line.

Fig. 35.11

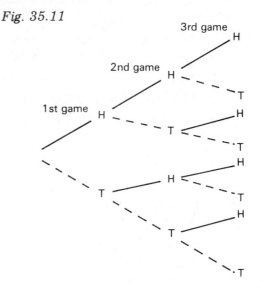

(a) The branch of the tree needed to find the probability of three heads occurring is shown in Fig. 35.12.

$$\text{Probability of three heads} = \frac{1}{2} \times \frac{1}{2} \times \frac{1}{2}$$

$$= \frac{1}{8}$$

Fig. 35.12

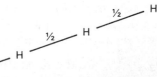

(b) The branches of the tree required to find the probability of only one head appearing are shown in Fig. 35.13.

Fig. 35.13

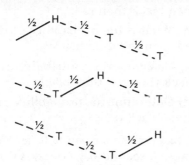

The diagram shows that there are three ways in which one head can occur: head, tail, tail; tail, head, tail; tail, tail, head.

Probability HTT $= \dfrac{1}{2} \times \dfrac{1}{2} \times \dfrac{1}{2} = \dfrac{1}{8}$

Probability THT $= \dfrac{1}{2} \times \dfrac{1}{2} \times \dfrac{1}{2} = \dfrac{1}{8}$

Probability TTH $= \dfrac{1}{2} \times \dfrac{1}{2} \times \dfrac{1}{2} = \dfrac{1}{8}$

Since the events HTT, THT, TTH are mutually exclusive

Probability of one head $= \dfrac{1}{8} + \dfrac{1}{8} + \dfrac{1}{8}$

$= \dfrac{3}{8}$

Example 21

A box contains 4 black and 6 red balls. A ball is drawn at random from the box and is not replaced. A second ball is then drawn. Find the probabilities of **(a)** red then red being drawn, **(b)** black then red being drawn, **(c)** red then black being drawn, **(d)** black then black being drawn.

The probability tree is shown in Fig. 35.14.

Fig. 35.14

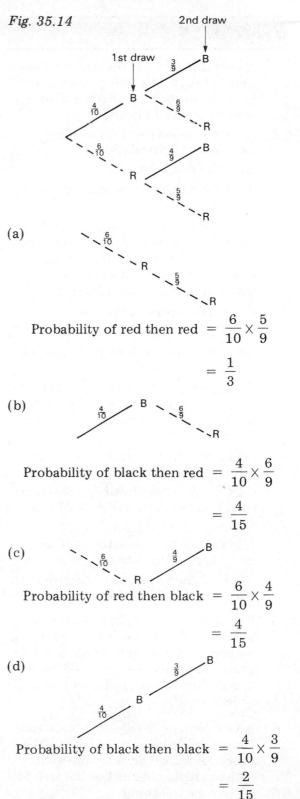

(a) Probability of red then red $= \dfrac{6}{10} \times \dfrac{5}{9}$

$= \dfrac{1}{3}$

(b) Probability of black then red $= \dfrac{4}{10} \times \dfrac{6}{9}$

$= \dfrac{4}{15}$

(c) Probability of red then black $= \dfrac{6}{10} \times \dfrac{4}{9}$

$= \dfrac{4}{15}$

(d) Probability of black then black $= \dfrac{4}{10} \times \dfrac{3}{9}$

$= \dfrac{2}{15}$

Exercise 35.5

1. A card is dealt from the top of a well-shuffled pack of playing cards. Determine the probability that it will be any ace or the king of hearts.

2. A coin is tossed and a die is rolled. Calculate the probability of

   (a) a tail and a 5 resulting

   (b) a head and an even number occurring

3. A box contains 8 red counters and 12 white ones. A counter is drawn at random from the box and then replaced. A second counter is then drawn. Determine the probabilities that

   (a) both counters will be red

   (b) both counters will be white

   (c) one counter will be white and the other red

4. A bag contains 3 red balls and 2 green ones. A ball is drawn from the bag but not replaced. A second ball is then drawn.

   (a) Draw a tree diagram showing the various probabilities.

   (b) Calculate the probability that a red ball followed by a red ball will be drawn.

   (c) What is the probability of two green balls being drawn?

   (d) Find the probability of a green ball being drawn followed by a red ball.

5. Six cards marked A, B, C, D, E and F are shuffled. Two cards are then drawn without replacement. What is the probability that the cards will be B and D although not necessarily in that order?

6. A box contains 3 red and 4 black balls. Draw a probability tree to show the probabilities of drawing three balls without replacement. Use the tree to answer the following

   (a) Find the probability of drawing red then black then red.

   (b) What is the probability of drawing red then red then black?

   (c) Calculate the probability of drawing black then red then black.

   (d) Determine the probability of drawing three black balls.

7. A bag contains 5 red balls and 7 blue balls. A second bag contains 6 red balls and 9 blue balls. If a ball is drawn at random from each bag, calculate the probability that each ball will be red.

Correlation

When two variables x and y, whose values have been obtained by experiment or statistical enquiry, are plotted, a diagram like Fig. 35.15 is obtained. This diagram is called a **scatter diagram**. Looking at the plotted points we see that a rough relationship, or **correlation**, is seen to exist. The straight line which approximates to the plotted data is called the **line of best fit**.

Fig. 35.15

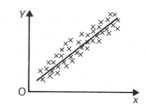

The correlation may be positive, precise or negative. For positive correlation (Fig. 35.16) the values of y increase as x increases. For the correlation to be precise all the points on the scatter diagram must lie on the straight line (Fig. 35.17). For negative correlation (Fig. 35.18) the values of y decrease as x increases. If no relationship is indicated by the points on the scatter diagram we say that there is no correlation between the variables x and y (Fig. 35.19).

Fig. 35.16   Fig. 35.17

Fig. 35.18   Fig. 35.19

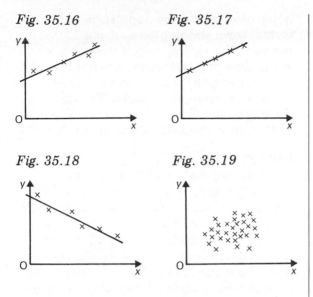

We can often see by looking at the scatter diagram how well the line of best fit describes the relationship between x and y. In Fig. 35.17 the straight line describes the relationship between x and y very well and we say that there is good correlation between x and y. We mean that the values of y depend upon the values of x. In Fig. 35.18 the relationship is not so good and we say that the correlation is poor. We mean that it is unlikely that the values of y depend upon the values of x. When no correlation exists as in Fig. 35.19 we mean that we are definitely sure that the values of y do not depend upon the values of x.

Example 22

The table below gives corresponding values of x and y. Plot a scatter diagram and draw the line of best fit.

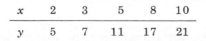

x	2	3	5	8	10
y	5	7	11	17	21

The scatter diagram is shown in Fig. 35.20 and we see that all the points lie on the line of best fit. Hence the correlation is precise and we can say that the values of y depend upon the values of x.

Example 23

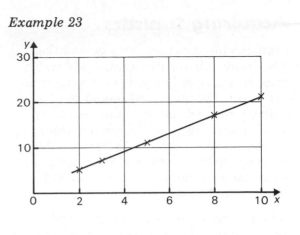

Example 23

The table below shows the weights (in kilograms) and the heights (in centimetres) of 12 adults. Plot a scatter diagram and draw the line of best fit.

Weight (kg)	74	70	55	58	51	56
Height (cm)	175	173	164	171	152	164
Weight (kg)	57	62	62	69	68	67
Height (cm)	155	169	168	178	186	175

The scatter diagram is shown in Fig. 35.21 and it shows that the correlation between height and weight is poor.

Fig. 35.21

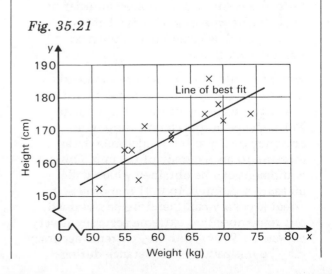

Acquiring Statistics

Statistics involves measuring or counting. The figures obtained as a result of a survey must be expressed in units such as metres, litres, etc. The units used should be of a suitable size. For instance the length of a small object might be measured in centimetres but the journey of a lorry would be measured in kilometres. Wherever possible standard units should be used and these should be of a constant size.

As an example consider the output of a small factory. The figures might be: January, 43 000 items; February, 37 000 items; March, 44 000 items; and April, 41 000 items. These figures are not directly comparable because the unit of size, the month, is not of constant size: January and March have 31 days, April has 30 days and February has 28 days. A better unit of time might be 28 days.

The data collected may be quantitative or qualitative. Quantitative data consists of facts in numerical form such as heights of individuals or the number of driving licences issued. Qualitative data consists of facts like colours of eyes or the taste of food. Qualitative data can often be in quantitative form. For instance, health can be measured by the number of days illness and intelligence by an IQ test. Qualities like taste and colour can be placed in order of rank. In the case of colour the lightest colour could be given rank 1, the next lightest rank 2, etc.

When the information is to be obtained by measuring, the type of equipment used can affect the accuracy of the measurement. For instance, a rule can measure to an accuracy of 2 mm but a micrometer can measure to an accuracy of 0.1 mm. Optical equipment can be obtained which will measure accurately to 0.01 mm or less. Great accuracy can therefore be obtained, but remember that extreme accuracy is very expensive. When counting, extreme accuracy may be misleading. For instance during a

population census the population of a certain town was quoted as being 73 058 persons. This figure, even if it was correct at the time of publication, was not likely to remain accurate for any length of time because of births and deaths. We might say, therefore, that the population was about 73 000, or with justification, about 70 000.

Surveys

A survey may be conducted by either a census or a sample taken from the population. In a census the whole population under study is used but in a sample only a proportion of the population is used. Problems arise with sampling because wrong conclusions about the population as a whole may be drawn by studying information obtained from only a fraction of the population.

If the population is small a census is preferable. For instance, in a survey of working conditions a census is better because the number of people working in the factory (the population) is small. However sampling has the advantages that it is cheaper and quicker than a census.

Information for a survey can be collected by direct observation (e.g. counting the number of vehicles passing a certain point in a given time). It may also be obtained by the face-to-face interview. This method has the advantages that many questions can be asked quickly and that high response rates are achieved. The disadvantage is that the interviewer can influence the answers and this may introduce a systematic bias into the survey.

There are no hard and fast rules which have to be obeyed when designing a questionnaire. However, the following points should be noted:

(1) The questions should be simple and unambiguous. Long questions should be broken up into sub-topics.

(2) Leading questions such as 'Don't you agree that all intelligent people read the *Times* newspaper?' should be

avoided because the answer is suggested. To the question posed, the answer 'yes' is strongly suggested.

(3) The questions should be as short as possible and they should be asked in a logical sequence because then the quality of the answers will be improved.

(4) Questions which allow the answer to be ticked are best.

Rules for Tabulation

Information given in narrative form is often difficult to understand. Consider the following statement:

'In 1951, 207 thousand persons received unemployment benefit, 906 thousand persons sickness benefit, 1437 thousand males retirement pensions, 2709 thousand females retirement pensions, 457 thousand received widows benefit. 217 thousand persons received other National Insurance benefits. In 1971 the corresponding figures for unemployment benefit was 457 thousand, for sickness benefit was 969 thousand, for male retirement pensions 2611 thousand, for female retirement pensions 5196 thousand, for widows benefit 448 thousand, for other National Insurance benefits 387 thousand.'

The narrative is difficult to read and very confusing. If the information is tabulated as below the confusion is eliminated.

NATIONAL INSURANCE BENEFITS

Type of benefit	Numbers obtaining benefits (thousands)	
	In 1951	In 1971
Unemployment	207	457
Sickness	906	969
Retirement pensions (males)	1437	2611
Retirement pensions (females)	2709	5196
Widow's pensions	457	448
Others	217	387

In this table we have only given the figures stated in the narrative. It may help the reader who is interested in the data if we add some secondary statistics such as the total number of people drawing benefits and percentages of the totals as shown in the next table.

NATIONAL INSURANCE BENEFITS

Type of benefit	1951		1971	
	Numbers drawing benefit (thousands)	% of total	Numbers drawing benefit (thousands)	% of total
Unemployment	207	3.49	457	4.54
Sickness	906	15.27	969	9.62
Retirement pensions (males)	1437	24.22	2611	25.93
Retirement pensions (females)	2709	45.66	5196	51.61
Widows pensions	457	7.70	448	4.45
Others	217	3.66	387	3.85
Totals	5933	100	10068	100

Testing Hypotheses

At the end of a statistical report we are often called upon to make a decision. To do this we make assumptions which are called **hypotheses.** Unless care is taken the wrong conclusions may be drawn from a set of figures.

Example 24

The death rate in Bournemouth is much higher than in some inland cities. Therefore we make the hypothesis that seaside resorts are not good for health.

On the face of it the hypothesis seems plausible but it goes against the fact that seaside resorts are usually healthy places to live in because there is less polution, etc. Why should the death rate be higher in Bournemouth? Is it anything to do with the type of population living there? We now realise that many elderly people retire to Bou mouth and because the death ra the elderly is higher than that population as a whole we wo expect the death rate for Bo to be high. It is therefore relate health and the dea hence we reject the hyp

Example 25

A school decided that they wanted to change the colour of its school blazer. Two girls decided to carry out a survey to see which colour pupils would prefer. They asked their classmates for their preference and noted their replies in a notebook as follows:

Black	Green	Black	Blue
Blue	Brown	Blue	Red
Black	Red	Grey	Green
Black	Red	Grey	Green
Blue	Red	Red	Blue

(a) Show a better way of noting the responses.

(b) Are the results as useful as they might be?

(c) Suggest a method whereby they could improve their survey.

(a) A tally chart as follows would be better:

Colour	Tally	Total
Black	I I I I	4
Red	┼┼┼┼	5
Grey	I I	2
Green	I I I	3
Blue	┼┼┼┼	5
Brown	I	1

(b) Since only classmates were interviewed only one age group contributed to the survey, and the sample is not representative of the population (i.e. pupils in the school).

The survey would be improved by taking a cross-section of the classes in the school thereby a more representative sample.

1. The results of ten candidates entered for examinations in both chemistry and physics are shown in the following table.

Physics mark	24	36	48	52	56
Chemistry mark	18	38	46	60	52
Physics mark	60	64	76	82	92
Chemistry mark	72	70	82	88	84

(a) Draw a scatter diagram for this information. Place the physics mark on the horizontal axis and use a scale of 2 cm to 20 marks on both axes.

(b) Draw the line of best fit on the diagram.

(c) One candidate took only the chemistry examination, for which he obtained 50 marks. Using your scatter diagram, estimate the mark this candidate was likely to have obtained in physics.

2. Twelve samples of home-made toffee were judged by two judges, A and B, who awarded marks out of 100. The results are shown in the following table:

Sample	1	2	3	4	5	6
Judge A	51	39	43	28	30	68
Judge B	60	45	50	35	35	80

Sample	7	8	9	10	11	12
Judge A	59	64	25	47	32	44
Judge B	70	75	30	55	40	55

(a) Draw a scatter diagram for this information with Judge A plotted on the horizontal axis and a scale of 2 cm to 20 marks on both axes.

(b) Draw the line of best fit on your scatter diagram. Two further samples were judged by one judge only. Judge A gave the thirteenth sample 65 marks, but judge B gave the fourteenth sample 71 marks. Use your graph to estimate the marks that would most likely have been awarded by the other judge.

3. The following values of P and Q were obtained in an experiment:

P	10	12	15	20	30
Q	80	76	70	60	40

Plot the scatter diagram placing P on the horizontal axis and using scales of 2 cm to 5 units on the horizontal axis and 2 cm to 10 units on the vertical axis. What kind of correlation exists between P and Q?

4. The marks obtained by 10 pupils studying languages were:

French	20	25	30	32	44	45	54	56	60	60
German	25	30	32	34	45	43	54	60	58	60

Plot a scatter diagram with French on the horizontal axis using a scale of 2 cm to 5 marks on both axes. Draw the line of best fit and hence state the type of correlation that exists between the two sets of marks.

5. In England in a certain year there were 16 070 000 dwellings of these 8 360 000 were owner occupied, 4 500 000 were rented from local authorities and 2 410 000 were rented from private owners, the remainder being held under other tenures. In Scotland the same year there were 1 800 000 dwellings: of these 540 000 were owner occupied, 940 000 were rented from local authorities and 200 000 were rented from private owners, the remainder being held under other tenures. Tabulate this information, calculate appropriate secondary statistics and include these in your tabulation.

6. 4629 students sat examinations in single-subject courses comprising Statistics, Mathematics and Computer Studies. The examinations were at GCSE, A-level and AS-level. 2216 students out of the 4629 took GCSE subjects and 1193 A-level. 1943 took the examinations in Statistics and 1147 the examinations in Mathematics. 984 students took the GCSE in Statistics and 765 the AS-level in that subject. 334 took A-level Mathematics and 665 A-level Computer Studies. 195 students sat the AS-level examinations in Mathematics. Draw up a suitable table to show the above information.

7. Criticise the following statements as evidence of statistical facts and explain why each one could be misleading as to its conclusions.

(a) Traffic fatalities increased from 15 to 45, an increase of 300%.

(b) The average age at which girls in Beeston begin to use lipstick is 11.768 years.

(c) A class in Mathematics, after listening to Brahms' Symphony No. 4 in E minor and Beethoven's No. 3 in E flat major, indicated the following preferences: Brahms 32; Beethoven 12. This shows clearly that people prefer Brahms to Beethoven.

8. Comment critically on the following statements showing (with reasons) whether you would accept them or consider them to be wrong:

(a) The illegitimacy rate in Birmingham is the highest in the country, which proves that large cities are immoral places.

(b) More people were killed in air accidents in 1978 than in 1928. Therefore it was more dangerous to fly in 1978 than in 1928.

(c) There is a greater incidence of rheumatism in Torquay per head of population than in London. Therefore the London environment is less likely to cause rheumatism than that of Torquay.

Miscellaneous Test 35

1. The marks of 5 students in an examination were as follows: 54, 63, 49, 78 and 61. What was the mean mark?

2. The goals scored by a football team in 60 matches were:

   4 1 3 2 0 1 1 1 1 0 2 5 0 0 4
   1 1 0 1 1 5 1 2 1 1 0 2 0 1 2
   3 1 1 1 0 1 2 0 1 0 2 0 1 3 0
   0 2 0 1 2 1 0 2 1 1 2 4 0 1 2

   Arrange this information in a frequency table using classes of 0, 1, 2, 3, 4 and 5 goals.

3. Draw a bar chart to depict the frequency distribution in question 2.

4. An investigation was carried out into reasons for absence or for late arrival at a school on a particular Monday morning. The result of the investigation was as follows:

Missed the bus	25%
Sickness	38%
Overslept	11%
Had a cold	21%
Bicycle had a puncture	5%

   On graph or squared paper, draw a bar chart to represent these statistics. Use a scale of 2 cm to represent 5% on the vertical axis.

5. Fig. 35.22 shows the numbers of children classified by weight in groups at intervals of 2 kg.

   (a) Calculate the number of children.

   (b) Calculate the mean weight of the children, correct to the nearest kilogram.

Fig. 35.22

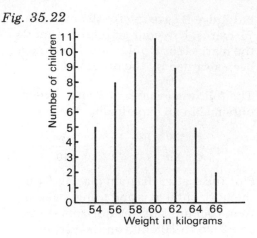

6. A letter is chosen from the word 'flagstaff'. Find the probability that it will be

   (a) l   (b) a   (c) f

7. A large transport firm had 70 lorries in different parts of the country. It wished to compile a table showing the distance travelled by each lorry in a particular period. The following information was obtained by inspectors who checked and recorded the distance, in kilometres, travelled by each lorry.

   10 12 30 56 57 36 27 32 57 20
   15 50 49 46 21 69 55 42 60 45
   35 47 22 45 44 31 67 43 17 54
   49 56 51 55 29 65 23 55 46 66
   40 63 24 33 62 58 15 57 64 37
   38 30 27 28 59 25 52 47 26 63
   48 31 31 34 65 41 16 28 39 46

   (a) Draw up a frequency distribution for the classes 0–9, 10–19, 20–29, etc.

   (b) Draw a histogram to represent this frequency distribution.

8. The information below gives the diameters of round steel bars.

Diameter (mm)	Frequency
14.96–14.98	3
14.99–15.01	8
15.02–15.04	12

(a) Give the upper and lower class boundaries for the second class.

(b) State the class width for each of the classes shown in the table.

(c) What is the class interval for the first class?

9. Classify each of the following as a continuous or discrete variable:

(a) the number of shirts sold per day

(b) the weight of packets of chemical

(c) the number of bunches of daffodils packed by a grower

(d) the daily temperature at noon

(e) the lifetime of electric light bulbs

(f) the number of telephone calls received by a police station per day

10. In a small firm five people earn £126 per week, three earn £114 per week and two earn £132 per week. Calculate the mean wage for the ten people.

11. 50 students had their height measured with the following results:

Height (cm)	160	161	162	163	164	165	166
Frequency	1	5	10	16	10	6	2

Calculate the mean height of the 50 students.

12. Given the set of numbers

$$3, 5, 2, 7, 5, 8, 5, 2, 7$$

find

(a) the mode

(b) the median

13. 100 pupils took a mathematics test. Their marks were as follows:

Mark	1	2	3	4	5	6	7	8	9
Frequency	2	8	20	32	18	9	6	3	2

Determine

(a) the median mark

(b) the modal mark

14. A fair die is rolled. Calculate the probability of scoring

(a) a four

(b) a number less than four

(c) a two or a three

(d) an odd number

15. A loaded die shows these probabilities:

Score	1	2	3	4	5	6
Probability	0.14	0.18	0.16	0.15	0.17	0.20

(a) If the die is thrown once what is the probability of a score less than 3?

(b) If it is thrown twice, what is the probability of scoring a 4 followed by a 6?

16. A bag contains 14 yellow counters and 10 red counters. Two counters are taken out in succession and not replaced.

(a) Copy and complete the tree diagram (Fig. 35.23) by writing the correct fractions in the boxes A, B, C and D.

(b) Calculate the probability that two red counters will be taken out.

Fig. 35.23

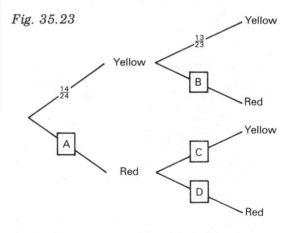

17. A survey was carried out by some students who recorded the first digit of the registration number of cars which passed their school gate. The results were as follows:

Digit	1	2	3	4	5	6	7	8	9
Frequency	38	27	27	22	22	34	20	7	3

Determine the probability that

(a) the next car that passes will have a registration number beginning with 5

(b) the next car that passes will have a registration number beginning with a digit greater than 6

18. The distribution of marks obtained by 300 candidates in an examination is given in the following grouped frequency table:

Mark	0–20	21–40	41–60	61–80	81–100
Frequency	24	56	116	74	30

(a) Complete the cumulative frequency table below:

Mark not exceeding	20	40	60	80	100
Cumulative frequency					

(b) Plot the points represented by the values in the cumulative frequency table and join them with a smooth curve.

(c) Use your graph to estimate
  (i) the number of candidates who obtained a mark less than 64
  (ii) the median mark.

19. The table below shows the number of university students attending lectures on psychology during a 50 week period.

Number of students	Mid-point of interval	Frequency
0–19	9.5	0
20–39	29.5	3
40–59	49.5	10
60–79	69.5	10
80–99	89.5	17
100–119	109.5	8
120–139	129.5	2

(a) Draw a frequency polygon to illustrate the data. Use the mid-points of the intervals. Suitable scales are 2 cm to 20 students on the horizontal axis and 2 cm to 5 units on the vertical axis.

(b) Calculate the mean of the number of students attending the lectures each week.

(c) State the interval in which the median lies.

Mental Test 35

1. The tally chart below shows the distribution of a sample of mixed nuts. How many of the sample were almonds and how many nuts were there in the sample altogether?

Brazils	++++
Walnuts	++++ I I
Almonds	++++ I I I
Chestnuts	++++ I I I
Cobnuts	++++ ++++ I I

2. What is the mode of the histogram shown in Fig. 35.24?

Fig. 35.24

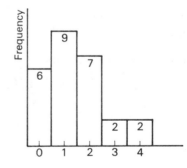

3. The probability of John being selected for the school football team is 7/25. What is the probability that he will not be selected?

4. Find the mean of £23, £42 and £25.

5. What is the median of the numbers 2, 4, 6, 7, 8, 9 and 10?

6. Find the mode of the numbers 3, 5, 2, 7, 5, 8, 2, 7, 2.

7. The following figures give the probability of an event occurring: 0, 0.6, 1 and 2. One of these figures cannot be correct. Which one is it?

8. A bag contains 7 red balls and 3 blue balls. One ball is drawn at random. What is the probability that it will be blue?

9. The table below shows the frequency distribution for the diameter of ball bearings:

Diameter (mm)	14–15	16–17	18–19
Frequency	8	11	14

(a) Write down the lower and higher class boundaries for the second class.

(b) What is the class width?

10. The mean of five numbers is 40. If one of the numbers is 4, what is the mean of the remaining four numbers?

11. One letter is chosen from the word 'emergency'. What is the probability that it will be an e?

36

Vectors

Vector Quantities

Scalar quantities are fully described by size or **magnitude**. Some examples are time, temperature and mass or weight.

Vector quantities need both direction and magnitude to describe them fully. Some examples of vector quantities are:

(1) A displacement from one point to another, e.g. 15 m due east.

(2) A velocity in a given direction, e.g. 30 km/h due north.

(3) A force of 15 newtons acting vertically downwards.

Representing Vector Quantities

If Susan walks 8 m due east, how can we represent this vector quantity in a diagram? We first choose a suitable scale to represent the magnitude. In Fig. 36.1 a scale of 1 cm to represent 1 m has been chosen. We then draw a straight line 8 cm long in an easterly direction. An arrow is placed on the line to make it clear that Susan walked to the east (and not to the west).

Fig. 36.1

Scale: 1 cm = 1 m

When we want to refer to a vector several times we need a shorthand way of doing this. The usual way is to name the end points of the vector. The vector described by Susan started at A and ended at B. If we now want to refer to the vector we write $\overrightarrow{AB}$ which means 'the vector from A to B'.

Exercise 36.1

Draw the following vectors using a convenient scale for each. Label each vector correctly.

1. $\overrightarrow{AB}$ 5 m due north.
2. $\overrightarrow{XY}$ 8 km due south.
3. $\overrightarrow{PQ}$ 7 km due west.
4. $\overrightarrow{MN}$ 10 miles due east.
5. $\overrightarrow{RS}$ 6 m south-east.

Resultant Vectors

In Fig. 36.2 A, B and C are three points marked out in a field. A man walks from A to B (i.e. he describes $\overrightarrow{AB}$) and then walks from B to C (i.e. he describes $\overrightarrow{BC}$). Instead the man could have walked from A to C thus describing the vector $\overrightarrow{AC}$.

Fig. 36.2

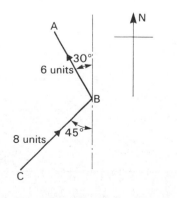

Now going direct from A to C has the same effect as going from A to C via B. We therefore call $\overrightarrow{AC}$ the **resultant** of the vectors $\overrightarrow{AB}$ and $\overrightarrow{BC}$. Note carefully that the arrows on the vectors $\overrightarrow{AB}$ and $\overrightarrow{BC}$ follow nose to tail. The resultant $\overrightarrow{AC}$ is marked with a double arrow which opposes the arrows on $\overrightarrow{AB}$ and $\overrightarrow{BC}$.

Triangle Law

The resultant of any two vectors is equal to the length and direction of the line needed to complete the triangle. This is called the **triangle law.**

Example 1

Two vectors act as shown in Fig. 36.3. Find the resultant of these two vectors.

Fig. 36.3

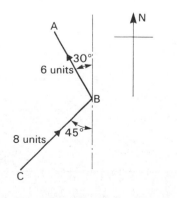

The vector triangle is drawn to scale in Fig. 36.4. The resultant is $\overrightarrow{CA}$. The length of $\overrightarrow{CA}$, written $|\overrightarrow{CA}|$, is 11.2 units. To state its direction we measure the angle ACX which is found to be 14°.

Fig. 36.4

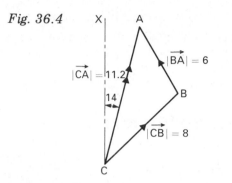

Equal Vectors

Two vectors are equal if they have the same magnitude and the same direction. Hence in Fig. 36.5, the vector $\overrightarrow{AB}$ is equal to the vector $\overrightarrow{CD}$ because the length AB is equal to that of CD and the lines AB and CD are parallel to each other.

Fig. 36.5

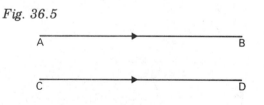

Inverse Vectors

In Fig. 36.6 the vector BA has the same magnitude as AB but its direction is reversed. Hence $\overrightarrow{BA} = -\overrightarrow{AB}$. $\overrightarrow{AB}$ is said to be the **inverse** of $\overrightarrow{BA}$.

Fig. 36.6

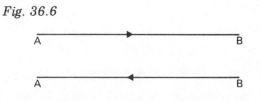

Adding Vectors

In Fig. 36.4 the arrows on the vectors $\overrightarrow{CB}$ and $\overrightarrow{BA}$ follow nose to tail. $\overrightarrow{CA}$ is the resultant of these two vectors and we write

$$\overrightarrow{CA} = \overrightarrow{BA} + \overrightarrow{CB}$$

The resultant $\overrightarrow{CA}$ is said to be the sum of the two vectors $\overrightarrow{BA}$ and $\overrightarrow{CB}$ and the sum of two vectors can always be found by making an accurate drawing.

Subtracting Vectors

Because $-\overrightarrow{CD} = +\overrightarrow{DC}$ we can treat $\overrightarrow{AB} - \overrightarrow{CD}$ as $\overrightarrow{AB} + \overrightarrow{DC}$.

Example 2

Fig. 36.7 shows two vectors v_1 and v_2. Find in magnitude and direction, the vector $v_1 - v_2$.

Fig. 36.7

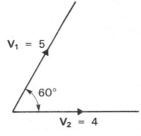

To find the vector $v_1 - v_2$ we treat $v_1 - v_2$ as $v_1 + (-v_2)$. This means that we reverse the arrow on the vector v_2 (Fig. 36.8) and then proceed in a similar way to that used in Example 1.

Fig. 36.8

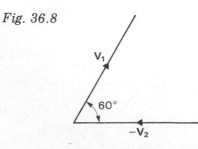

The vector triangle is drawn in Fig. 36.9. The resultant of v_1 and $-v_2$ is CA which is found to be 4.6 units. To state its direction, we measure the angle ACB which is found to be 71°.

Fig. 36.9

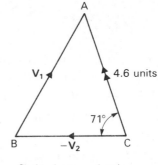

Scale: 1 cm = 1 unit

Multiplying Vectors by a Scalar

If the vector $\overrightarrow{AB}$ is 5 m due west then the vector $3\overrightarrow{AB}$ is 15 m due west (Fig. 36.10). In general if k is any number, the vector $k\overrightarrow{AB}$ is a vector in the same direction as $\overrightarrow{AB}$ but k times its magnitude.

Fig. 36.10

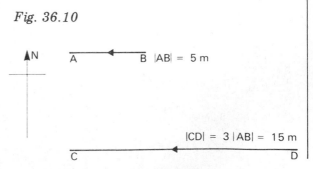

Exercise 36.2

Find by accurate drawing, the resultant in magnitude and direction of each of the pairs of vectors shown in Fig. 36.11.

Fig. 36.11

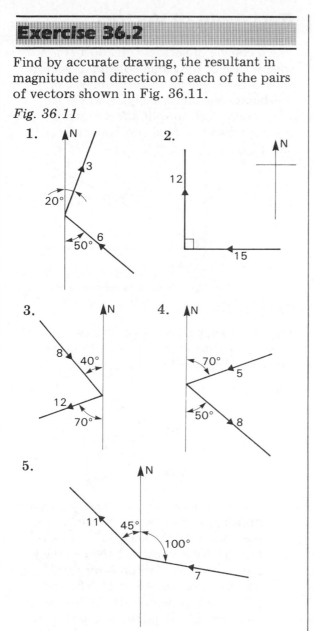

1. 2.

3. 4.

5.

Find by accurate drawing the vectors $v_1 - v_2$ in magnitude and direction for each of the systems shown in Fig. 36.12.

Fig. 36.12

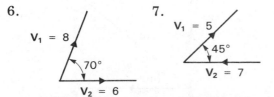

6. 7.

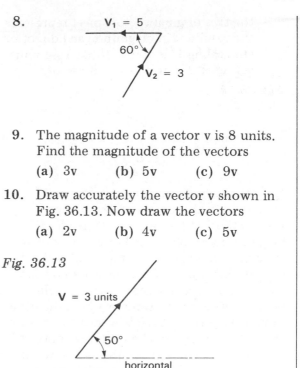

8.

9. The magnitude of a vector v is 8 units. Find the magnitude of the vectors

    (a) 3v   (b) 5v   (c) 9v

10. Draw accurately the vector v shown in Fig. 36.13. Now draw the vectors

    (a) 2v   (b) 4v   (c) 5v

Fig. 36.13

V = 3 units

50°

horizontal

Parallelogram of Vectors

Two vectors may be added by drawing a parallelogram of vectors. This is an alternative to the triangle of vectors discussed on page 377.

Example 3

Draw the parallelogram of vectors for the two vectors shown in Fig. 36.14. Hence find their resultant in magnitude and direction.

Fig. 36.14

5

60°

7

The two vectors are drawn to scale with the angle between them accurately made by using a protractor. The parallelogram is then completed as shown by the dotted lines in Fig. 36.15. The diagonal drawn from the point where the two original vectors meet represents the resultant in magnitude and direction. On scaling it is found to be 10.4 units acting at 25° to the 7 unit vector.

Fig. 36.15

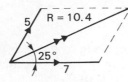

Note that when using a parallelogram of vectors both the arrows on the original vectors must either point away from the point of intersection of the two vectors or they must point towards the point of intersection. In Fig. 36.16(a) the arrows on the two vectors do not conform to this rule. One arrow points towards A whilst the other points away from A. However, the horizontal vector remains unchanged if we draw it as shown in diagram (b). Alternatively the parallelogram may be drawn as shown in diagram (c) with both arrows pointing towards A.

Note that if both arrows point away from A (diagram (b)) then the arrows on the resultant also point away from A. If both arrows point towards A then arrows on the resultant also point towards A.

Fig. 36.16

(a)

(b)

(c)

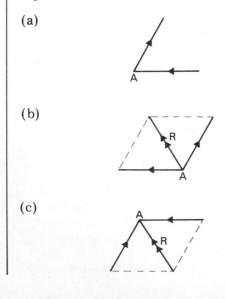

Rectangle of Vectors

The rectangle of vectors (Fig. 36.17) is a special case of the parallelogram of vectors. It is often used to solve problems in mechanics and stress analysis, where it is generally used to split a force into horizontal and vertical components.

Fig. 36.17

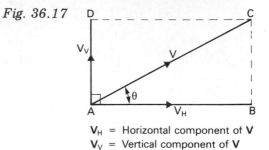

V_H = Horizontal component of **V**
V_V = Vertical component of **V**

Example 4

Fig. 36.18 shows a force of 20 newtons acting at 40° to the horizontal. Find its horizontal and vertical components.

Fig. 36.18

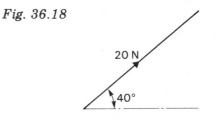

To find the two components by accurate drawing, first draw the vector to scale measuring the angle of 40° with a protractor. Next construct the rectangle ABCD with the vector as diagonal. Finally measure AB = 15.3 N and AD = 12.9 N which are the horizontal and vertical components respectively (Fig. 36.19).

Fig. 36.19   Scale: 1 cm = 5 N

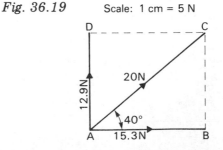

Exercise 36.3

By drawing a parallelogram of vectors, find the resultant vector for each of the cases shown in Fig. 36.20.

Fig. 36.20

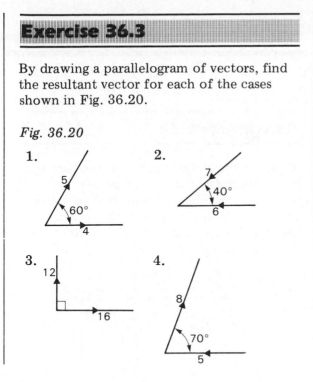

5.

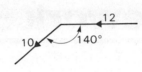

By drawing a rectangle of vectors find the vertical and horizontal components of the following vectors:

6. 5 N acting at $60°$ to the horizontal.

7. 30 km/h in a direction of $40°$ due east.

8. 80 N in a direction of $70°$ to the horizontal.

Coursework 3

Spacefiller Ltd manufactures blocks having six faces, which may be squares or rectangles. There is an economy drive organised to save the paint needed to cover the surface of the blocks. Customers order blocks according to a particular volume.

Investigate the surface areas of various blocks.

One customer required a Spacefiller block with a volume of 36 cubic centimetres. This was the first block the firm produced:

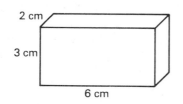

Painter's calculation:

$2(6 \times 2) = 24 \, cm^2$

$2(6 \times 3) = 36 \, cm^2$

$2(3 \times 2) = 12 \, cm^2$ TOTAL: $72 \, cm^2$

The firm's present policy is to use whole-number measurements.

Investigate the block with a volume of $36 \, cm^3$ in the light of the firm's economy drive.

Investigate blocks with other volumes such as $30 \, cm^3$, $48 \, cm^3$ and so on. If the firm decides to include half-centimetres in the measurements, will it affect the conclusions you have reached so far? Could other fractional lengths be included?

In the rectangle, there are nine numbers and six mathematical signs. There are two addition signs, one multiplication sign, two equality signs and one sign which means 'less than'. Using that sign for instance we could write:

$$4 \times 2 < 16$$

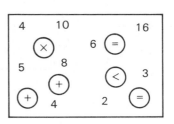

All the numbers and all the signs must be used to make three correct statements. Here is one example in which all the signs are correctly used:

$$5 + 3 = 8$$
$$4 \times 4 = 16$$
$$6 + 2 < 10$$

Find at least ten other ways of producing three accurate statements.

This assignment is concerned with the application of mathematics to planet Earth.

(a) Find out the diameter of the Earth at its equator. Assume that the Earth is a perfect sphere and calculate its circumference, its total surface area and its volume. How many significant figures can be trusted in your answers? Express each answer in standard form.

The Earth rotates once on its axis in 24 hours. Calculate the speed of rotation of a town located at the equator. Will London spin along at the same speed? Give reasons. Does it make any difference whether a place is north or south of the equator?

To calculate the speed of rotation of a town 'T' on the Earth's surface, we need to find out its latitude which is the angle θ subtended at the centre of the Earth:

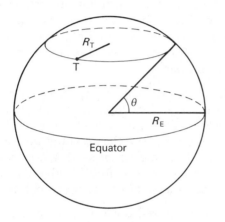

The radius of the circle of latitude where T is located (R_T) can be calculated using this formula:

$$R_T = R_E \times \cos \theta$$

where R_E is the Earth's equatorial radius and θ is the latitude of T. Can you show why this formula works? Use the formula to calculate the speed of rotation of major cities such as London, New York, Melbourne, etc.

At what latitude must a town be located to have a rotational speed exactly half that of a town on the equator? Describe what happens at the Poles. Plot a graph to show how rotational speeds vary with latitude.

(b) The Earth's axis of rotation is tilted at an angle of about $23\frac{1}{2}°$ from the perpendicular to the plane of its orbit around the Sun:

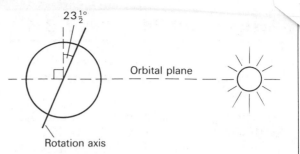

What has this to do with the four seasons and the Tropics of Cancer and Capricorn? Find out the meaning of equinoxes and solstices, particularly with regard to the Sun's position. Calculate the maximum angle of elevation the Sun can reach above the horizon in London (this will occur at midday on the summer solstice). Draw diagrams to show why the Poles experience six months of daylight followed by six months of night each year.

(c) A Space Shuttle orbiting at a circular altitude of 300 km above the Earth's surface takes 90.5 minutes to complete one revolution. Calculate its orbital speed in km/h.

Telecommunications satellites, which relay live TV pictures of events from all over the world, are placed in special **geosynchronous** orbits. A satellite in such an orbit takes 24 hours to complete one circuit of the Earth. Why are these orbits so useful?

A geosynchronous satellite moves round its circular orbit at 11 040 km/h. How high above the Earth's surface must such a satellite be positioned?

(d) In June 1992, over 170 of the world's nations gathered for the Earth Summit in Rio de Janeiro, Brazil, and signed treaties to safeguard the world's environment and wildlife. Seek out statistics on such concerns as air pollution, acid rain, ozone depletion, the 'greenhouse effect', drought, population growth, famine, etc. (modern atlases and fact-finders are sources of reference for such data). Collate the information and analyse it statistically. Draw relevant conclusions from your analysis.

Assignment 24   Ma 2, 3, 4, 5

We will now expand our horizons and embark on a tour of the larger Universe beyond our home planet.

(a) The Moon follows an orbit around the Earth which takes 27 days 7 hours 43 minutes 11.5 seconds to complete. Express this period in days correct to 4 decimal places.

The Moon's orbit is roughly circular with a mean radius of 3.844×10^5 km. Calculate its average orbital speed around the Earth in km/h. How many degrees of arc will the Moon travel across our skies in 24 hours?

(b) The Moon has a radius of 1738 km. Calculate its angular diameter in degrees as seen from the Earth. How do its surface area and volume compare with those of the Earth? It has a 'captured' rotation, which means it takes exactly the same amount of time to spin once on its axis as it does to circle the Earth once. Calculate the speed of rotation experienced by an astronaut standing in a crater at the Moon's equator.

The diameter of the Sun is 1.392×10^6 km. Here is a comparison of the relative sizes of the Sun and Moon:

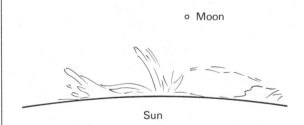

o Moon

Sun

Express the diameter of the Sun compared with that of the Moon as a ratio in the form $n:1$. At the time of a total solar eclipse, the disc of the Moon completely hides the Sun's disc. Explain how this is possible. Use the given statistics to calculate an approximate value for the Earth–Sun distance.

(c) The Sun loses roughly 4 million tonnes of its mass every second through radiation. Calculate its mass loss in an hour, a day, a

year, etc. It has a total mass of 2×10^{27} tonnes. How long would it be before the Sun sent out its last sunbeam at this rate of loss? (Actually, it will be bankrupt long before this time, but there is no cause for alarm!)

(d) Look up sets of data for the planets – diameters, masses, distances from the Sun, surface gravity, etc. – and prepare diagrammatic representations to show the true scale of the Solar System and different characteristics of the planets. Compare their sizes, the apparent diameter of the Sun's disc as seen from each planet, etc. How much would an average human weigh on the surface of each world? Is there any obvious relationship between the size of a planet and the number of moons circling it? Look for other comparisons.

Build your own scale model of the Solar System using everyday objects. For instance, if the Sun was represented by a beach ball two feet across, the Earth would be a pea and the Moon a grain of sand. Would you be able to place your 'planets' at the correct scaled distance from your beach ball Sun?

(e) Light and radio waves travel at a speed of 299 792 km/s. During the Apollo Moon flights, Mission Control in Houston had to make allowances for this when talking to their astronauts on the lunar surface. How far away is the Moon, reckoned in light seconds (a light second is the distance travelled by light in one second)? If you said 'hello' over the radio to a man on the Moon, what is the minimum time you would have to wait for a reply?

If the Sun exploded at this instant, no one on Earth would notice. The Sun would still appear whole and perfect to holidaymakers basking on the beach. Why is this? How long would it be before we knew about the catastrophe? What about someone taking a holiday on Mercury or Pluto?

(f) To 'boldly go' beyond the Solar System at warp speed, we need larger units of measurement. A **light year** is the distance travelled by light in one year. After the Sun, the nearest star to us is Proxima Centauri at a distance of 4.28 light years. How far is this in kilometres?

Another unit employed for measuring astronomical distances is the **parsec**. Find out how it is defined and how it compares with the light year.

On our scaled-down Solar System, with the Sun as a beach ball, Proxima Centauri would be about the size of a melon. To get the distance scale to match, how far away would you have to place it from the beach ball?

(g) A supernova is a star which explodes cataclysmically. The Crab Nebula in Taurus is the remnant of a supernova explosion recorded by the Chinese in AD 1054. The Crab lies at a distance of about 5.7×10^{16} km. When did the supernova really explode?

Our Solar System lies in a vast, rotating system of stars called the Galaxy. The Sun is situated 28 000 light years from its centre and has a rotational speed about the Galactic centre of about 235 km/s. How far do we travel through the Galaxy each day? The time taken by the Sun to complete one circuit of the Galaxy is called the 'cosmic year'. Calculate its length.

Assignment 25 Ma 5

Many surveys involve the use of a questionnaire. When television companies, newspapers or manufacturers of certain products wish to conduct a survey among members of the public, they produce a set of questions. The questions are either of the type which require a simple answer of 'Yes' or 'No', or a range of choices from which each person taking part in the survey can select an appropriate answer.

The 'Yes' and 'No' questions would be of the following type:

Do you use this particular product?
Did you watch this particular TV programme?
Did you vote for this candidate in the election?

Choices can be offered as follows:

Do you rise in the morning:

(a) Before 6.00 a.m.?
(b) 6.00 to 6.30 a.m.?
(c) 6.30 to 7.00 a.m,?
(d) 7.00 to 7.30 a.m.?
(e) 7.30 to 8.00 a.m.?
(f) After 8.00 a.m.?

Questions that could have answers not appearing on the question sheet should not be asked.

Test the questionnaire on yourself or a friend before putting it into use, and make whatever improvements to the questions you consider necessary.

Analysis of the results of a survey is simple if the questions are well constructed and prepared.

Make some questionnaires for surveys on the following subjects:

(a) The belief in superstitutions among the general public
(b) Weekly purchases required for a variety of households
(c) TV viewing among young people
(d) Daily habits of people aged from 11 to 16 years
(e) The extent of active or passive sport participation
(f) People and their pets
(g) Young people and their tastes

Test one of your questionnaires on as many people as possible. Analyse and interpet the results and make a display using graphics to bring out the main conclusions.

Assignment 26   Ma 2, 4

Obtain or construct a die in the shape of a cube.

This investigation concerns the turning of the die on an edge to show a different number on top.

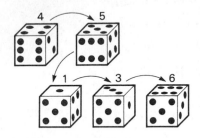

Is it possible to show the numbers in proper sequence as the die is turned over on an edge in any direction?

What kind of sequences are possible?

Work the die around inside this grid:

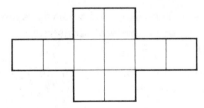

Is it possible to start and finish on the same square showing the same number on top?

Use this grid of nine squares. Always start on the centre square marked X.

H	A	B
G	X	C
F	E	D

You can leave the centre in four ways, by turning the die on its edge into squares A, C, E or G. Travel clockwise and note the sequences of numbers which appear on the top of the die until it arrives back at the centre again. Each sequence has eleven numbers.

Comment on the number patterns formed on the four separate routes.

Does the starting number affect the sequence?

How are the four routes related?

What is the effect of reversing the direction and moving anticlockwise?

Investigate 13-figure sequences produced by a figure-of-eight route such as XABCDEXAHGFEX.

Assignment 27   Ma 2, 5

This investigation centres around the calendar.

(a) In order to pursue this study you must first become familiar with the routine for calculating days of the week. The procedure involves some basic arithmetical deductions which will enable you to work out the day upon which a particular date falls in the past, present or future.

The following sequence is tailor-made for the computer programmer. A suitable set of program subroutines based on this sequence would permit days to be identified with great speed, reducing the donkey work.

Firstly, we need to note this list of digits for the twelve months:

> January 1 (0 in leap years); February 4 (3 in leap years); March 4; April 0; May 2; June 5; July 0; August 3; September 6; October 1; November 4; December 6

Century digits follow a recurring sequence (4, 2, 0, 6, ...):

18th	19th	20th	21st	22nd	23rd
4	2	0	6	4	2

Taking the date 13 June 1947, this is the routine:

Year number:

Last two digits of the year: 47
Divide by 4 (ignore remainders): 11

Century number (20th Century): 0

Month number (find June in the list): 5

Date in the month: 13

Total: $47 + 11 + 0 + 5 + 13 = 76$

Divide by 7: $76 \div 7 = 10$ remainder 6

The remainder fixes the day in this list:
Sun 1, Mon 2, Tue 3, Wed 4, Thu 5,
Fri 6, Sat 0

Our remainder of 6 tells us that 13 June 1947 was a FRIDAY.

(b) For the superstitious, Friday 13th is an unlucky day. How can you quickly prove that 13 June 1947 was the only Friday 13th in that year? What is the maximum number of Friday 13ths which can occur in a particular year and which months will they always appear in?

(c) Work out a few trial dates to familiarise yourself with the routines. Most people do not realise that the first day of the 21st Century will be 1 January 2001, NOT 1 January 2000. Can you explain this anomaly? You must discover how the routines handle this to be sure that your calculations will be accurate in these special cases. Check 1800 and 1900 as well.

There is a limit to how far we can venture into the past with our day calculations. The earliest date for which the procedure operates in the English-speaking countries is Thursday, 14 September 1752. Find out why.

(d) A whole range of interesting surveys can now be conducted. The one essential requirement for all such studies is a sizeable collection of dates for analysis. Birthdays offer a wide scope and the admission rolls at your own centre will provide a good source of data here. Names are not necessary for this study. You could investigate the frequency of people born in each month of the year, on each day of the week, etc. and follow up with a statistical study to highlight any significant trends.

(e) You can gather the birthdays of famous people, past and present, from books of dates and anniversaries which are now easily obtainable. You should find some interesting coincidences along the way. For instance, the American President, Abraham Lincoln and the famous English naturalist, Charles Darwin, were both born on Sunday, 12 February 1809.

Astrologers claim that people who pursue certain careers tend to be born under particular signs of the Zodiac. Are sports people more likely to be born in a particular month or season? Is there any pattern in the birthdays of scientists, artists, musicians, pop stars, politicians, show business personalities, etc?

Look up important dates in history. Do more battles take place on a particular day of the week? Are more new discoveries made at certain times of the year? What were 'Bloody Sunday', 'Ruby Tuesday' and 'Black Friday'?

Assignment 28   Ma 4

On the next page is an aerial view of the popular holiday resort of Amity Bay.

Tiger Rock is south east of Arrow Point and 0.92 km from Jaws Tower. The Coastguard's hut is 430 m due south of Jaws Tower; the bearing of Arrow Point from Jaws Tower is 042° and they are 0.62 km apart. The bearing of Falcon Stack from Tiger Rock is 197°. The straight-line distance from the Coastguard's hut to Tiger Rock is the same as from the hut to Falcon Stack.

(a) Construct an accurate drawing of the resort using a suitable scale factor. Work out the representative fraction of your map in the form $1:n$.

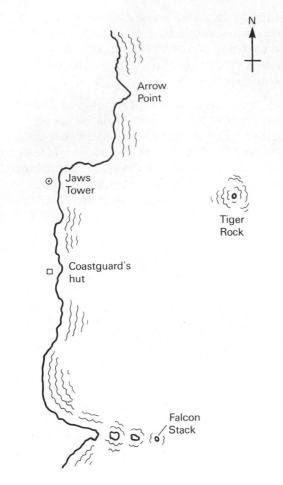

The nearest lighthouse lies on a line equidistant from Arrow Point and the Coastguard's hut, 0.9 km from Tiger Rock. Locate the lighthouse and measure its distance from Jaws Tower.

For safety purposes, local trawlers must ensure that they are sufficiently distant from the coast that the angle subtended by Arrow Point and Falcon Stack as measured from their vessel's location is always less than 90°. Show the limit of this fishing zone on your diagram. What is the closest that any trawler is allowed to approach Tiger Rock?

(b) The Coastguard and Amity Bay police went on high alert at the start of the holiday season this year when a report came in from a local trawlerman fishing off the

coast. His message was broken and garbled: 'position. . . . . . bearing. . . . 234° to . . . . ack and 340° to Ti . . . . sighted whale shark due south of us . . . . Boy, what a fish supper! . . . . distance estimated 60 . . . . uh oh! . . . . holy mackerel! . . . . tangled in our nets . . . . . going down . . . . .' This was the last recorded message.

Try to interpret the message and show where the shark was probably sighted. Was the trawler fishing illegally?

A warning notice was immediately posted:

ATTENTION!

KEEP A SHARK LOOKOUT!

BATHERS ARE STRONGLY ADVISED NOT TO SWIM IN THE SEA UNLESS ARMED WITH A HARPOON. SMALL CRAFT ARE RESTRICTED TO A ZONE BOUNDED BY A LINE JOINING ARROW POINT TO FALCON STACK.

Amity Bay Coastguard

Based on these new restrictions, what is the closest a small craft is allowed to sail to the lighthouse?

Assignment 29   Ma 4

Fashion designers sometimes use miniature models to test their creations. Draw up patterns for the design of a modern suit of clothes. This may be either a coat and trousers or a female two-piece outfit. You must not consult prepared commercial patterns.

You may simply use your imagination here or you could study the way in which various modern styles of clothing are constructed. Study the stitching to decide on the number of individual pieces of cloth that will be required. Strips of paper placed on the clothing, for instance, will help you to discover the shapes you need.

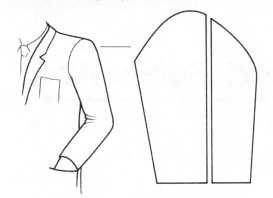

When you have obtained the necessary shapes, you then need to decide on a suitable scale according to your own choice of model. Test your patterns by attaching tabs so that the various pieces can be assembled to form the two items of clothing.

Assignment 30 Ma 2

This problem is based on a puzzle from a mathematical magazine published in Czechoslovakia several years ago.

Four circles form two interlocking paths and at the start numbers 1 to 5 occupy the positions as shown in the first diagram. By moving the numbers, you must arrange them to appear as they are in the second drawing.

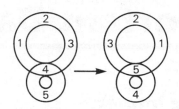

These are your rules:

(a) In the larger rings, the numbers can all move one position clockwise or anti-clockwise but must not change their order.

(b) In the smaller ring, the two numbers there at any time can change places.

(c) Moves must alternate in the large and small circles. A move in the large circle must be followed by an exchange in the small circle and vice versa. You may start in either circle.

You may take as many moves as you wish but the quickest method requires 14 moves only.

Appendix
Important Data and Formulae

Simple Interest

$$I = \frac{PRT}{100}$$

I = simple interest

P = principal (i.e. amount invested or borrowed)

R = rate per cent

T = time in years of investment or loan

Indices

$$a^m \times a^n = a^{m+n}$$

$$a^m \div a^n = a^{m-n}$$

$$(a^m)^n = a^{mn}$$

$$a^0 = 1$$

$$a^{-m} = \frac{1}{a^m}$$

Mensuration

Circumference of a circle $= 2\pi r$ or πd

Area of a circle $= \pi r^2$

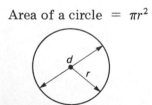

Area of a triangle $= \frac{1}{2} \times$ base $\times$ height (or altitude)

Area of a trapezium $= \frac{1}{2}(a + b)h$

Area of a rectangle $= a \times b$

Area of a square $= a^2$

Area of a parallelogram $= ah$

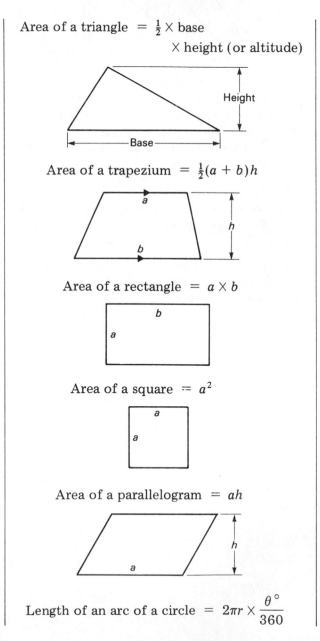

Length of an arc of a circle $= 2\pi r \times \dfrac{\theta^\circ}{360}$

Area of a sector of a circle $= \pi r^2 \times \dfrac{\theta^\circ}{360}$

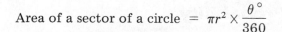

Volume of any solid having a uniform cross section

$\quad = $ Cross-sectional area $\times$ Length of solid

Volume of a cylinder $= \pi r^2 h$

Surface area of a cylinder $= 2\pi r(h + r)$

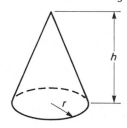

Volume of a cone $= \frac{1}{3}\pi r^2 h$

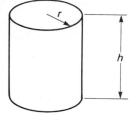

Volume of a sphere $= \frac{4}{3}\pi r^3$

Surface area of a sphere $= 4\pi r^2$

Volume of a pyramid $= \frac{1}{3}Ah$

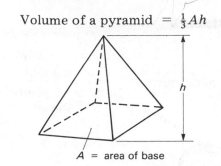

A = area of base

Equation of a Straight Line

$$y = mx + c$$

(m is the gradient, c is the intercept on the y-axis).

Profit and Loss

$$\text{Profit \%} = \frac{\text{Selling price} - \text{Cost price}}{\text{Cost price}} \times 100$$

$$\text{Loss \%} = \frac{\text{Cost price} - \text{Selling price}}{\text{Cost price}} \times 100$$

Pythagoras' Theorem

$$b^2 = a^2 + c^2 \quad \text{(see triangle below)}$$

Trigonometry

$$\sin A = \frac{\text{Side opposite to } \angle A}{\text{Hypotenuse}} = \frac{a}{b}$$

$$\cos A = \frac{\text{Side adjacent to } \angle A}{\text{Hypotenuse}} = \frac{c}{b}$$

$$\tan A = \frac{\text{Side opposite to } \angle A}{\text{Side adjacent to } \angle A} = \frac{a}{c}$$

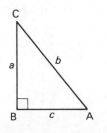

Statistics

$$\text{Arithmetic mean} = \frac{\text{Sum of the quantities in the set}}{\text{Number of quantities in the set}}$$

Probability

$$\text{Probability} = \frac{\text{Number of equiprobable events which produces a favourable outcome}}{\text{Total number of equiprobable events}}$$

Answers

Answers to Chapter 1

Exercise 1.1

1.

```
 -8   -6   -4   -2    0    3    5    7
```

2. $+3$ 3. $+2, 3, +5, 7$
4. $0, +2, 3, 5, 7$ 5. $-9, -5, -4, -3, -1$
6. $+25, 18, 12, 3, -2, -6, -8, -15$
7. (a) 12   (b) -10
   (c) $+5, +7, 8, 12$
8. -5
9. $+5, 7, +11$
10. Below the freezing point of water

Exercise 1.2

1. 40	2. 60	3. 90
4. 90	5. 80	6. 700
7. 700	8. 900	9. 400
10. 900	11. 3000	12. 5000
13. 8000	14. 9000	15. 6000

Exercise 1.3

1. 16	2. 18	3. 27
4. 20	5. 30	6. 462
7. 4710	8. 58 225	9. 8 599 293
10. 140 700		

Exercise 1.4

1. 13	2. -9	3. -32
4. -16	5. -4	6. -8
7. -14	8. 4	9. 3
10. -3	11. 6	12. 194
13. -227	14. -98	15. 181

Exercise 1.5

1. 4	2. 5	3. 3
4. 6	5. 5	6. 158
7. 896	8. 1429	9. 2037
10. 135 895	11. 1	12. 1
13. 8	14. 1	15. -6

Exercise 1.6

1. (a) 30   (b) 140
2. 40
3. (a) 60   (b) 280
4. 210   5. 1000
6. 3150   7. 9165   8. 17 010

9. 31 680	10. 110 880	11. -24
12. -10	13. -18	14. 8
15. -60	16. 24	17. -40
18. 60		

Exercise 1.7

1. 133	2. 15	3. 27
4. 16	5. 43	6. 59
7. 18	8. 36	9. 21
10. 350	11. -4	12. -3
13. 4	14. 2	15. -2
16. 3	17. -3	18. 3
19. -2	20. 4	21. -3
22. $+2$	23. 54	24. 28

25. 1 825 740   26. 779 471
27. 7 remainder 1   28. 7 remainder 2
29. 4 remainder 3   30. 1 remainder 5
31. 6 remainder 3   32. 5 remainder 7
33. 27 remainder 5   34. 31 remainder 12
35. 268 remainder 3

Exercise 1.8

1. 19	2. 9	3. 7
4. 44	5. 33	6. 5
7. 23	8. 25	9. 7
10. 2	11. -5	12. -12
13. 56	14. 34	15. -20

Exercise 1.9

1. 768	2. 256	3. 0
4. 0	5. 18	6. 0
7. 13	8. 8	9. 0
10. 0	11. -42	12. -5

Exercise 1.10

1. 250, 1250	2. 17, 21	3. 27, 33
4. 22, 11	5. 15, 21	6. 21, 34
7. 78, 158	8. 165, 489	9. $-6, -8$
10. 3, 5	11. 6, 8	12. 21, 89
13. 82, 244	14. 245, 2189	15. 9, 1

Miscellaneous Exercise 1

1. 27	2. 18	3. 0
4. 3	5. 6, 8, 16	6. 185 min
7. 9, 5, 1	8. $8°C$	

9. (a) 9, 5, 1   (b) 16, 8, 0
10. 900 000   11. 45   12. 36
13. 36, 49
14. (a) 40   (b) −13   (c) −4   (d) −300
15. −5, −3, 0, 3
16. 7866
17. $(5 + 2) \times (7 − 4) = 21$; $5 + 2(7 − 4) = 11$; 10
18. 3, 6, 96
19. (a) 44   (b) −26
20. (a) $27 + (5 − 12)$ (b) $13 − (4 − 9)$

Mental Test 1

1. 20	2. 20	3. 11
4. 9	5. 70	6. 90
7. 9	8. 11	9. 13
10. 12	11. −9	12. −30
13. −2	14. 3	15. −60
16. 4	17. −6	18. 54
19. 2	20. −2	

Answers to Chapter 2

Exercise 2.1

1. Odd	2. Even	3. Even
4. Odd	5. Odd	6. Even
7. Even	8. Odd	

9. (a) 17, 59, 121, 259   (b) 36, 98, 136
10. 19, 35, 89, 137   11. 36
12. 125
13. (a) 49   (b) 64
14. 8   15. 81
16. (a) 4   (b) 12
17. 18   18. 64   19. 90
20. 336
21. (a) 81   (b) 32   (c) 4096
22. 100   23. 432   24. 7
25. (a) 13   (b) 25   (c) 32   (d) 87
    (e) 125
26. (a) 7   (b) 11   (c) 25

Exercise 2.2

1. 8	2. 9	3. 9
4. 4	5. 1	6. 7
7. 9	8. 2	9. Yes
10. No	11. Yes	12. Yes
13. Yes	14. No	15. Yes
16. Yes	17. Yes	18. Yes
19. No	20. No	21. Yes
22. No	23. No	24. Yes
25. Yes	26. Yes	27. No
28. Yes	29. Yes	30. Yes
31. Yes	32. No	33. Yes
34. Yes	35. Yes	36. No
37. Yes	38. Yes	39. No
40. Yes	41. No	42. Yes
43. No	44. Yes	45. Yes

Exercise 2.3

1. (a) Yes   (b) No   (c) Yes   (d) Yes
   (e) No
2. (a) No   (b) No   (c) Yes   (d) No
   (e) Yes
3. (a) No   (b) Yes   (c) Yes   (d) No
   (e) Yes
4. (a) Yes   (b) Yes   (c) No   (d) No
   (e) Yes   (f) Yes   (g) No   (h) No
5. (a) Yes   (b) Yes   (c) Yes   (d) No
   (e) Yes   (f) No   (g) Yes   (h) Yes
   (i) No   (j) Yes   (k) No   (l) No
6. 7

Exercise 2.4

1. (a) 1, 3, 5, 15   (b) 1, 2, 4, 8, 16, 32, 64
   (c) 1, 2, 3, 4, 6, 8, 12, 16, 24, 48
2. 23, 29
3. (a) 24   (b) 33, 45, 61, 49 or 27
   (c) 61   (d) 49
4. (a) 21   (b) 7
5. (a) $2^3 \times 3$ (b) $2^3 \times 3^2$ (c) $3^2 \times 5$
6. (a) 30   (b) 12   (c) 30   (d) 60
   (e) 45
7. 1, 2, 4, 5, 10, 20; 1, 2, 3, 4, 6, 8, 12, 24;
   1, 2, 4; 4
8. 6, 12, 18, 24, 30, 36, 42, 48, 54, 60;
   10, 20, 30, 40, 50, 60; 30, 6; 30
9. (a) 10   (b) 3   (c) 2
10. (a) 2, 3, 4, 6, 12   (b) 6, 12, 18, 24
11. (a) 1, 2, 3, 4, 6, 12
    (b) 1, 2, 3, 4, 6, 9, 12, 18, 36
    (c) 1, 2, 3, 4, 5, 6, 10, 12, 15, 20, 30, 60
    (d) 1, 2, 4, 5, 10, 20, 25, 50, 100
12. 1, 2, 3, 4, 6, 7, 12, 14, 21, 28, 42, 84

Miscellaneous Exercise 2

1. 5 and 11
2. (a) 8, 16, 24, 32, 40   (b) 29, 31, 37
3. 36   4. 25   5. 28
6. 60   7. 36   8. 12
9. (a) 55, 60   (b) 60, 63, 81
   (c) 60, 122
10. $2^3 \times 3^2 \times 5$
11. (a) 30   (b) 60
12. (a) 1, 2, 3, 6, 7, 14, 21, 42
    (b) 15, 20, 25
13. (a) 61; yes   (b) 5291; no
14. (a) 57   (b) 72
15. (a) 2, 8, 34   (b) 2, 3, 5, 13
    (c) 123   (d) 89
16. (a) $2^2 \times 5^2$
    (b) 6, 12, 18, 24, 30, 36, 42, 48, 54
    (c) 29, 31, 37

Mental Test 2

1. Yes 2. No 3. 64
4. 16 5. No 6. 6
7. 40 8. 1 9. Yes
10. $1, 2, 3, 4, 6, 8, 12, 24$
11. $7, 14, 21, 28, 35$ 12. 2×3^2
13. 10 14. 5 15. 33

Answers to Chapter 3

Exercise 3.1

1. $\frac{1}{6}$ 2. $\frac{3}{4}$ 3. $\frac{3}{8}$
4. $\frac{3}{5}$ 5. $\frac{5}{7}$

Exercise 3.2

6. $\frac{15}{20}$ 7. $\frac{8}{12}$ 8. $\frac{12}{28}$
9. $\frac{12}{32}$ 10. $\frac{4}{24}$ 11. $\frac{27}{36}$
12. $\frac{2}{4}$ 13. $\frac{3}{5}$ 14. $\frac{5}{6}$
15. $\frac{3}{7}$

Exercise 3.3

1. $\frac{1}{2}$ 2. $\frac{3}{5}$ 3. $\frac{3}{5}$
4. $\frac{1}{8}$ 5. $\frac{2}{3}$ 6. $\frac{7}{8}$
7. $\frac{2}{3}$ 8. $\frac{2}{3}$ 9. $\frac{7}{8}$
10. $\frac{5}{8}$

Exercise 3.4

1. $3\frac{1}{2}$ 2. $2\frac{3}{8}$ 3. $2\frac{4}{9}$
4. $1\frac{1}{11}$ 5. $2\frac{5}{8}$ 6. $\frac{19}{8}$
7. $\frac{51}{10}$ 8. $\frac{26}{3}$ 9. $\frac{47}{7}$
10. $\frac{38}{9}$

Exercise 3.5

1. $\frac{1}{2}, \frac{7}{12}, \frac{2}{3}, \frac{5}{6}$ 2. $\frac{3}{5}, \frac{3}{4}, \frac{7}{8}, \frac{9}{10}$
3. $\frac{17}{32}, \frac{9}{16}, \frac{5}{8}, \frac{3}{4}$ 4. $\frac{1}{5}, \frac{1}{4}, \frac{1}{3}, \frac{1}{2}$
5. $\frac{5}{18}, \frac{2}{6}, \frac{3}{8}, \frac{5}{9}$

Exercise 3.6

1. $\frac{5}{6}$ 2. $1\frac{3}{10}$ 3. $1\frac{1}{8}$ 4. $\frac{11}{20}$
5. $2\frac{1}{8}$ 6. $1\frac{5}{24}$ 7. $4\frac{15}{16}$ 8. $14\frac{4}{15}$
9. $13\frac{23}{56}$ 10. $10\frac{2}{3}$ 11. $11\frac{5}{16}$ 12. $10\frac{13}{15}$

Exercise 3.7

1. $\frac{1}{6}$ 2. $\frac{1}{24}$ 3. $\frac{2}{15}$ 4. $\frac{1}{6}$
5. $2\frac{1}{8}$ 6. $2\frac{2}{7}$ 7. $2\frac{19}{80}$ 8. $2\frac{19}{40}$
9. $\frac{39}{40}$ 10. $\frac{11}{12}$

Exercise 3.8

1. $1\frac{3}{8}$ 2. $\frac{7}{20}$ 3. $6\frac{7}{8}$ 4. $\frac{2}{3}$
5. $8\frac{13}{80}$ 6. $8\frac{9}{16}$ 7. $1\frac{3}{8}$ 8. $1\frac{1}{24}$

Exercise 3.9

1. $\frac{8}{15}$ 2. $\frac{15}{28}$ 3. $\frac{10}{27}$ 4. $1\frac{19}{36}$
5. $4\frac{9}{10}$ 6. $5\frac{5}{6}$ 7. $1\frac{32}{45}$ 8. $2\frac{53}{56}$

Exercise 3.10

1. $1\frac{1}{3}$ 2. 4 3. $\frac{7}{16}$ 4. $1\frac{1}{2}$
5. $\frac{1}{24}$ 6. 4 7. $6\frac{3}{4}$ 8. $8\frac{1}{4}$
9. 12 10. 100 11. 3 12. 2

Exercise 3.11

1. $\frac{3}{5}$ 2. 8 3. $1\frac{1}{3}$ 4. $1\frac{1}{2}$
5. $\frac{2}{3}$ 6. $\frac{25}{26}$ 7. $1\frac{1}{5}$ 8. $3\frac{5}{6}$

Exercise 3.12

1. $\frac{5}{9}$ 2. $2\frac{1}{2}$ 3. $\frac{5}{6}$ 4. $\frac{2}{3}$
5. $2\frac{1}{2}$ 6. $\frac{2}{3}$ 7. $1\frac{1}{3}$ 8. $1\frac{1}{2}$

Exercise 3.13

1. £150 2. 4
3. (a) $\frac{13}{20}$ (b) $\frac{7}{20}$
4. $262\frac{1}{2}$ litres 5. 330 6. £1.60
7. 30 min 8. £15 950

Miscellaneous Exercise 3

1. $\frac{1}{12}$ 2. $\frac{2}{9}$ 3. $2\frac{1}{2}, 2\frac{3}{4}, 3$
4. $4\frac{1}{8}$ 5. $4\frac{33}{40}$ 6. $\frac{4}{7}$
7. $\frac{2}{3}$ 8. 40 9. $1, \frac{2}{3}$
10. $\frac{27}{16}$
11. (a) $\frac{3}{4}$ (b) $\frac{3}{20}$ (c) $\frac{9}{20}$ (d) $\frac{12}{13}$
12. (a) $\frac{2}{7}$ (b) $\frac{5}{22}$ (c) $\frac{5}{24}$ (d) $4\frac{1}{12}$
13. $\frac{1}{2}$ 14. $3\frac{3}{8}, 3\frac{3}{4}, 4\frac{1}{8}$ 15. 20
16. $\frac{3}{16}$ 17. $\frac{17}{32}$ 18. $1\frac{1}{2}$
19. £2500 20. $5\frac{1}{2}$

Mental Test 3

1. $\frac{7}{8}$ 2. $\frac{13}{20}$ 3. $\frac{1}{10}$
4. $\frac{7}{8}$ 5. $\frac{1}{10}$ 6. $\frac{2}{7}$
7. $\frac{1}{4}$ 8. 5 9. $\frac{1}{10}$
10. $\frac{1}{2}$ 11. $\frac{4}{5}$ 12. $\frac{37}{7}$

Answers to Chapter 4

Exercise 4.1

1. (a) $\frac{9}{100}$ (b) $\frac{9}{1000}$ (c) $\frac{9}{10}$ (d) 9
2. (a) 500   (b) $\frac{5}{100}$ (c) $\frac{5}{10}$ (d) $\frac{5}{100}$
3. (a) 8000   (b) $\frac{8}{1000}$ (c) $\frac{8}{10}$ (d) $\frac{8}{100}$
4. 0.396
5. 89.1
6. (a) 0.7   (b) 0.37   (c) 0.589
   (d) 0.009   (e) 0.03   (f) 0.017
7. (a) $\frac{2}{10}$ (b) $4\frac{6}{10}$ (c) $3\frac{58}{100}$
   (d) $\frac{256}{1000}$ (e) $\frac{4}{1000}$ (f) $\frac{36}{1000}$
   (g) $80\frac{29}{100}$ (h) $\frac{32}{1000}$

Exercise 4.2

1.

$$-1.8 \ -0.8 \quad 0.7 \qquad\qquad 3.7 \ 4.2 \qquad\qquad 7.2$$

$$-2 \ -1 \quad 0 \quad 1 \quad 2 \quad 3 \quad 4 \quad 5 \quad 6 \quad 7 \quad 8 \quad 9$$

2. (a) 0.8   (b) 4.9   (c) 2.0
   (d) 0.3   (e) 1.2
3. $-3, -9, -29$
4. $-6, -4, 0, 3$
5. $-52.05, -5.052, 5.025, 5.205, 5.502$

Exercise 4.3

1. 3.005   2. 24.176   3. 61.75
4. 55.355   5. 8.98   6. 0.828
7. 5.4109   8. 18.69   9. 60.43
10. -79.8

Exercise 4.4

1. (a) 4.5   (b) 45   (c) 450
2. (a) 78.93   (b) 789.3   (c) 7893
3. (a) 0.58   (b) 5.8   (c) 58
4. (a) 892.346   (b) 8923.46   (c) 89 234.6
5. (a) 82.643   (b) 826.43   (c) 8264.3
6. (a) 0.63   (b) 6.3   (c) 63
7. (a) 0.0028   (b) 0.028   (c) 0.28
8. (a) 0.08   (b) 0.8   (c) 8
9. (a) 28.9   (b) 2.89   (c) 0.289
10. (a) 2.817   (b) 0.2817   (c) 0.028 17
11. (a) 82.734   (b) 8.2734   (c) 0.827 34
12. (a) 0.004   (b) 0.0004   (c) 0.000 04
13. (a) 0.0625   (b) 0.006 25   (c) 0.000 625
14. (a) 0.000 38   (b) 0.000 038 (c) 0.000 003 8
15. (a) 0.652   (b) 0.0652   (c) 0.006 52

Exercise 4.5

1. 743.0266   2. 0.949 43   3. 0.2888
4. 7.411 25   5. 0.013 76   6. 3.5

7. 13.4   8. 1.9   9. 4.25
10. 0.045   11. 48.36   12. 896.98
13. 5.32   14. 0.918   15. 0.1395

Exercise 4.6

1. (a) 19.37   (b) 19.4
2. (a) 0.007 52   (b) 0.008   (c) 0.01
3. (a) 4.970   (b) 4.97
4. 1.33   5. 0.016
6. 189.74   7. 0.1380
8. 0.000 018 9

Exercise 4.7

1. (a) 24.9358   (b) 24.94   (c) 25
2. (a) 0.008 357   (b) 0.008 36   (c) 0.0084
   (d) 0.008
3. (a) 17 359 000   (b) 17 000 000
   (c) 20 000 000
4. (a) 0.0780   (b) 0.078   (c) 0.08
5. 48.9   6. 4
7. 0.5   8. 600
9. 108 070   10. 308.9 (4 s.f.)
11. 130 (2 s.f.)   12. 33 (2 s.f.)
13. 47 (2 s.f.)   14. 13 (2 s.f.)

Exercise 4.8

1. $20 + 40 + 430 = 490;\ 487.35$
2. $80 - 40 - 10 - 30 = 0;\ 1.685$
3. $20 \times 0.5 = 10;\ 13$
4. $40 \times 0.25 = 10;\ 11$
5. $0.7 \times 0.1 \times 2 = 0.14;\ 0.16$
6. $90 \div 30 = 3;\ 2.931$
7. $0.09 \div 0.03 = 3;\ 2.6$
8. $30 \times 30 \times 0.030 = 27;\ 24$
9. $\dfrac{1.5 \times 0.01}{0.05} = 0.3;\ 0.342$
10. $\dfrac{30 \times 30}{10 \times 3} = 30;\ 29.2$

Exercise 4.9

1. 0.75   2. 0.2   3. 0.875
4. 0.8125   5. 1.625   6. 2.593 75
7. 3.234 375   8. 0.555 56   9. 0.888 89
10. 0.177 78   11. 0.454 55   12. 0.353 54
13. 0.212 12   14. 0.428 43   15. 0.567 16
16. $0.\dot{2}$ 17. $0.1\dot{8}$ 18. $0.\dot{7}$
19. $0.4\dot{5}$ 20. $0.\dot{4}$ 21. $\frac{3}{10}$
22. $\frac{13}{20}$ 23. $\frac{3}{8}$ 24. $\frac{7}{16}$
25. $2\frac{31}{50}$ 26. $1\frac{3}{4}$ 27. $9\frac{37}{200}$
28. $7\frac{9}{25}$ 29. 0.0175   30. 1.748
31. 0.0895   32. 0.0001   33. 0.001 875

Exercise 4.10

1. 700	**2.** 700	**3.** 900
4. 3600	**5.** 4900	**6.** 50
7. 250	**8.** 500	**9.** 900
10. 17 400	**11.** 90	**12.** 380
13. 690	**14.** 590	**15.** 600

Exercise 4.11

1. 5, 198

2. -9

3. $1.57, \frac{1}{7}, -5.625, \sqrt{16}, 6.76$ and $-3\frac{1}{2}$ are all rational; $\sqrt{15}$ is irrational.

4. 0, 6

5. $-4.7, -3\frac{3}{4}, -3.4, 0, \frac{5}{8}, 2.5, 3, 7$

Exercise 4.12

1. 1.31	**2.** 2.93	**3.** 1.96
4. 3.55	**5.** 8.73	**6.** 9.51
7. 11.31	**8.** 18.49	

Miscellaneous Exercise 4

1. 0.63	**2.** 0.082	**3.** 9
4. 0.8i	**5.** 0.9	**6.** 49.95

7. (a) 738.06   (b) 738.1

8. (a) 0.9375   (b) 9.16

9. 0.02   **10.** 56.798

11. 2.52

12. (a) 0.0495   (b) 0.049

13. (a) 15.33   (b) 16.3

14. 62

15. 0.167

16. 0.027

17. (a) $\frac{1}{3}$ (b) 0.333

18. (a) 87 400   (b) 0.0326

19. -0.68

20. $1.6, \frac{2}{3}, 0.3, -0.7, -4.2$

21. 277.155

22. (a) 0.4375   (b) 0.05

23. (a) $\frac{19}{20}$ (b) $\frac{8}{25}$

24. 9.291

Mental Test 4

1. 2.55	**2.** 0.262	**3.** 1.12
4. $\frac{16}{}$	**5.** 0.06	**6.** $\frac{9}{100}$
7. $\frac{2}{5}$	**8.** 0.53	**9.** 4.3
10. 7.8	**11.** 357	**12.** 0.5321
13. 2.47	**14.** 54 000	**15.** 17.5
16. 0.8	**17.** 0.178	**18.** 4.75
19. 3600	**20.** Yes	

Answers to Chapter 5

Exercise 5.1

1. 50	**2.** 48	**3.** 9
4. 2000	**5.** 8300	**6.** 28 100
7. 700	**8.** 1940	**9.** 800
10. 4800	**11.** 80	**12.** 280
13. 32.5	**14.** 80	**15.** 896.4
16. 0.54	**17.** 8	**18.** 25.634
19. 0.0543	**20.** 3	**21.** 378.64
22. 0.0935	**23.** 18 000	**24.** 6200
25. 9	**26.** 0.46	**27.** 0.058
28. 0.018	**29.** 0.0075	**30.** 0.470

Exercise 5.2

1. 4	**2.** 2.5	**3.** 3.25
4. 6.2	**5.** 24	**6.** 27.6
7. 2439	**8.** 118.2	**9.** 6
10. 4.4	**11.** 16.3	**12.** 56.15
13. 12	**14.** 2.9	**15.** 22.5
16. 403	**17.** 96	**18.** 276
19. 189.6	**20.** 382.8	**21.** 252
22. 183.6	**23.** 244.8	**24.** 1951.2
25. 3	**26.** 6.4	**27.** 5
28. 9.3		

Exercise 5.3

1. 5	**2.** 6.98	**3.** 8000
4. 11 200	**5.** 9000	**6.** 39 000
7. 750	**8.** 45	**9.** 1.83
10. 0.085	**11.** 7000	**12.** 4900
13. 300	**14.** 9	**15.** 50
16. 8.57	**17.** 0.453	**18.** 0.045

Exercise 5.4

1. 4	**2.** 80	**3.** 72
4. 132.8	**5.** 7	**6.** 0.25
7. 0.75	**8.** 3.25	**9.** 448
10. 582.4	**11.** 13 440	**12.** 16 352
13. 4	**14.** 3.9	**15.** 160
16. 248	**17.** 6	**18.** 0.6
19. 4	**20.** 9.2	

Exercise 5.5

1. 600	**2.** 430	**3.** 25
4. 50	**5.** 800	**6.** 2000
7. 300	**8.** 5	**9.** 500
10. 30	**11.** 5.5	**12.** 0.4
13. 0.08	**14.** 3.25	**15.** 0.09

Exercise 5.6

1. 3	**2.** 1.25	**3.** 2.25
4. 72	**5.** 50	**6.** 80

7. 26   **8.** 9   **9.** 320
10. 48   **11.** 3   **12.** 0.65
13. 95   **14.** 2   **15.** 25

Exercise 5.7

1. 144 lb   **2.** 6 fl oz   **3.** 20 miles
4. 40 km   **5.** 4.5 lb   **6.** 2.27 kg
7. 4.92 tons   **8.** 28 litres

Exercise 5.8

1. 6.315 m   **2.** 93 462.05 m
3. 0.2668 m   **4.** 2.072 kg
5. 98.05 kg   **6.** 68.7 cm
7. 628 mm   **8.** 5 cm

Exercise 5.9

1. 120.65 m   **2.** 17.85 m
3. 47; 9 cm   **4.** 83
5. 32 000   **6.** 468 complete jars
7. 48   **8.** 0.041

Exercise 5.10

1. £15.97   **2.** £2.37   **3.** £100.64
4. £7.13   **5.** £2.48   **6.** £190.89
7. £701.69   **8.** £0.78

Exercise 5.11

1. £1.44   **2.** £5.22   **3.** £217.58
4. £21.00   **5.** 52 p   **6.** £2.49
7. £5.46   **8.** £5.72

Miscellaneous Exercise 5

1. £109.92   **2.** £59.85   **3.** £1.32
4. (a) 12 340   (b) 1.234   (c) 1.234
5. 14   **6.** 6.75 lb   **7.** 870 g
8. 40   **9.** 100 miles
10. (a) 2000   (b) 2056   (c) 6.72
     (d) 28.4
11. 60
12. (a) 3000   (b) 900   (c) 3.5
13. 13; 25 cm
14. $\frac{9}{20}$ **15.** 0.8 kg   **16.** 17.3 kg
17. £1203.75   **18.** £90.10   **19.** £3.68

Mental Test 5

1. 400   **2.** 5000   **3.** 2
4. 70   **5.** 6   **6.** 9200
7. 3   **8.** 2   **9.** 5280
10. 4   **11.** 8000   **12.** 7.5

13. 9200   **14.** 320   **15.** 336
16. 4   **17.** 100   **18.** 400
19. 48   **20.** 10   **21.** 3
22. 13   **23.** 12   **24.** 1950
25. 200   **26.** £3.75   **27.** £1.23
28. £5.70   **29.** 50 p

Answers to Chapter 6
Exercise 6.1

1. (a) 800 g butter, 640 g castor sugar, 2 eggs,
     1600 g self-raising flour, 400 g currants,
     120 g mixed peel, 2 pinches of salt.
     (b) 960 g   (c) 240 g   (d) 3200 g
2. 160 kg, 320 kg
3. 40 kg
4. (a) 15 kg   (b) 20 kg
5. (a) 240 g flour, 2 eggs, 500 ml milk
     (b) 3   (c) 480 g

Exercise 6.2

1. 1 : 4   **2.** 1 : 3   **3.** 6 : 7
4. 5 : 6   **5.** 8 : 7   **6.** 5 : 4
7. 3 : 2   **8.** 5 : 6 : 7   **9.** 5 : 6 : 9
10. 4 : 5 : 6 : 7

Exercise 6.3

1. $\frac{9}{7}$ **2.** $\frac{1}{2}$ **3.** $\frac{2}{1}$
4. $\frac{4}{5}$ **5.** $\frac{2}{3}$ **6.** $\frac{4}{5}$
7. $\frac{1}{3}$ **8.** $\frac{5}{6}$ **9.** $\frac{7}{8}$
10. $\frac{5}{9}$

Exercise 6.4

1. 9 : 4   **2.** 4 : 3   **3.** 4 : 3
4. 2 : 5   **5.** 9 : 1   **6.** 7 : 3
7. 1 : 2   **8.** 11 : 6   **9.** 16 : 39
10. 48 : 103

Exercise 6.5

1. 4 : 1   **2.** 5 : 1   **3.** 3 : 100
4. 3 : 1   **5.** 20 : 1   **6.** 1 : 10
7. 5 : 1   **8.** 10 : 3   **9.** 32 : 1
10. 100 : 3

Exercise 6.6

1. £500, £300   **2.** 112 kg, 48 kg
3. 24 m, 36 m, 60 m
4. 168 mm, 588 mm, 924 mm
5. £280
6. 15 kg, 22.5 kg; 37.5 kg
7. £88   **8.** £3.60

Exercise 6.7

1. £8.48
2. £48.125
3. £1.75
4. 36 litres
5. £224
6. 7 hours
7. 90
8. 4 kg

Exercise 6.8

1. 16 days
2. $13\frac{1}{2}$ days
3. 6
4. 6
5. $4\frac{1}{2}$ days
6. $10\frac{1}{2}$ min
7. 24

Exercise 6.9

1. 44 km
2. 20 min
3. 80 min
4. (a) 40 miles　(b) 200 miles
5. £1.32
6. 10 hours
7. 40 min
8. (a) £1　(b) £8

Exercise 6.10

1. (a) 6 p per kg　(b) 72 p
2. (a) 12 litres per min　(b) 3.75 min
   (c) 84 litres
3. (a) 30 miles per gallon　(b) 150 miles
   (c) 4 gallons
4. (a) 150 km　(b) 3.5 hours
5. 13 500 kg
6. 7
7. (a) 2 m/s　(b) 175 s

Exercise 6.11

1. 42.45
2. 4425
3. 14.88
4. 120.97
5. 5722.85
6. 4.18
7. 97.84
8. 15.63

Exercise 6.12

1. 8 m
2. 1600 m
3. (a) 2 m　(b) 6 cm
4. 25 cm
5. (a) 2.16 m　(b) 0.36 m
6. (a) 3 km　(b) 1.5 cm
7. 405 km
8. 5.76 in

Miscellaneous Exercise 6

1. £1.08
2. 31.5 km
3. 4.5 km
4. (a) 15　(b) £43.50
5. 7 : 8
6. 4 : 1
7. $\frac{1}{8}$
8. Yes
9. 30
10. 27.3 m
11. £178
12. 9
13. £3.48
14. 24
15. 277.50
16. 4 hours
17. 8
18. 58 p
19. 56.25
20. £34
21. 15
22. 36 ℓ
23. 46.06

Mental Test 6

1. 10 kg
2. 2 : 5
3. $\frac{3}{4}$
4. 4 : 1
5. £80, £120
6. 18 p
7. 1 day
8. 5
9. 500
10. 4 cm

Answers to Chapter 7

Exercise 7.1

1. 70%
2. 80%
3. 55%
4. 36%
5. 62%
6. 25%
7. 90%
8. 20%
9. 20%
10. 34%
11. 73%
12. 68%
13. 81.3%
14. 92.7%
15. 33.3%

Exercise 7.2

1. $\frac{8}{25}$
2. $\frac{6}{25}$
3. $\frac{3}{10}$
4. $\frac{9}{20}$
5. $\frac{3}{50}$
6. $\frac{3}{8}$
7. $\frac{2}{3}$
8. $\frac{29}{400}$
9. $\frac{1}{12}$
10. $\frac{5}{8}$
11. 0.78
12. 0.31
13. 0.482
14. 0.025
15. 0.0125
16. 0.0395
17. 0.201
18. 0.0196

	Fraction	Decimal	Percentage
19.	$\frac{11}{20}$	0.55	55
20.	$\frac{17}{50}$	0.34	34
21.	$\frac{7}{8}$	0.875	$87\frac{1}{2}$
22.	$\frac{19}{25}$	0.76	76
23.	$\frac{2}{25}$	0.08	8
24.	$\frac{3}{8}$	0.375	$37\frac{1}{2}$
25.	$\frac{3}{20}$	0.15	15
26.	$\frac{27}{100}$	0.27	27
27.	$\frac{9}{20}$	0.45	45
28.	$\frac{21}{400}$	0.0525	$5\frac{1}{4}$

Exercise 7.3

1. (a) 10　(b) 24　(c) 6
   (d) 2.4　(e) 21.315　(f) 2.516
2. (a) $12\frac{1}{2}$　(b) 20　(c) 16
   (d) 16.3　(e) 45.5
3. (a) 60%　(b) 27
4. 115 cm
5. 88.667 cm
6. (a) £7.20　(b) £13.20　(c) £187.50
7. 584 kg
8. 39 643

Exercise 7.4

1. £800   2. £20   3. £1000
4. 800   5. £1600   6. 2000 kg
7. £8000

Miscellaneous Exercise 7

1. (a) $\frac{12}{25}$ (b) $\frac{3}{50}$
2. (a) 0.64   (b) 0.07
3. $a = \frac{17}{20}$; $b = 85$; $c = 0.35$; $d = 35$; $e = \frac{3}{25}$; $f = 0.12$
4. (a) £1.80   (b) £26.60
5. (a) £12   (b) £68
6. (a) 12   (b) 60%
7. 88%   8. £13   9. £65.25
10. £144   11. 1100   12. 28%
13. £12   14. £6.30   15. 55%; 54
16. (a) £1.20   (b) £120   (c) £110.40

Mental Test 7

1. 30%   2. 40%   3. $\frac{7}{10}$
4. 0.24   5. 5   6. 16%
7. £20   8. 40 cm   9. 14
10. $\frac{2}{3}$

Answers to Chapter 8

Exercise 8.1

1. £175   2. £5
3. (a) £140   (b) £192.50   (c) £280
4. £7.00   5. £204.40

Exercise 8.2

1. (a) £6.25   (b) £7.50   (c) £10
2. (a) £4.50   (b) £6.00   (c) £10.50
3. £15.84   4. £138.60
5. (a) £7.52   (b) £52.64   (c) £184.24

Exercise 8.3

1. £38.00   2. £32.40   3. £57.00
4. £38.00   5. £15.25

Exercise 8.4

1. (a) £2.00   (b) £6.00   (c) £40.00
    (d) £100.00
2. £70.00   3. £240.00   4. £97.00
5. £168.00

Exercise 8.5

1. £443.33   2. £500.00   3. £700.00
4. £508.00   5. £620.00

Exercise 8.6

1. £765   2. £350   3. £66.28
4. £1800   5. £1200
6. (a) £4975   (b) £414.58
7. (a) £4555   (b) £1038.75
8. (a) £15 835   (b) £3858.75
9. (a) £9145   (b) £31 950
10. £9117.75

Miscellaneous Exercise 8

1. £116   2. £750   3. £4750
4. £5   5. £3.75   6. £240
7. £90   8. 12%
9. (a) £4000   (b) £1200
10. (a) £6.00   (b) £9.00
    (c) £303.00
11. £1600   12. £19.00
13. (a) £3020   (b) £4980   (c) £415
14. £13 545
15. (a) £140   (b) £5.39
    (c) £116.31
16. £138

Mental Test 8

1. £3   2. £160   3. £3
4. £15   5. £10   6. £60
7. £450   8. £25   9. £155
10. £500

Answers to Chapter 9

Exercise 9.1

1. £25   2. £110   3. £600
4. £1600   5. £82.50   6. £180
7. £480   8. £36   9. £119
10. £945

Exercise 9.2

1. £280   2. £600   3. £200
4. £90   5. £768   6. 4 years
7. 3 years   8. 4 years   9. 7%
10. 12%   11. 10%   12. £4000
13. £200   14. £1200   15. £850
16. 8%   17. 4 years

Exercise 9.3

1. £30.75   2. £331   3. £161.25
4. £1036.02   5. £2508.80   6. £26 450
7. £1983.75
8. £10 400; £10 077.70; £322.30;
    10% p.a. simple interest
9. £2742   10. £6395   11. £750.50
12. £328.30   13. £3432   14. £8338.50

Exercise 9.4

1. £1166.40   2. £26 622.32   3. £5242.88
4. £16 875

Miscellaneous Exercise 9

1. £80   2. £20
3. (a) £24   (b) £48
4. (a) £20   (b) £42   (c) £242
5. £500   6. £80   7. £5832
8. (a) £4318   (b) £4283.02
   (c) savings bond, £34.98
9. £2178   10. £5000   11. £3973.75

Mental Test 9

1. £10   2. £20   3. £1200
4. £146.90   5. £900

Answers to Chapter 10

Exercise 10.1

1. 25%   2. 20%   3. 50%
4. 75%   5. 20%   6. $12\frac{1}{2}$%
7. (a) £42   (b) 25%
8. (a) £10   (b) $33\frac{1}{3}$%

Exercise 10.2

1. £36   2. £280   3. £100
4. £15 000   5. £200   6. £17.28
7. 9p

Exercise 10.3

1. £180   2. £56   3. £77
4. £8.50   5. £0.91

Exercise 10.4

1. £282   2. £45.96   3. £376
4. £47   5. £117.45   6. £195.75
7. £23.83

Exercise 10.5

1. £216   2. £120
3. £0.90 in the £   4. 25p in the £
5. £4 644 000   6. 9.1p in the £
7. £85 000
8. (a) £87 960   (b) 55p in the £
9. £39 100   10. 2.9p in the £

Exercise 10.6

1. £37.50   2. £50.00
3. £96   4. £32
5. £5.25   6. £60.00
7. £9.54
8. (a) £6000   (b) £5241
9. £600
10. £973.38; 8.1p per km
11. 5.98p per km

Miscellaneous Exercise 10

1. (a) £80   (b) $66\frac{2}{3}$%
2. (a) £150   (b) 50%
3. £60
4. (a) £8.75   (b) £58.75
5. £500 per annum
6. £90   7. £84   8. £27 000
9. £260.40   10. £48   11. £15.75
12. £4.05   13. £17.50   14. £2700
15. (a) £835 000   (b) £7 181 000
16. £120   17. £64.84   18. £1.88

Mental Test 10

1. £2   2. £500   3. 20%
4. 20%   5. £160   6. £7.50
7. £450   8. £5000   9. £10
10. £160

Answers to Chapter 11

Exercise 11.1

1. £25 200   2. £47.77   3. £43.24
4. £292.80   5. £260.53   6. £267.50
7. £16.67

Exercise 11.2

1. £120
2. (a) £120   (b) £12   (c) £132
   (d) £13.20
3. £24.75
4. (a) £128   (b) £512.00   (c) £61.44
   (d) £573.44   (e) £143.36
5. (a) £260   (b) £60

Exercise 11.3

1. £60   2. £144
3. (a) £93.15   (b) £11 178
4. £28.01   5. 20%

Exercise 11.4

1.

Charge per kWh	Number of kWh used	Cost of gas used in pence
1.256	3263	4098
1.492	2531	3776
2.063	7235	14 926
2.412	12 500	30 150

2. £50.35   3. (a) 6397   (b) £114.49
4. 1.563   5. 8350

Exercise 11.5

1.

Charge per unit	Number of units used	Cost of electricity used	
		In pence	In pounds
7.71	100	771	7.71
2.63	600	1578	15.78
9.93	450	4469	44.69
8.17	800	6536	65.36
12.95	750	9713	97.13

2. £76.44   3. £216.65   4. 116; £17.66
5. £145.16

6.

(a)

(b)

(c)

7. (a) 23 968   (b) 84 947   (c) 46 936
8. 1200

Exercise 11.6

1. £19.07   2. £24.79   3. £24.25
4. (a) £28.94   (b) £45.09   (c) £52.99
5. (a) 1252   (b) £61.35   (c) £78.64
   (d) £13.76   (e) £92.40

Miscellaneous Exercise 11

1. £6
2. (a) £52.50   (b) £352.50   (c) £39
   (d) £315
3. (a) 24 612   (b) 25 372

(c)

   (d) £67.40
4. Quarterly rental charge   £9.75
   Dialed units 600 at 4.7 p per unit   £28.20
   Total exclusive of VAT   £37.95
   VAT at 17.5%   £6.64
   Total payable   £44.59

5. (a) £495.72   (b) £384.80
   (c) Economy; £110.92
6. (a) £375   (b) £2875
   (c) £239.58
7. (a) £2340   (b) £53.10
   (c) £248.10
8. (a) £27 000   (b) £313.20
9. (a) £230.40   (b) £49.60
10. (a) £175   (b) £836.50
   (c) £136.50

Mental Test 11

1. £40   2. £200   3. £720
4. £650   5. £60   6. £60
7. £100

Answers to Chapter 12

Exercise 12.1

1. (a) 240 min   (b) 210 min
2. (a) 540 s   (b) 135 s
3. 5 hours 50 min   4. 18 hours 7 min
5. 17 hours 15 min   6. 12 hours 39 min
7. 11 hours 47 min   8. 12 hours 6 min
9. 12 hours 39 min   10. 12 hours 47 min

Exercise 12.2

1. 0730; 15 min
2. 0937; 2 hours 12 min
3. 0835; 3 min
4. 1518; 1607; 49 min
5. 1040 from Paddington; 29 min
   1202 from Stroud

Exercise 12.3

1. 2 hours
2. 4 hours
3. 4 s
4. 60 miles
5. 140 km
6. 480 m
7. 15 mile/h
8. 40 km/h
9. 16 m/s
10. 300 km
11. 9 km/h
12. 35.2 mile/h
13. 26 km/h
14. 75 km/h
15. 80 km/h

Miscellaneous Exercise 12

1. (a) 2015   (b) 30 min
2. 1815 hours
3. (a) 2 hours 30 min   (b) 50 km/h
4. (a) 1 hour 30 min   (b) 8 mile/h
5. (a) 2.35 p.m.   (b) 30 mile/h
6. (a) 4 hours   (b) 64 km/h
7. (a) 1210   (b) 40 mile/h
8. (a) 36 min   (b) 2 hours 24 min
   (c) 3 hours   (d) 39 km/h
9. (a) 360 km   (b) 4 hours 30 min
10. (a) 8 m/s   (b) 28.8 km/h

Mental Test 12

1. 4 min
2. 240 s
3. 3 h
4. 300 min
5. 5 days
6. 2 h 15 min
7. 3 h 20 min
8. 3 h 8 min
9. 40 km/h
10. 50 mile/h
11. 150 km
12. 2 h
13. 1500 miles
14. 48 km/h
15. 3 h
16. 29 min
17. 1222
18. 426
19. 27 min
20. Yes

Coursework 1

1. 12×84, 12×63, 12×42, 13×62, 13×93, 14×82, 23×64, 23×96, 24×84, 24×63, 26×93, 34×86, 36×84, 46×96
4. (b) If you have struggled to make 73 without success, here is a clue: you will need 4!, $\sqrt{4}$ and $\sqrt{.4}$ twice – but it's still tough!
   (d) The largest number possible is $(4! \times 4! \times 4! \times 4!)$ or 331 776.
   (e) Using the same rules applied to the four fours, all the numbers from 1 to 18 can be made with three threes except 11. If you broaden the rules to allow the threes to be 'glued together', 11 can be made by the division $33 \div 3$.
7. The month is February 1984. Under normal circumstances, the day/date arrangement for Februarys in leap years repeats once every 28 years – but there is a catch! Since 1899, only the years 1928 and 1956 had Februarys identical with that of 1984: 1900 did not match

perfectly because it was not a leap year. The pattern will not be repeated until February 2012.

8. Two examples for each number:

```
    2           9           3           2
  1 7         5 3         7 8         7 5
 9 0 6       1 0 4       5 1 2       3 1 6
8 3 4 5     8 6 2 7     4 0 9 6     9 4 0 8

    0           0           2           5
  7 6         9 7         5 4         6 7
 5 2 9       5 3 6       7 3 9       2 4 0
4 8 3 1     1 4 8 2     1 6 8 0     8 1 3 9

    8           4           0           0
  3 9         7 2         9 4         7 4
 1 5 0       3 5 9       1 6 8       5 6 9
7 4 6 2     1 6 8 0     7 2 3 5     2 3 8 1

    3           5           0           1
  6 2         6 9         9 4         6 5
 4 7 9       1 8 0       1 8 7       4 9 7
1 8 5 0     4 3 7 2     6 2 3 5     2 3 8 0

    9           3           5           8
  1 0         5 2         4 0         4 6
 7 2 6       7 4 9       3 7 6       0 9 1
3 8 4 5     0 6 8 1     8 1 2 9     5 3 7 2
```

9. Four numbers: Line of 0's; Three numbers: 101, 110 and 011 will repeat; Line of 0's is always possible using 2^x numbers.
10. (a) Break four links in one length and use the four to join the other four lengths in a loop. Cost: £8.
    (b) Break each link in the 3 and 4 lengths. Use the seven separate links to join the remaining seven pieces in a loop. Cost: £14.

Answers to Chapter 13

Exercise 13.1

1. $7a$
2. $5b - 7$
3. $5x + 3y$
4. xyz
5. $8mn$
6. $\dfrac{2p}{3q}$
7. $9r + 3s$
8. $\dfrac{xy}{p}$

Exercise 13.2

1. 21
2. -15
3. 18
4. -2
5. 13
6. 1
7. 3
8. 2
9. 9
10. 0
11. 19
12. 15
13. -21
14. -42
15. -126

16. 90   17. 4   18. 2
19. 49   20. -71 21. $-\frac{7}{3}$
22. $-\frac{6}{7}$

Exercise 13.3

1. 8   2. 81   3. 25
4. 162   5. 48   6. -48
7. 200   8. -750 9. 360
10. 13   11. -16 12. 43

Exercise 13.4

1. $8x$ 2. $4y$ 3. $-3p$
4. $-3q$ 5. $-2x$ 6. $10p$
7. $-6x$ 8. $-2m$
9. $13x - 9y + 15c$ 10. $5a - 2b - 5c$

Exercise 13.5

1. $6py$ 2. $-6mp$ 3. $-20ab$
4. $30abc$ 5. $12xyz$ 6. $-14mnp$
7. $-6abc$ 8. $-120qyz$ 9. $-6m^2$
10. $-24p^2q^2$

Exercise 13.6

1. $3a + 6b$ 2. $10x - 15y$ 3. $20p - 15q$
4. $-p + q$ 5. $-6x + 8y$ 6. $-20m + 15n$
7. $13a - 21b$ 8. $-4p - 7q$ 9. $25n$
10. $7x - 9y$

Exercise 13.7

1. $2x$ 2. $-\dfrac{a}{b}$ 3. $\dfrac{x}{y}$

4. $\dfrac{2a}{b}$ 5. $2b$ 6. $3xy$

7. $-2ab$ 8. $2ab$ 9. $7ab$
10. $2xyz^2$

Exercise 13.8

1. $3(x + 2)$ 2. $4(y - 2)$ 3. $2(p - 2)$
4. $5(m + 2)$ 5. $2(x + y)$ 6. $3(p - q)$
7. $x(ax + b)$ 8. $mx(1 - x)$
9. $(a + b)(x + y)$ 10. $(x - y)(a - b)$

Exercise 13.9

1. $\dfrac{2}{ab}$ 2. $\dfrac{3y}{2x}$ 3. $\dfrac{4qs}{r}$

4. $\dfrac{6a^2d^3}{bc}$ 5. $\dfrac{b^2c}{a}$ 6. $\dfrac{px}{2qy}$

7. $\dfrac{by}{ax}$ 8. $\dfrac{pq}{mn}$

Exercise 13.10

1. $\dfrac{9a}{20}$ 2. $\dfrac{3p - 2q}{6}$ 3. $\dfrac{11a}{20}$

4. $\dfrac{13x}{12}$ 5. $\dfrac{13}{6x}$ 6. $\dfrac{7m}{6x}$

7. $\dfrac{14x}{15y}$ 8. $\dfrac{x + 5}{3}$ 9. $\dfrac{10x - 3}{12}$

10. $\dfrac{9m + 2}{2}$ 11. $\dfrac{-9 - x}{12}$ 12. $\dfrac{9}{4x}$

13. $\dfrac{x + 6}{6}$ 14. $\dfrac{a + 12b}{6}$ 15. $\dfrac{25 - 7x}{10}$

Miscellaneous Exercise 13

1. 26   2. $\frac{1}{4}$ 3. 74
4. 30   5. $54°F$ 6. 270
7. 2
8. (a) $6a$ (b) $16a^2$ (c) 4
9. (a) $5x + 7y$ (b) $2x + 5y$ (c) $-2y$
10. 2
11. (a) $30a^2b$ (b) $5x - 6y$
12. (a) 0   (b) 0   (c) $\frac{3}{4}$
13. (a) 4   (b) -11
14. $\dfrac{x}{2}$ 15. $\dfrac{x + 8}{6}$ 16. $\dfrac{3ab^2c^2d}{2}$

17. $\dfrac{21b^2}{10ac}$

Mental Test 13

1. $11x$ 2. $4a$ 3. $3b$
4. $4p$ 5. $6ab$ 6. p^2
7. $12m^2$ 8. $5a$ 9. $4a$
10. $4p$ 11. $3p + 3q$ 12. $-x + y$
13. $-2m - 2n$ 14. $\dfrac{ap}{bq}$ 15. $-2x$

Answers to Chapter 14

Exercise 14.1

1. 5^6 2. a^7 3. p^9
4. y^{23} 5. 2^8 6. $24a^6$
7. $15y^7$ 8. $18p^7$

Exercise 14.2

1. 3^2 2. p^2 3. d
4. q^4 5. m 6. t^5
7. m 8. $4p^3$

Exercise 14.3

1. y^8
2. 7^6
3. $5^3 c^6$
4. $5^5 a^5 b^{10} c^{15}$
5. $\dfrac{3^4 a^8}{2^4 b^{12}}$
6. m^{15}
7. $3^4 b^4$
8. $2^3 x^6 y^3$
9. $\dfrac{3^4}{4^4}$
10. $\dfrac{1}{2^5}$

Exercise 14.4

1. 5^{-1}
2. 3^{-2}
3. a^{-3}
4. b^{-5}
5. $2x^{-3}$
6. $5x^{-1}$
7. m^{-6}
8. $8y^{-7}$
9. $\frac{1}{3}$
10. $\dfrac{1}{5^2}$
11. $\dfrac{1}{a^3}$
12. $\dfrac{1}{m^5}$
13. $\dfrac{3}{b}$
14. $\dfrac{7}{q^3}$
15. $\frac{1}{10}$
16. $\frac{1}{9}$
17. $\frac{1}{16}$
18. $\frac{1}{25}$
19. $\frac{1}{27}$
20. $\frac{1}{1000}$
21. 1
22. 3
23. 1
24. 1
25. (a) $x^{1/5}$ (b) $x^{1/4}$ (c) $x^{1/7}$
26. (a) 2 (b) 2 (c) 2
27. 30
28. $\sqrt{2} = 1.414$

Exercise 14.5

	10^6	10^5	10^4	10^3	10^2	10^1	10^0
1.					9	3	6
2.				7	2	5	4
3.				8	6	1	9
4.			1	8	3	4	2
5.		2	7	8	9	4	8
6.		5	6	3	7	4	0
7.	8	6	2	1	9	7	5
8.				7	3	0	0
9.			8	3	5	0	7
10.			9	0	0	0	0	7

11. $7 \times 10^2 + 5 \times 10^1 + 3 \times 10^0$
12. $3 \times 10^5 + 8 \times 10^4 + 4 \times 10^3 + 6 \times 10^2 + 8 \times 10^1 + 5 \times 10^0$
13. $1 \times 10^5 + 7 \times 10^4 + 3 \times 10^3 + 2 \times 10^2 + 7 \times 10^1 + 8 \times 10^0$
14. $2 \times 10^5 + 3 \times 10^4 + 5 \times 10^0$
15. $3 \times 10^6 + 9 \times 10^5 + 7 \times 10^3 + 6 \times 10^2$

Exercise 14.6

	10^4	10^3	10^2	10^1	10^0	10^{-1}	10^{-2}	10^{-3}	10^{-4}
1.						5	2		
2.						3	7	5	
3.						9	1	3	4
4.	5	6	3	2	7	5	8		
5.				9	6	3			
6.			1	8	7	2	5		
7.		3	0	5	9	2	7	3	
8.					0	6	3	5	
9.		7	3	0	0	2	7	6	3
10.	8	3	5	0	7	0	0	3	8

11. (a) $5 \times 10^{-1} + 2 \times 10^{-2}$
   (b) $3 \times 10^{-1} + 7 \times 10^{-2} + 5 \times 10^{-3}$
   (c) $9 \times 10^{-1} + 1 \times 10^{-2} + 3 \times 10^{-3} + 4 \times 10^{-4}$
   (d) $5 \times 10^4 + 6 \times 10^3 + 3 \times 10^2 + 2 \times 10^1 + 7 \times 10^0 + 5 \times 10^{-1} + 8 \times 10^{-2}$
   (e) $9 \times 10^1 + 6 \times 10^0 + 3 \times 10^{-1}$
   (f) $1 \times 10^2 + 8 \times 10^1 + 7 \times 10^0 + 2 \times 10^{-1} + 5 \times 10^{-2}$
   (g) $3 \times 10^3 + 5 \times 10^1 + 9 \times 10^0 + 2 \times 10^{-1} + 7 \times 10^{-2} + 3 \times 10^{-3}$
   (h) $6 \times 10^{-2} + 3 \times 10^{-3} + 5 \times 10^{-4}$
   (i) $7 \times 10^3 + 3 \times 10^2 + 2 \times 10^{-1} + 7 \times 10^{-2} + 6 \times 10^{-3} + 3 \times 10^{-4}$
   (j) $8 \times 10^4 + 3 \times 10^3 + 5 \times 10^2 + 7 \times 10^0 + 3 \times 10^{-3} + 8 \times 10^{-4}$

12. 2753
13. $84\,605$
14. 800.073
15. $9\,076\,083.26$
16. 62.754

Exercise 14.7

1. 8.27×10^2
2. 1.734×10^3
3. 1.7632×10^4
4. $8.937\,62 \times 10^5$
5. 8.036×10^6
6. 3200
7. 500
8. $1\,870\,000$
9. $43\,200$
10. $876\,200$

Exercise 14.8

1. 3×10^{-1}
2. 5×10^{-2}
3. 5.6×10^{-3}
4. 7×10^{-6}
5. 6.7×10^{-5}
6. 0.002
7. 0.567
8. 0.032
9. $0.000\,005$
10. $0.000\,41$

Exercise 14.9

1. 2.1×10^3 by 1520
2. 3.95×10^4 by $30\,100$
3. 8.58×10^4 by $75\,930$
4. 2.1×10^{-2} by 0.0156

5. 1.26×10^{-1} by 0.11727
6. 1.2489×10^4
7. 1.858×10^{-1}
8. 1.04×10^1
9. 4×10^{-5}
10. 2.758×10^3
11. 8.3228×10^4
12. 8.1368×10^{-2}
13. 8.322×10^{-2}
14. 4.55×10^6
15. 1.5×10^{-2}
16. 4.22×10^{10}
17. 3.4×10^{-2}
18. 5.93×10^0

Miscellaneous Exercise 14

1. 1.2×10^6
2. 3.397×10^3
3. $0.005\,01$
4. 8.93×10^{-4}
5. (a) 3.2×10^3 (b) 6×10^5
   (c) 2.8×10^3 (d) 1.5×10^1
6. (a) $\frac{1}{9}$ (b) 1 (c) $\frac{1}{10}$
7. 1.2×10^3 by 327 8. 10^{15}
9. $x = 4$ 10. $2^6 = 64$ 11. 0.08
12. (a) $n = 4$ (b) $n = 1$ (c) $n = 4$
13. 3×10^{-13} 14. $0.000\,067$ 15. 3.57×10^{-2}
16. (a) $\frac{1}{8}$ (b) 1

Mental Test 14

1. 8 2. $\frac{1}{3}$ 3. $9a^4$
4. a^6 5. a^2 6. a^{10}
7. $\dfrac{a^6}{b^3}$ 8. 32 9. 8×10^3
10. 3×10^{-3}

Answers to Chapter 15

Exercise 15.1

1. 2 2. 2 3. 2
4. 6 5. 6 6. 4
7. 4 8. 3 9. 12
10. 9 11. 10 12. 12
13. 8 14. 15 15. 21
16. 24 17. 10

Exercise 15.2

1. 8 2. 4 3. 2
4. -1 5. -2 6. 5
7. 6 8. 6 9. 2
10. 9 11. 2 12. -4

Exercise 15.3

1. 15 2. 21 3. 7
4. 12 5. 15 6. $\frac{6}{5}$
7. $\frac{15}{4}$ 8. -15 9. $\frac{50}{47}$
10. $\frac{15}{28}$ 11. $\frac{1}{2}$ 12. 13

Exercise 15.4

1. $3x + 5$ 2. $\dfrac{sP}{r}$
3. $x - 5$ 4. $(5a + 7b)$ pence
5. (a) $5x + y + z$ (b) $q(5x + y + z)$
6. Philip £$(p - n)$; Richard £$(q + n)$
7. £$Y + \dfrac{Nx}{100}$ 8. $6a + 4b + c$

Exercise 15.5

1. 18 and 19 2. 15
3. 15 4. 9
5. (a) $8, 16, 32$ (b) (i) $y = 2x$ (ii) 32
6. (b) $n + 1, n + 2$ (c) $3(n + 1)$
   (d) 24 (e) $3(n + 1) = 39$
   (f) $12, 13, 14$
7. (a) 316
   (b) (i) $n, n + 1, n + 5, n + 6$
       (ii) $t = 4n + 12$
   (c) 77
8. 15

Exercise 15.6

1. 32 2. 80 3. 360
4. 144 5. 704 6. 21
7. 6

Exercise 15.7

1. 0.5 2. 5 3. 3
4. 4 5. 100 6. 8.59
7. 2.55 8. 6

Exercise 15.8

1. $V = \dfrac{c}{P}$ 2. $R = \dfrac{I}{PT}$
3. $h = \dfrac{E}{mg}$ 4. $a = xy$
5. $R = \dfrac{E}{I}$ 6. $T = \dfrac{PV}{R}$
7. $t = \dfrac{ST}{s}$ 8. $p = v - 3$
9. $q = 3 - p$ 10. $a = \dfrac{v - u}{t}$
11. $r = \dfrac{n - p}{3}$ 12. $x = \dfrac{b - a}{c}$
13. $m = \dfrac{y - c}{x}$ 14. $D = TL + d$
15. $L = \dfrac{S(C - F)}{P}$

Exercise 15.9

1. $2 < 5$
2. $\frac{1}{2} > \frac{1}{3}$
3. $10\,mm = 1\,cm$
4. $£1 > 80p$
5. $9 > 1$
6. $\frac{2}{3} > \frac{4}{9}$
7. $-3 > -7$
8. $7 < 9$
9. $x^2 > x^2 - 3$
10. $2 < 4$

Exercise 15.11

1. $5, 6, 7, 8$
2. $4, 5, 6, 7, 8, 9$
3. $-2, -1, 0, 1, 2$
4. $5, 6, 7, 8$
5. $-3, -2, -1, 0, 1, 2, 3, 4$

Exercise 15.12

1. $x \geqslant 4$
2. $x < 5$
3. $x < -2$
4. $x \leqslant 8$
5. $x < 6$
6. $x \geqslant 5$
7. $x \leqslant 12$
8. $x < -2$

Exercise 15.13

1. 45
2. 43
3. 256
4. 6561
5. (a) $2, 6, 14, 26, 42, 62, 86, 114$
6. (a) n^2 (b) 81
7. (b) 33
8. (a) $G4 = 40;\ G5 = 60;\ G6 = 84$
  (b) $2n(n + 1)$ (c) 760 (d) $G13$
9. (a) $4n^2$ (b) 400
10. (a) $8 + 2(n - 1)$

(b)

Number of black tiles	1	2	3	4	5	6	7	8	9
Number of white tiles	8	10	12	14	16	18	20	22	24

(c) 31

Miscellaneous Exercise 15

1. 70
2. 28
3. 62.8
4. 78.5
5. 628
6. 11
7. (a) $5^2 - 3^2 = 4^2$ (b) $9^2 - 3^2 > 3^2$
  (c) $\frac{5}{9} - \frac{1}{3} < \frac{2}{3}$ (d) $\sqrt{14.4} > 1.2$
  (e) $0.5^2 > 0.5^3$
8. $x = -3$
9. (a) $x = 5$ (b) $x = 3$
10. $x = 3$
11. (a) $x = \frac{1}{2}$ (b) $x = 4$
12. (a) $p(p - 5)$ (b) $x^2(2x + y)$
13. 1080
14. 2
15. $t = \dfrac{v - u}{a}$
16. 22
17. $N = 3(T - 20)$

Mental Test 15

1. 3
2. 8
3. 3
4. 6
5. 10
6. $N = \dfrac{P}{M}$
7. $a = xy$
8. 4

Answers to Chapter 16

Exercise 16.1

1. $x = 5,\ y = 3$
2. $x = 4,\ y = 2$
3. $x = 3,\ y = 4$
4. $x = 1,\ y = 2$
5. $x = 2,\ y = 1$
6. $x = 5,\ y = 2$
7. $x = 4,\ y = 3$
8. $x = 2,\ y = 7$

Exercise 16.2

1. $x = 2,\ y = 3$
2. $x = 5,\ y = 3$
3. $x = 3,\ y = 2$
4. $x = 2,\ y = 5$
5. $x = 1,\ y = 2$
6. $x = 1,\ y = 5$
7. $x = 5,\ y = -1$
8. $x = 4,\ y = 5$

Exercise 16.3

1. $x^2 + 4x + 3$
2. $x^2 + 6x + 8$
3. $x^2 + 6x + 5$
4. $x^2 + 9x + 18$
5. $x^2 - 3x + 2$
6. $x^2 - 7x + 12$
7. $3x^2 - 17x + 10$
8. $10x^2 - 23x + 12$
9. $6x^2 - 7x - 3$
10. $x^2 - 6x + 9$
11. $25x^2 + 30xy + 9y^2$
12. $4x^2 - 25$

Exercise 16.4

1. 3.9
2. 5.74
3. 31
4. 3.4
5. 6.3

Exercise 16.5

1. 8
2. 44
3.

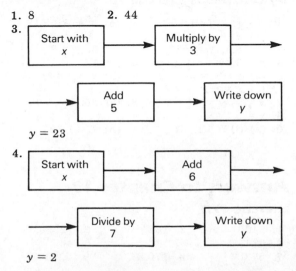

$y = 23$

4.

$y = 2$

5.

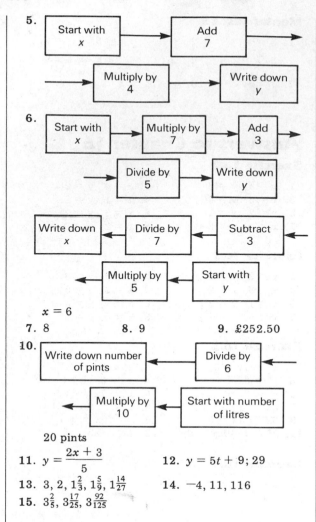

6.

$x = 6$

7. 8   **8.** 9   **9.** £252.50

10.

20 pints

11. $y = \dfrac{2x + 3}{5}$ **12.** $y = 5t + 9; 29$

13. $3, 2, 1\frac{2}{3}, 1\frac{5}{9}, 1\frac{14}{27}$ **14.** $-4, 11, 116$

15. $3\frac{2}{5}, 3\frac{17}{25}, 3\frac{92}{125}$

Miscellaneous Exercise 16

1. (a) $x^2 + 7x + 10$ **(b)** $x^2 - 9x + 18$
(c) $x^2 - 4x - 21$
2. $x = 0.5, y = -3$ **3.** $x = 3, y = 2$
4. $x = 3, y = -1$ **5.** $p = 12$
6. (a) $x + y = 5$ **(b)** $x - y = -7$
(c) $x = -1, y = 6$
7. $x = 7, y = -1$ **8. (c)** 20
9. $x = 9.1$
10. (a) (i) 7   **(ii)** 21   **(b)** $T = 1.62$

Answers to Chapter 17

Exercise 17.1

1. $240°$ **2.** $135°$ **3.** $72°$
4. $324°$ **5.** $54°$ **6.** $46.5°$
7. $57.0°$ **8.** $170.3°$ **9.** $52°42'$

10. $80°6'5''$ **11.** $36.3°$ **12.** $8.2°$
13. $23°48'$ **14.** $17°56'54''$
15. (a) $12.3833°$ **(b)** $35.6619°$
(c) $71.9017°$
16. (a) $37°10'58''$ **(b)** $71°33'24''$
(c) $12°28'4''$

Exercise 17.2

1. A acute; B reflex; C acute; D obtuse;
E reflex; F acute; G obtuse
2. (a) $60°$ **(b)** $22.5°$ **(c)** $180°$
(d) $135°$ **(e)** $54°$ **(f)** $31.5°$
3. (a) $45°$ **(b)** $120°$ **(c)** $72°$
(d) $60°$
4. (a) $50°$ **(b)** $133°$ **(c)** $52°$
(d) $125°$ **(e)** $270°$
5. (a) $33°$ **(b)** $56°$ **(c)** $153°$
(d) $B = 82°$

Exercise 17.3

1. $a = 103°$
2. $x = 101°, y = 101°$
3. $m = 40°, n = 40°, p = 130°, q = 310°$
4. $y = 117°$
5. $p = 90°, q = 80°, r = 60°, s = 120°$
6. $x = 72°, y = 134°$
7. $a = 85°$
8. $u = 62°, v = 72°, w = 53°, x = 55°,$
$y = 118°, z = 108°$

Mental Test 17

1. $\frac{1}{360}$ **2.** $60°$ **3.** $108°$
4. Acute   **5.** $30°$ **6.** $50°$
7. (a) $x = 140°$ **(b)** $x = 130°$ **(c)** $x = 270°$
8. $A = 50°, B = 130°$
9. $b = 110°, c = 110°, d = 70°, e = 70°,$
$f = 110°, g = 70°, h = 110°$
10. $80°$

Answers to Chapter 18
Exercise 18.1

1. (a)   **(b)**

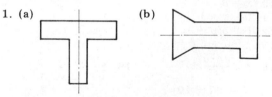

(c) (d)

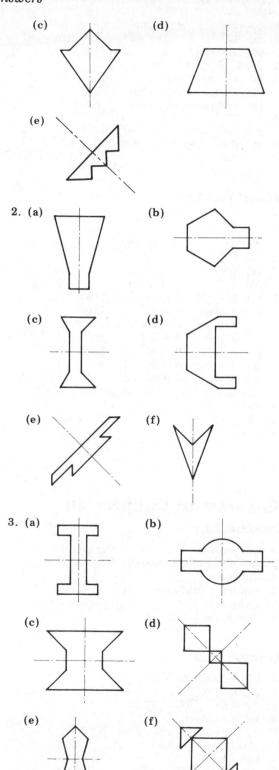

(e)

2. (a) (b)

(c) (d)

(e) (f)

3. (a) (b)

(c) (d)

(e) (f)

4.

Exercise 18.2

1. (a) 2   (b) 2   (c) Yes
2. (a) 5   (b) 5   (c) No
3. (a) 2   (b) 2   (c) Yes
4. (a) 8   (b) 8   (c) Yes
5. (a) 5   (b) 5   (c) No
6. (a) 2   (b) 4   (c) Yes
7. (a) 1   (b) 1   (c) No
8. (a) 3   (b) 3   (c) No
9. (a) 9   (b) ∞ (c) ∞
   (d) ∞ (e) 4

Answers to Chapter 19
Exercise 19.1

1. *g, i* and *k*   2. *d, h* and *j*
3. *a, b, d, e, h* and *j*   4. *c, f* and *l*
5. *c* and *l*

Exercise 19.2

1.

Triangle	Angles		
A	28°	67°	85°
B	37°	61°	82°
C	80°	80°	20°
D	62°	90°	28°
E	104°	13°	63°
F	90°	60°	30°

2.

Triangle	$\angle A$	$\angle B$	x
R	49°	54°	103°
S	39°	87°	126°
T	68°	94°	162°
U	107°	35°	142°
V	27°	54°	81°

3.

Triangle	$\angle A$	$\angle B$	x
G	$39°$	$73°$	$112°$
H	$41°$	$67°$	$108°$
I	$65°$	$24°$	$89°$
J	$54°$	$111°$	$165°$
K	$78°$	$90°$	$168°$
L	$11°$	$86°$	$97°$

4. (a) $x = 59°$, $y = 121°$
   (b) $x = 85°$, $y = 43°$
   (c) $x = 11°$, $y = 47°$
   (d) $x = 79°$, $y = 84°$
   (e) $x = 111°$, $y = 29°$

Exercise 19.3

1. $b = 5.8$, $c = 9.8$, $\angle C = 71°$
2. $a = 3.71$, $b = 9.44$, $\angle B = 56°$
3. $b = 6.80$, $c = 2.92$, $\angle B = 46°$
4. $b = 7$, $\angle B = 50°$, $\angle C = 80°$
5. $\angle A = 38.2°$, $\angle B = 81.7°$, $\angle C = 60.1°$
6. $\angle B = 61.3°$, $\angle C = 75.7°$, $c = 9.95$
7. $\angle B = 70.8°$, $\angle C = 47.2°$, $a = 8.42$
8. $\angle A = 56°$, $\angle C = 40°$, $c = 3.88$
9. $\angle A = 44°$, $\angle B = 36°$, $\angle C = 100°$
10. $a = 6.30$, $\angle A = 44.2°$, $\angle B = 50.8°$
11. $\angle A = 50°$, $b = 9.0$, $c = 9.8$
12. $a = 13.2$, $b = 9.0$, $\angle C = 30°$

Exercise 19.4

1. (a) AC   (b) BC   (c) AC
   (d) AB   (e) AC
2. (a) 13   (b) 11.1   (c) 10.2
   (d) 16.6   (e) 9.14
3. (a) $a = 6.40$ (b) $c = 13.6$ (c) $a = 12.7$
   (d) $b = 14.4$ (e) $c = 5.20$

Exercise 19.5

1. 11.9 cm   2. 12.3 cm   3. 13.1 cm
4. (a) $A = B = 58°$
   (b) $B = 48°$, $A = 84°$
   (c) $A = 33°$, $B = 57°$
   (d) $A = 27°$, $B = 63°$
5. (a) $x = 44°$, $y = 31°$
   (b) $x = 72°$, $y = 36°$
6. 10.4 cm   7. 10.9 cm   8. 7.79 cm

Miscellaneous Exercise 19

1. 9 cm   2. $a = 59°$, $b = 57°$ 3. 9 cm
4. (a) $x = 70°$ (b) $y = 50°$
   (c) $\angle DAC = 40°$
5. (a) $52°$ (b) $74°$ (c) $52°$
   (d) $54°$ (e) $52°$
6. 10.8 cm   7. 9.17 cm

8. 13.4 cm   9. 53.2 cm
10. (a) $42°$ (b) $42°$ (c) $138°$
    (d) isosceles
11. 4 cm   12. 300 m
13. (a) 4, 4 and 4 cm
    (b) 2 cm, 5 cm and 5 cm
    (c) 3 cm, 4 cm and 5 cm
14. $x = 103°$, $y = 52°$, $z = 51°$
15. $p = 70°$, $q = 20°$, $r = 60°$, $s = 120°$
16. $m = 40°$, $n = 40°$, $p = 50°$, $q = 310°$
17. 11.3 cm

Mental Test 19

1. $80°$ 2. $80°$ 3. $110°$
4. $70°$
5. (a) $60°$ (b) $120°$ (c) $60°$
   (d) $70°$
6. (a) 10   (b) 16   (c) 12
   (d) 39   (e) 24   (f) 75
7. (a) $B = 70°$, $C = 40°$, $x = 110°$
   (b) $A = 60°$, $B = 60°$, $x = 120°$
   (c) $A = 30°$, $C = 120°$, $x = 150°$
   (d) $A = 80°$, $B = 80°$, $C = 20°$
   (e) $A = 50°$, $B = 50°$, $x = 130°$
   (f) $A = 60°$, $B = 60°$, $C = 60°$
   (g) $B = 45°$, $C = 90°$, $x = 135°$
8. Equilateral   9. $x = 50°$, $y = 80°$
10. $30°$ 11. $110°$
12. $60°$ 13. Hypotenuse
14. 13 cm

Answers to Chapter 20

Exercise 20.1

1. Congruent   2. Congruent
3. Not congruent necessarily
4. Congruent
5. Not congruent necessarily
6. Congruent   7. Congruent
8. Congruent

Exercise 20.2

1. CD = 8 cm, EC = 5 cm
2. AFD, FDE, FEB, DEC; FGH, GHJ, GJD, HJE
3. TP = 8 cm, TQ = 6 cm
4. BC = 8 cm
5. ST = 8 cm, LX = 11.3 cm, RY = 11.3 cm, $\angle$MLP = 40°, $\angle$SRY = 20°
6. CD = 5 cm
7. AE = 4 cm, CD = 5 cm
8. (a) $\angle$ADE = 80°   (b) $\angle$FEC = 40°
   (c) EF = 6 cm

Answers to Chapter 21

Exercise 21.1

1. $x = 40°$, $y = 100°$, $z = 71°$
2. $x = 109°$, $y = 127°$
3. $a = 116°$, $b = 64°$
4. $\angle DAC = 65°$, $\angle BCD = 130°$, $\angle ABC = 50°$, $\angle DOC = 90°$
5. 11.7 cm   6. 11.3 cm
7. 10.2 cm   8. 10.5 m
9. (a) $\angle AXD = 90°$ (b) $AX = 3$ cm
   (c) trapezium
10. (a) trapezium
    (b) $\angle XDC = 61°$, $\angle DXB = 119°$
11. (a) $x = 42°$ (b) $y = 42°$
    (c) $\angle ACB = 96°$ (d) isosceles
12. $AO = 12$ cm, $BO = 5$ cm, $AB = 13$ cm
13. Kite and isosceles trapezium
14. Rectangle and square
15. $a = 35°$, $p = 65°$
16. $\angle ABE = 64°$, $\angle BED = 134°$
17. $\angle ABC = 100°$, $\angle ECD = 20°$
18. $y = 61°$

Mental Test 21

1. $50°$ 2. $40°$
3. $\angle BEA = 90°$, $\angle ABC = 60°$
4. 5 cm   5. 4
6.

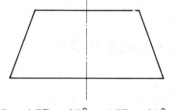

7. $\angle ACD = 25°$, $\angle ACB = 80°$
8. $150°$

Answers to Chapter 22

Exercise 22.1

1. (a) $720°$ (b) $1260°$ (c) $1620°$
2. (a) $45°$ (b) $135°$
3. Each interior angle is $120°$; the hexagon has 6 axes of symmetry
4. A seven sided polygon has 7 axes of symmetry; each interior angle is $128.6°$
5. $150°$ 6. $108°$
7. 6   8. $115°$
9. (a) $20°$ (b) 18
10. $\angle AOB = 72°$, $\angle ABC = 108°$
11. 15   12. $\angle UVO = 60°$
13. $x = 45°$, $y = 45°$

14. (a) $36°$ (b) $126°$
15. $\angle AFX = 6°$ 16. 7
17. $\angle BCD = 132°$, $\angle AED = 75°$

Mental Test 22

1. $140°$ 2. 6   3. 8
4. $36°$ 5. 6   6. Hexagon
7. 8   8. $45°$ 9. $135°$

Answers to Chapter 23

Exercise 23.1

1. $\angle ACB = 47°$, $\angle ADB = 47°$, $\angle BDC = 43°$
2. 5 cm
3. $\angle ABO = 70°$, $\angle OAC = 20°$
4. $\angle BAC = 54°$, $\angle BYC = 54°$, $\angle ACY = 27°$, $\angle YBC = 36°$
5. $\angle CBA = 90°$, $\angle ACB = 21°$, $\angle CAD = 21°$
6. $\angle AOB = 84°$, $\angle ADB = 42°$
7. $x = 63°$, $y = 48°$, $\angle BDA = 42°$
8. $\angle C = 110°$, $\angle D = 75°$, $\angle BCX = 70°$

Exercise 23.2

1. $OP = 17.9$ cm   2. $a = 78°$, $b = 12°$
3. $CB = 22.9$ cm
4. $\angle AOB = 66°$, $\angle OBD = 33°$, $\angle CBD = 57°$
5. 11.3 cm   6. $x = 59.89$ cm
7. $h = 24$ 8. $h = 57.0$ cm

Mental Test 23

1. $60°$ 2. $80°$ 3. 10 cm
4. $40°$ 5. $80°$ 6. $10°$
7. $60°$ 8. 13 cm   9. $20°$
10. $AB = 6$ cm, $AC = 5$ cm, $BC = 5$ cm

Answers to Chapter 26

Exercise 26.1

1. 32 cm   2. 20 cm   3. 30 cm
4. 20.2 cm   5. 26.8 cm   6. 31.4 cm
7. 24.2 cm   8. 22.2 cm

Exercise 26.2

1. 35 cm^2 2. 220 mm^2 3. 280 ft^2
4. 43.16 cm^2 5. 1820 mm^2 6. 1.11 m^2
7. 0.96 m^2 8. 28.42 ft^2 9. 44 m^2
10. 174 m^2
11. (a) 52.29 m^2 (b) 36.21 m^2
    (c) 16.08 m^2
12. (a) 29 cm^2 (b) 103 m^2 (c) 64 cm^2
    (d) 52 mm^2 (e) 72 mm^2 (f) 68 cm^2
    (g) 64 cm^2 (h) 48 cm^2
13. 6 ft   14. 12 m   15. 6 cm
16. 64 cm^2 17. 144 cm^2; 12 cm

Exercise 26.3

1. 35 cm^2 2. 180 mm^2 3. 15 ft^2
4. 4 m 5. 8 in 6. 52 cm^2
7. 4.24 cm
8. (a) 81 cm^2 (b) 13.5 cm

Exercise 26.4

1. 9 cm^2 2. 28 cm^2 3. 88 in^2
4. 6.88 m^2 5. 3 cm^2
6. (a) 6 cm^2 (b) 54 cm^2 (c) 30 cm^2
   (d) 120 cm^2
7. 10 yd 8. 32 mm 9. 9.8 cm^2
10. 7.5 ft^2 11. 55.4 cm^2 12. 143 cm^2
13. 30 cm^2 14. 9.61 cm^2

Exercise 26.5

1. 60 in^2 2. 40 cm^2 3. 6 cm
4. 600 in^2 5. 198 mm^2

Exercise 26.6

1. 72 m^2 2. 36 m^2 3. 266.7 mm^2
4. 3075 mm^2 5. $EF = 12 \text{ m}$ 6. 12 cm

Miscellaneous Exercise 26

1. (a) 32 cm (b) 52 cm^2
2. 25 in^2 3. 36 cm^2
4. (a) 9 cm (b) 28 cm^2
5. 5.76 cm^2 6. 3 m^2 7. 84 cm^2
8. 36 cm 9. $\frac{15}{16}$ 10. 48 cm^2

Mental Test 26

1. 21 cm^2 2. 5 in 3. 40 cm^2
4. 8 cm 5. 16 ft^2 6. 8 cm
7. 50 cm^2 8. 8 cm^2

Answers to Chapter 27

Exercise 27.1

1. 132 cm 2. 2200 cm 3. 88 ft
4. 88 cm 5. 1760 ft 6. 270.2 cm
7. 19.9 m 8. 267.1 in 9. 134.2 mm
10. 270.7 ft 11. 201.1 m 12. 2.0 m
13. 35.0 cm 14. 263.9 m^2 15. 691.2 in

Exercise 27.2

1. 154 cm^2 2. $61\,600 \text{ mm}^2$ 3. 2464 in^2
4. 1386 m^2 5. 962.5 ft^2 6. 50.3 cm^2
7. $15\,396 \text{ in}^2$ 8. 154.0 m^2 9. 1963.8 ft^2
10. $13\,275 \text{ mm}^2$ 11. 141.4 in^2 12. 581.3 mm^2
13. 116.3 m^2 14. 3.39 in^2 15. 3.4 cm

Exercise 27.3

1. $l = 3.142 \text{ cm}$; $A = 6.284 \text{ cm}^2$
2. $l = 15.71 \text{ cm}$; $A = 78.55 \text{ cm}^2$
3. $l = 6.284 \text{ in}$; $A = 9.426 \text{ in}^2$
4. $r = 8.91 \text{ in}$
5. $l = 7.96 \text{ m}$; $A = 31.8 \text{ m}^2$
6. $A = 9.05 \text{ yd}^2$
7. $A = 288.7 \text{ mm}^2$

Exercise 27.4

1. 14.284 cm^2 2. 76.3 cm^2 3. 292.3 cm^2
4. 5.36 cm^2 5. 13.73 cm^2 6. 23.43 cm^2
7. 57.72 cm^2 8. 42.04 cm^2

Miscellaneous Exercise 27

1. 220 cm 2. 22 yd 3. 90.28 cm^2
4. 1386 in^2 5. 33.2 cm^2
6. 28.26 cm^2 7. 14 in
8. (a) 1000 cm^2 (b) 73 cm
9. 308 cm^2 10. 28 in 11. 476.1 cm^2
12. 5.5 cm 13. 14 cm 14. 62.8 in^2
15. (a) 18.85 cm (b) 28.28 cm^2 (c) 36 cm
    (d) 111.4 cm

Mental Test 27

1. 44 cm 2. 88 in 3. 154 cm^2
4. 4 in 5. 16 cm^2 6. $48\pi \text{ cm}^2$
7. $2\pi \text{ cm}$ 8. $12\pi \text{ cm}^2$ 9. 50 cm^2
10. 100 cm

Answers to Chapter 28

Exercise 28.1

1. 6 2. 8 3. 4
4. 6 5. 5 6. 5
7. 12 8. 8 9. 9
10. 6

Exercise 28.2

6. Slant height $= 8.73 \text{ cm}$
7. (a) cylinder   (b) rectangular pyramid
   (c) rectangular pyramid

Exercise 28.3

1. 54 m^2 2. 148 cm^2 3. 3871 cm^2
4. (a) 440 cm^2 (b) 484 cm^2
5. 176 cm^2 6. 216 cm^2 7. 179 m^2
8. 204 cm^2

Exercise 28.4

1. 160 in^3 2. 0.0072 m^3 3. 308 m^3
4. $62\,840 \text{ mm}^3$ 5. 60 m^3 6. 73.28 cm^3
7. $13\,856 \text{ in}^3$ 8. 300 cm^3

Exercise 28.5

1. $8210 \, cm^3$ 2. $209 \, in^3$ 3. $24 \, cm^3$
4. $106.7 \, ft^3$
5. $179.6 \, cm^3$; $154.0 \, cm^2$
6. $3.8 \, m^3$ 7. $20 \, m^3$ 8. $4295 \, cm^3$

Exercise 28.6

1. $1.232 \, \ell$ 2. $1500 \, \ell$ 3. $5.75 \, \ell$
4. $42 \, 417 \, \ell$ 5. $38.4 \, c\ell$
6. $31.8 \, m\ell$; 22 glasses
7. $11.3 \, \ell$ 8. $9000 \, \ell$
9. Formula 3
10. Formula 2 and formula 6

Exercise 28.7

1. (a) $50 \, cm^2$ (b) $280 \, cm^3$
2. 6 in   3. $20.25 \, cm^2$ 4. $60.9 \, in^3$
5. $1.728 \, pt$ 6. $3 \, m$; $112 \, m^2$; $128 \, m^3$
7. $113.1 \, ft^2$; $113.1 \, ft^3$; $314.2 \, ft^2$; $523.7 \, ft^3$
8. $46.1 \, cm^2$

Miscellaneous Exercise 28

1. 6
2. (a) $48 \, m^2$ (b) $576 \, m^3$ (c) $26.9 \, m$
   (d) $323 \, m^2$
3. (a) $250 \, cm^3$ (b) $250 \, cm^2$ (c) $132 \, cm^2$
4. (a) cube   (b) none   (c) none
   (d) none   (e) tetrahedron   (f) none
   (g) none
5. (a) $29.68 \, in^3$ (b) $237.44 \, in^3$
   (c) 2.4 in by 5.25 in by 7.95 in
6. (a) $81 \, 000 \, cm^3$ (b) $25 \, cm$
7. $28 \, m^3$
8. (a) $5600 \, cm^2$ (b) $4400 \, cm^3$ (c) 19.6; 15.4
9. The 35-cm high packet

Mental Test 28

1. $40 \, in^3$ 2. $100 \, cm^3$ 3. $210 \, cm^3$
4. $250 \, in^3$ 5. $200 \, in^2$ 6. $96 \, cm^3$
7. $110 \, cm^2$ 8. $100\pi \, cm^3$ 9. $28 \, cm^3$
10. $120 \, cm^3$ 11. $36\pi \, cm^2$ 12. $1:2$

Coursework 2

11. XBCAEDF, XBCADEF, XBDECAF,
    XBDACEF, XBEACDF, XBEDCAF,
    XCDABEF, XCDEBAF, XCBAEDF,
    XCBADEF, XCEABDF, XCEDBAF,
    XAECBDF, XAEBCDF, XADCBEF,
    XADBCEF

13. (b) The flow chart tests for divisibility by 7;
    if a number is divisible by 7, the final result
    is always zero.
    (c) The numbers must have no common
    factors other than 1; the HCF is 1.
    (d) The LCM must be a multiple of 4; the
    smallest number will be the HCF; the largest
    number will be the LCM; the middle number
    will be twice the HCF and half the LCM; the
    sum $p + q + r$ will always be divisible by 7:
    $q = 2p$ and $r = 4p$, so $p + q + r = 7p$.

15. Diagrams produce this table:

Lines	Squares				
4	1				
5	2				
6	3	5			
7	4	8			
8	5	11	14		
9	6	14	20		
10	7	17	26	30	
11	8	20	32	40	
12	9	23	38	50	55

16. Tile G has limited use. Its reflection would be
    a useful inclusion.

17. Maximum 9-path including a return to X.
    Tetrahedron 5-path, Triangular prism 7-path,
    Octahedron 12-path, Pyramid 6-path

18.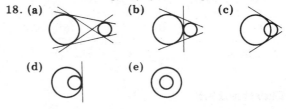

Answers to Chapter 29

Exercise 29.1

1. (a) 0.707   (b) 0.866
2. (a) $18.4°$ (b) $38.7042°$
   (c) $79.9147°$
3. (a) $43°12'$ (b) $54°45'7''$
   (c) $43°11'46''$
4. (a) 0.2419   (b) 0.9563   (c) 0.4258
   (d) 0.8643   (e) 0.8111   (f) 0.9839
   (g) 0.3404   (h) 0.9713
5. (a) $65°59'37''$ (b) $78°59'31''$
   (c) $47°17'56''$ (d) $83°52'46''$
   (e) $54°12'48''$ (f) $8°20'56''$

Exercise 29.2

1. (a) $11.47 \, cm$ (b) $12.04 \, cm$
   (c) $18.25 \, cm$

2. (a) $36.9°$ (b) $32.2°$ (c) $53.1°$
3. 5.95 cm   4. 3.63 cm   5. 23.1 cm
6. 20.0 cm
7. $41.8°$, $41.8°$ and $96.4°$
8. $\angle B = 61.4°$, $\angle C = 28.6°$

Exercise 29.3

1. (a) 0.9511   (b) 0.2419   (c) 0.7955
   (d) 0.1461   (e) 0.5818   (f) 0.9880
2. (a) 2.74 cm   (b) 10.20 cm   (c) 5.73 cm
3. (a) $63.3°$ (b) $77.2°$ (c) $66.4°$
4. 7.47 cm
5. $47.16°$; $66.42°$; $66.42°$
6. (a) $25°0'4''$ (b) $11°0'29''$
   (c) $59°18'10''$ (d) $78°11'59''$
   (e) $37°22'57''$ (f) $68°11'8''$
7. 11.1 cm
8. $\angle BAC = 60°$; $BD = 6.93$ cm

Exercise 29.4

1. (a) 0.5095   (b) 4.705   (c) 21.20
   (d) 0.2754   (e) 0.1456   (f) 1.5092
2. (a) $16°59'54''$ (b) $79°0'0''$
   (c) $12°17'53''$ (d) $43°48'4''$
   (e) $27°16'0''$ (f) $59°37'21''$
3. (a) 8.40 cm   (b) 9.00 cm   (c) 2.92 cm
4. (a) $33.7°$ (b) $38.7°$ (c) $60.3°$
5. 8.26 cm   6. 3.96 cm
7. 2.02 cm   8. 1.46 cm

Exercise 29.5

1. 4.82 m   2. 27.7 m   3. $42°$
4. 96.2 m   5. 8.29 m   6. 23.0 m
7. 575.2 m   8. $39.8°$ 9. 5.87 m
10. 14.9 m

Exercise 29.6

1. 22.13 cm^2 2. 31.9 cm^2 3. 6 cm
4. 540.4 cm^2 5. 737.5 m^2 6. 8.07 m^2
7. 14.0 m^2 8. 3870 m^2 9. 13.9 cm^2
10. 173.8 m^2

Exercise 29.7

1. (a) $323°$ (b) $190°$
2. (a) $323°$ (b) $277°$ (c) $350°$
3. $335°$ 4. $248°$
5. (a) $300°$ (b) $\angle ABC = 75°$
6. $140°$
7. (a) 264 km   (b) $062°$
8. (a) 69.3 km   (b) 40 km

Exercise 29.8

1. (a) $FH = 13$ cm   (b) $BH = 15.3$ cm
2. (a) $EF = 4$ cm   (c) $71.6°$
   (d) 12.6 cm
   (e) Area $\triangle VAD = 50.6 \text{ cm}^2$
   (f) Area of pyramid $= 266.4 \text{ cm}^2$
3. 12.7 cm   4. $54.5°$ 5. 4.53 cm
6. $EA = 14.4$ cm; $q = 28.7°$
7. (a) $62°$ (b) 1700 m^2
   (c) $12\,000 \text{ m}^3$

Miscellaneous Exercise 29

1. (a) $11.5°$ (b) 4.90 m
2. (a) $80°$ (b) 12.86 cm
   (c) 7.66 cm
3. (a) 39.61 cm   (b) 0.3051   (c) $17.0°$
4. (a) 60.8 m   (b) $37.2°$
5. (a) 7.75 cm   (b) 51.6 cm^2 (c) $31.8°$
6. (a) $22.9°$ (b) $66.4°$
7. 373 m
8. (a) $60°$ (b) 17.32 m
9. $027°$ 10. 78 km; $190°$

Answers to Chapter 30

Exercise 30.1

1. 1600 m   2. 25 cm
3. (a) 1:200   (b) 4 m by 6 m
4. (a) 1:50 000   (b) 5.4 km
5. 405 km
6. (a) 14 km   (b) 7.25 cm; 34.8 km

Exercise 30.2

1. (a) 15 m by 12.5 m   (b) 187.5 m^2
2. 640 m^2

3.
Name of room	Length (m)	Width (m)	Area (m²)
Lounge	5	5	25
Dining room	5	3	15
Kitchen	5	3	15
Bathroom	4	3	12

Area of carpet required is 13 m^2.

4.
Name of room	Length (m)	Width (m)	Area (m²)
Lounge	7	6	42
Hall	3	2	6
Bedroom	5	4	20
Kitchen	4	2	8
Bathroom	4	3	12

5. (a) 3   (b) 3   (c) 25.5 m
(d) 13.5 m   (e) 344.25 m^2 (f) 51.64 m^3
6. 11875 m^2
7. (a) 48 m^2 (b) 1120

Exercise 30.3

1. $40\,000 \text{ m}^2$ **2.** 12 m^2 **3.** 20 km^2
4. 10.4 km^2
5. (a) 13.4 cm^2 (b) $134\,000 \text{ m}^2$
6. 35.5 hectares

Exercise 30.4

1. $24\,000 \text{ m}^3$ **2.** 1.536 m^3
3. 0.6 cm^3 **4.** 7812 m^3
5. 840 m^3

Exercise 30.5

1. 1 in 25   **2.** 1 in 40   **3.** 100 m
4. 30 m   **5.** $7.2°$
6. (a) $1:4.8$ (b) 20.8%
7. 15 m

Exercise 30.6

1. $62°$
2. (a) 1 in 2.85   (b) 35%
3. $17.4°$
4. (a) 150 m   (b) 2100 m   (c) 7.1%
5. (a) 100 m   (b) 236 m
   (c) 1.55 km; 2.75 km   (d) $11°19'$

Miscellaneous Exercise 30

1. 500 m
2. $1.75 \text{ cm} \times 2.25 \text{ cm} \times 1.25 \text{ cm}$
3. (a) 10 m   (b) 14 cm   (c) 20 m^2
4. (a) 80 m   (b) 5 m^2 (c) $288\,000 \text{ m}^3$
5. (a) 50 cm   (b) 648 m^2 (c) 19.44 m^3

6.

Name of room	Length (m)	Breadth (m)	Area (m^2)
Lounge	7	6	42
Dining room	7	4	28
Bedroom 1	7	4	28
Bedroom 2	4	3	12
Hall			30
Bathroom	3	3	9
WC	2	2	4
Kitchen	5	3	15

7 windows and 8 doors
7. (a) $11.3°$ (b) 1 in 5   (c) 20%

Answers to Chapter 31

Exercise 31.1

1. (a) £6.24   (b) £26.52
   (c) £215.02   (d) £174.72
   (e) £1.69   (f) £2.42
   (g) £21.37
2. (a) $28.3°C$ (b) $45.9°C$
   (c) $131°F$ (d) $155.7°F$
3. (a) 488 miles   (b) 130 miles
   (c) 63 miles
4. (a) 14.3%   (b) $1:14$
5. (a) 24.13 km/h   (b) 30 mile/h
   (c) 93.33 km/h   (d) 914.12 km/h
6. (a) $37.8°C$ (b) $176°F$
   (c) $17.8°C$ (d) $118°F$
7. (a) 1.5 kg   (b) 258 kg
   (c) 4.0 lb
8. (a) 1.54 kg/cm^2 (b) 30 lb/in^2
   (c) 43 lb/in^2

Exercise 31.2

1. (a) 40   (b) 11

2.

0.60 cm	160 — Other
1.54 cm	420 — Private car
2.86 cm	780 — Bus

3.

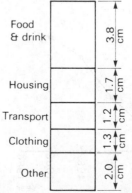

Food & drink	3.8 cm
Housing	1.7 cm
Transport	1.2 cm
Clothing	1.3 cm
Other	2.0 cm

4.

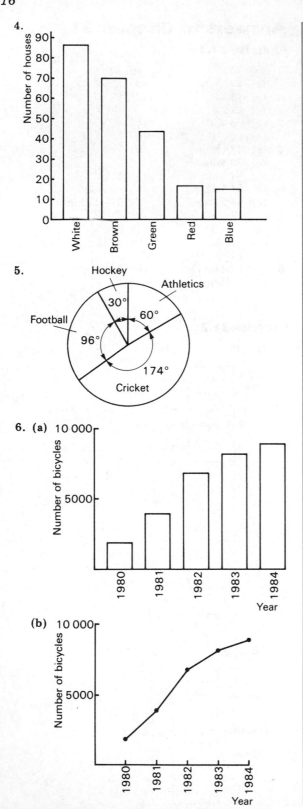

5.

6. (a)

(b)

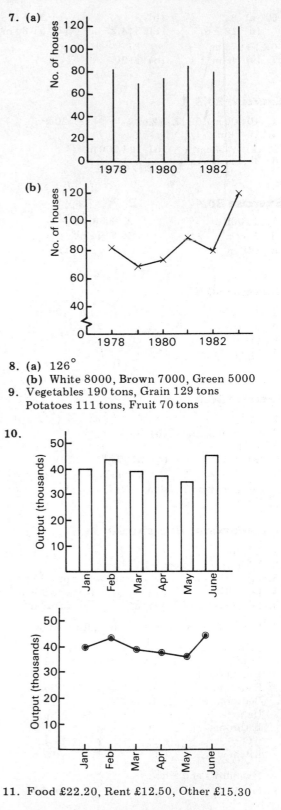

7. (a)

(b)

8. (a) 126°
   (b) White 8000, Brown 7000, Green 5000
9. Vegetables 190 tons, Grain 129 tons
   Potatoes 111 tons, Fruit 70 tons

10.

11. Food £22.20, Rent £12.50, Other £15.30

Mental Test 31

1. (a) 110 miles (b) 240 miles
   (c) 110 miles (d) 72 miles
2. (a) $-40\,^{\circ}$F (b) $68\,^{\circ}$F (c) $38\,^{\circ}$C
3. (a) 1600 (b) 880 (c) 375
4. 90; 150
5. (a) £60 000 (b) £200 000
6. (a) 1978 (b) 50 000
   (c) 35 000 (d) 200 000
7. 72°

Answers to Chapter 32

Exercise 32.1

1.

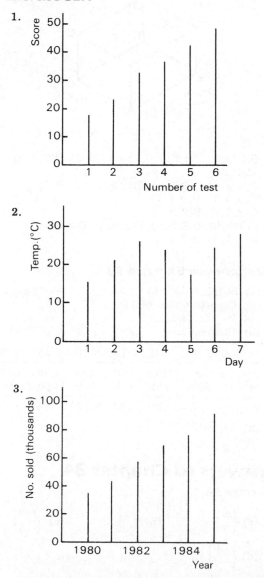

2.

3.

4.

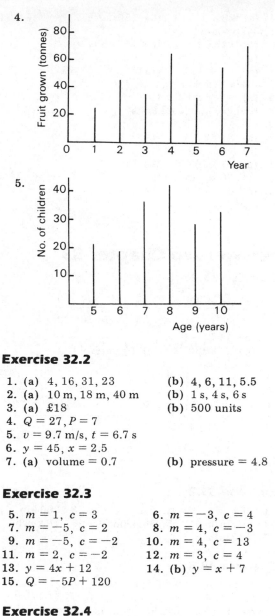

5.

Exercise 32.2

1. (a) 4, 16, 31, 23 (b) 4, 6, 11, 5.5
2. (a) 10 m, 18 m, 40 m (b) 1 s, 4 s, 6 s
3. (a) £18 (b) 500 units
4. $Q = 27, P = 7$
5. $v = 9.7$ m/s, $t = 6.7$ s
6. $y = 45$, $x = 2.5$
7. (a) volume $= 0.7$ (b) pressure $= 4.8$

Exercise 32.3

5. $m = 1, c = 3$ 6. $m = -3, c = 4$
7. $m = -5, c = 2$ 8. $m = 4, c = -3$
9. $m = -5, c = -2$ 10. $m = 4, c = 13$
11. $m = 2, c = -2$ 12. $m = 3, c = 4$
13. $y = 4x + 12$ 14. (b) $y = x + 7$
15. $Q = -5P + 120$

Exercise 32.4

1. $a = 5$, $b = 1$ (both approx.)
2. $a = 0.5$, $b = 3$ (both approx.)
3. $m = 1.3$, $c = 20$
4. $E = 4I$
5. $a = 0.03$, $b = 20$

Miscellaneous Exercise 32

1. (a) £18.75 (b) 232 units (c) £7.50
2. (a) A$(-2, 8)$, B$(3, -4)$, C$(0, 3.2)$
3. (b) (i) 38.5 miles per gallon
   (ii) 15 mile/h, 78 mile/h
   (iii) 50 mile/h

4. (a) 26.4 lb   (b) 15.9 kg
5. Trapezium
6. (a) £29.50   (b) 150 units
   (c) £8   (d) 6 p
7. (b) (0, 4)   (c) 1
   (d) $y = x + 4$
8. (a) $y = 4x + 3$ (b) -5
9. 9 units; 18 square units
10. $c = 4$
11. (a) A($-2, 8$)   (b) -2
    (c) $y = -2x + 4$
12. (a) A(0, 6)   (b) B($-12, 0$)
    (c) 36 square units
13. (a) $c = 5$ (b) OB $= 10$

Answers to Chapter 33
Exercise 33.1

6. (a)

T (hours)	1	2	3	4	5
N		2	4	8	16	32

(b) 11.3
7. (b) £12 280   (c) $5\frac{1}{2}$ years
8. 14 millions
9. $-5, 19$
10. $-4, 8$
11.

x	-2	$-1\frac{1}{2}$	-1	0	$\frac{1}{2}$	1	$1\frac{1}{2}$	2	3
y	-9	-4		0	5	6	6	5	3	-4

$x = 1.25$
12. $x = 8$

Exercise 33.2
1. 60 km/h   2. 5 h   3. 300 km
4. 50 km/h   5. 36.4 km/h   6. 10 km/h
7. 22 km/h

Exercise 33.3
1. 10.20 a.m.; 20 km from A
2. (b) 2124 hours
   (d) 1800 hours, 195 km from London
3. (a) 6 km/h   (b) 20 min
   (d) 6 km/h
4. (a) 48 km/h   (b) 344 km   (c) 95 km
   (d) 3.5 hours
5. (a) 1910 hours
   (b) 69.5 km   (c) 24.4 km/h
   (d) 1854 and 1924 hours

Exercise 33.4
1. (a) 60 m   (b) 6 m   (c) 80 m
   (d) 120 m   (e) 300 m
2. (a) 5 m/s²   (b) 0.5 m/s²   (c) -1.5 m/s²
   (d) -2.5 m/s²

3. 150 m
4. (a) 0.2 m/s²   (b) 4 m/s   (c) 210 m
5. (a) 1 m/s²   (b) 3 m/s²   (c) 5 m/s
   (d) 100 m
6. (a) 1 m/s²   (b) 2 m/s²   (c) 0.5 m/s²
   (d) 30 m/s   (e) 875 m
7. (b) 2 m/s²   (c) 1.6 m/s²   (d) 384 m
8. 239 m
9. (b) (i) 2.5 m/s²   (ii) 10 m/s²
10. 9 m/s

Exercise 33.5
1. (a)

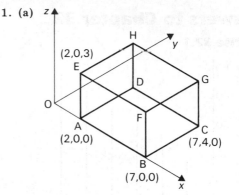

(b) D(2, 4, 0), F(7, 0, 3), G(7, 4, 3), H(2, 4, 3)
2. (2, 2, 3)
3. C(9, 5, 0), D(3, 5, 0), E(3, 0, 7), F(3, 5, 7)
   G(9, 5, 7), H(9, 0, 7)
4. O(2, 2, 0), E(2, 2, 5)
5. C(7, 5, 0), D(2, 5, 0), E(2, 0, 5), F(2, 5, 5),
   G(7, 5, 5), H(7, 0, 5)

Miscellaneous Exercise 33
1. (a) 50 min   (b) 33 km   (c) 72 km/h
   (d) Departure time 0830
2. $-4, 2$
3. (b) 13.15 million
4.

t	0	5	10	15	20	25	30
20t	0	100	200	300	400	500	600
$8t^2$	0	200	800	1800	3200	5000	7200
v	0	300	1000	2100	3600	5500	7800

(b) 340 m/s²

Answers to Chapter 34
Exercise 34.1
1. (a) $\begin{pmatrix} 4 \\ 3 \end{pmatrix}$ (b) $\begin{pmatrix} -5 \\ 2 \end{pmatrix}$ (c) $\begin{pmatrix} -6 \\ -6 \end{pmatrix}$
   (d) $\begin{pmatrix} 0 \\ -6 \end{pmatrix}$

2. (a) $\begin{pmatrix} 6 \\ 3 \end{pmatrix}$ (b) $\begin{pmatrix} 3 \\ 0 \end{pmatrix}$ (c) $\begin{pmatrix} 1 \\ 2 \end{pmatrix}$

(d) $\begin{pmatrix} 1 \\ -2 \end{pmatrix}$

3. $\begin{pmatrix} -4 \\ -1 \end{pmatrix}$

4. A'(5, 5), B'(7, 5), C'(7, 7), D'(5, 7)
5. A'(0, -3), B'(1, -5), C'(1, -2)

Exercise 34.2

1.
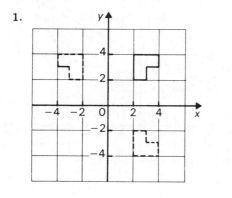

2.
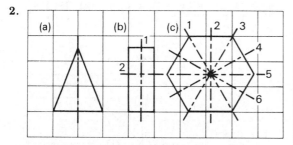

3.
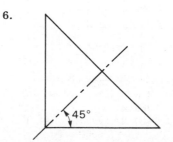

4.
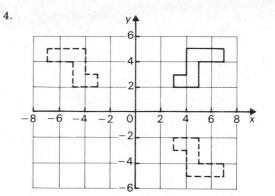

5.
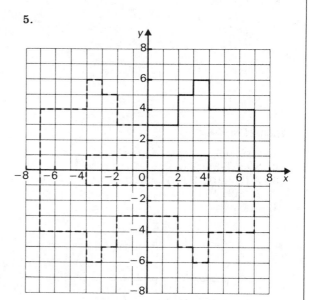

6.

7. (a) S(2, 6)
  (e) P'(2, 2), Q'(2, 6), R'(6, 6), S'(6, 2)

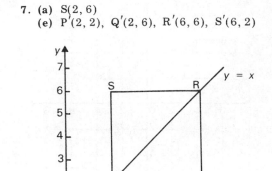

8. $y = 2$

Exercise 34.3

1. P'(5, 0.6)
2. X'(−3, 3), Y'(−3, 6), Z'(−5, 6)
3. A'(2.8, 0), B'(7.1, −4.2), C'(9.2, −2.1), D'(4.9, 2.1)
4.

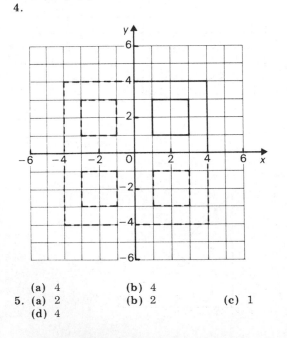

(a) 4	(b) 4	
5. (a) 2	**(b)** 2	**(c)** 1
(d) 4		

Exercise 34.4

1. (a) 3   **(b)** (−2, 1)
2. (a) 3   **(b)** (0, 0)
3. A'(4, 0), B'(4, 4), C'(8, 6), D'(8, 0);
  4 times as large
4. (a) 2; (−3, −4) **(b)** 3; (3, −3)
  (c) 2; (−1, 0.5)
5. A'(−2.5, 2.5), B'(−2.5, 4), C'(0.5, 5.5), D'(−1, 2.5)

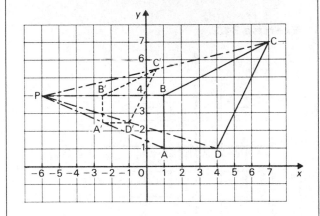

6. A'(4, 2), B'(4, −2), C'(−4, −2), D'(2, 4)

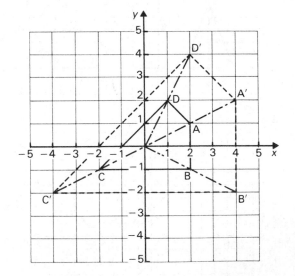

7. $\frac{1}{3}$; 8 cm
8. EF = 2 cm; CD = 12 cm

Miscellaneous Exercise 34

1. A$'(-2, 4)$
2. (a) A$(4, -3)$ (b) B$(-4, 3)$
3. $(6, 3)$
4. P$'(5.1, 0.4)$
5. $(2, 7)$
6. A$'(0, 0)$, B$'(4, 0)$, C$'(4, 2)$

7.

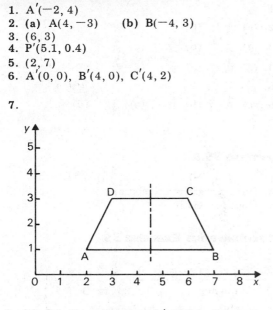

8. (a) R is the reflection of △ABC in the y-axis
   (b) T is the translation of △ABC under the
       vector $\begin{pmatrix} -5 \\ 3 \end{pmatrix}$

Answers to Chapter 35

Exercise 35.1

2.
Mark	1-10	11-20	21-30	31-40	41-50
Frequency	2	3	6	6	3

3. (b) $20.355, 20.455, 0.1$
5. (a) $39.5, 44.5, 5$ (c) continuous

6.
No. of letters	1-2	3-4	5-6	7-8	9-10	11-12
Frequency	15	35	5	13	3	1

7. (a) cont.   (b) disc.   (c) cont.
   (d) disc.   (e) cont.   (f) cont.
   (g) cont.   (h) disc.
8. (a) 50   (b) continuous

9. (a)
Noise level (dBA)	Frequency
81–85	4
86–90	12
91–95	9
96–100	7
101–105	4

(c)
Less than	Number of discos
81	0
86	4
91	16
96	25
101	32
106	36

(d) 15 discos had a noise level exceeding 94 dBA.

10. (b)
Less than	Cumulative frequency
10	5
20	38
30	120
40	268
50	538
60	753
70	840
80	884
90	896
100	900

(d) (i) 162   (ii) 324

11. £32

12. (a)
Weight (kg)	Cumulative frequency
Less than 47.9	3
Less than 51.9	20
Less than 55.9	70
Less than 57.9	115
Less than 59.9	161
Less than 63.9	218
Less than 67.9	241
Less than 71.9	250

(b) Semi-interquartile range = 3 kg

13. (a)
Estimated length (cm)	Cumulative frequency
Less than 35	0
Less than 36	1
Less than 37	4
Less than 38	8
Less than 39	16
Less than 40	22
Less than 41	27
Less than 42	30
Less than 43	32

(b) Median = 38.9 cm; semi-interquartile range = 1.2 cm

14.

Score	Cumulative frequency
Less than 40	1
Less than 45	3
Less than 50	8
Less than 55	18
Less than 60	34
Less than 65	62
Less than 70	81
Less than 75	92
Less than 80	97
Less than 85	99
Less than 90	100

Median score = 63; semi-interquartile range = 5.5

15. £154

16. 29.76 cm

17. 4 mm

Exercise 35.2

1. 176.7 cm   **2.** 8   **3.** 14
4. 5   **5.** 5.5   **6.** 5
7. (a) $2\frac{1}{2}$ kg   (b) $3\frac{1}{2}$ kg   (c) $2\frac{1}{2}$ kg
8. 4
9. (a) 78 kg   (b) 77 kg
10. (a) 40   (b) 3   (c) 3.3
   (d) 3
11. (a) 4.34   (b) 4   (c) 4
12. (a) 200   (b) 23   (c) 23
   (d) 22.6

Exercise 35.3

1. (a) $\frac{1}{6}$ (b) $\frac{1}{2}$ (c) $\frac{1}{2}$
2. (a) $\frac{1}{8}$ (b) $\frac{1}{4}$ (c) $\frac{3}{8}$ (d) $\frac{5}{8}$
3. (a) $\frac{1}{5}$ (b) $\frac{1}{2}$ (c) $\frac{3}{10}$ (d) $\frac{7}{10}$
4. $\frac{4}{7}$
5. (a) $\frac{1}{2}$ (b) $\frac{3}{10}$ (c) $\frac{3}{5}$ (d) $\frac{7}{20}$
6. (a) $\frac{1}{52}$ (b) $\frac{1}{13}$ (c) $\frac{4}{13}$ (d) $\frac{1}{26}$
7. (a) $\frac{1}{6}$ (b) $\frac{1}{6}$ (c) $\frac{1}{6}$
8. (a) $\frac{1}{13}$ (b) $\frac{1}{17}$
9. $\frac{12}{23}$
10. $\frac{1}{3}$

Exercise 35.4

1. (a) $\frac{7}{20}$ (b) $\frac{13}{20}$
2. (a) $\frac{3}{20}$ (b) $\frac{1}{4}$ (c) $\frac{1}{20}$
3. (a) $\frac{8}{15}$ (b) $\frac{18}{25}$ (c) $\frac{1}{15}$
4. (a) $\frac{7}{25}$ (b) $\frac{19}{50}$ (c) $\frac{31}{50}$
5. (a) $\frac{4}{25}$ (b) $\frac{14}{25}$ (c) $\frac{2}{25}$

Exercise 35.5

1. $\frac{5}{52}$
2. (a) $\frac{1}{12}$ (b) $\frac{1}{4}$
3. (a) $\frac{4}{25}$ (b) $\frac{9}{25}$ (c) $\frac{6}{25}$
4. (b) $\frac{3}{10}$ (c) $\frac{1}{10}$ (d) $\frac{3}{10}$
5. $\frac{1}{30}$
6. (a) $\frac{4}{35}$ (b) $\frac{4}{35}$ (c) $\frac{6}{35}$ (d) $\frac{4}{35}$
7. $\frac{1}{6}$

Exercise 35.6

1. (c) 48 to 50   **2.** (b) 77 and 60
3. Perfect negative correlation
4. Good correlation

Miscellaneous Exercise 35

1. 61
2.

Number of goals	0	1	2	3	4	5
Frequency	16	24	12	3	3	2

5. (a) 50   (b) 59 kg
6. (a) $\frac{1}{9}$ (b) $\frac{2}{9}$ (c) $\frac{1}{3}$
7.

Distance	0–9	10–19	20–29	30–39
Frequency	0	6	12	13

Distance	40–49	50–59	60–69
Frequency	15	14	10

8. (a) 14.985 mm; 15.015 mm
   (b) 0.03 mm   (c) 14.96–14.98 mm
9. (a) disc.   (b) cont.
   (c) disc.   (d) cont.
   (e) cont.   (f) disc.
10. £123.60
11. 163.1 cm
12. (a) 5   (b) 5
13. (a) 4   (b) 4
14. (a) $\frac{1}{6}$ (b) $\frac{1}{2}$ (c) $\frac{1}{3}$ (d) $\frac{1}{2}$
15. (a) 0.32   (b) 0.03
16. (a) A $= \frac{10}{24}$ (or $\frac{5}{12}$); B $= \frac{10}{23}$; C $= \frac{14}{23}$; D $= \frac{9}{23}$;
   (b) $\frac{15}{92}$
17. (a) $\frac{11}{100}$ (b) $\frac{3}{20}$
18. (c) (i) 218   (ii) 53
19. (b) 78.7   (c) 80–99

Mental Test 35

1. 8, 40   **2.** 1   **3.** $\frac{18}{25}$
4. £30   **5.** 7   **6.** 2
7. 2   **8.** $\frac{3}{10}$
9. (a) 15.5, 17.5, (b) 2 mm
10. 49   **11.** $\frac{1}{3}$

Answers to Chapter 36

Exercise 36.1

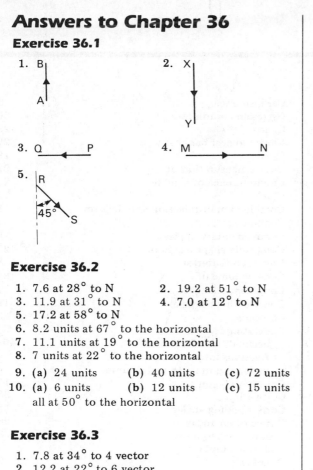

1. B
   A

2. X
   Y

3. Q ⟵ P

4. M ⟶ N

5. R
   45° S

Exercise 36.2

1. 7.6 at 28° to N
2. 19.2 at 51° to N
3. 11.9 at 31° to N
4. 7.0 at 12° to N
5. 17.2 at 58° to N
6. 8.2 units at 67° to the horizontal
7. 11.1 units at 19° to the horizontal
8. 7 units at 22° to the horizontal
9. (a) 24 units   (b) 40 units   (c) 72 units
10. (a) 6 units   (b) 12 units   (c) 15 units
    all at 50° to the horizontal

Exercise 36.3

1. 7.8 at 34° to 4 vector
2. 12.2 at 22° to 6 vector
3. 20.0 at 37° to 16 vector
4. 7.8 at 73° to 5 vector
5. 20.7 at 18° to 12 vector
6. 4.3 N and 2.5 N
7. 19.3 km/h and 23.0 km/h
8. 75.2 N and 27.4 N

Coursework 3

22. A small sample of the possible combinations:

    $4 \times 4 = 16$ $5 \times 2 = 10$
    $6 + 2 = 8$ $4 + 4 = 8$
    $5 + 3 < 10$ $6 + 3 < 16$

    $6 + 4 = 10$ $6 + 10 = 16$
    $3 + 2 = 5$ $5 + 3 = 8$
    $16 < 4 \times 8$ $4 < 4 \times 2$

There are 18 basic arrangements using the equality signs. The 'less than' sign then gives a total of 54 solutions.

23. (a) Equatorial diameter: 12 756 km; circumference: 40 074 km; surface area: $5.11 \times 10^8 \, \text{km}^2$; volume: $1.09 \times 10^{12} \, \text{km}^3$; equatorial rotation speed: 1670 km/h; latitude of town: 60°
    (b) The latitude of London is $51\frac{1}{2}°$N, so the maximum angle of elevation of the Sun here is $(90° - 51\frac{1}{2}°) + 23\frac{1}{2}°$ or 62°.
    (c) Shuttle speed: 27 800 km/h; geosynchronous altitude: 35 800 km

24. (a) 27.3217 days; 3680 km/h; 13.2°
    (b) Moon's mean angular diameter: 0.52° (31′ of arc);
    Earth–Moon ratios: surface area, 13.5 : 1; volume, 49.4 : 1; lunar equatorial rotation speed: 16.7 km/h;
    Sun–Moon diameters are in the ratio 400 : 1;
    Earth–Sun distance: 1.539×10^8 km (approx.; mean actual distance is 1.496×10^8 km)
    (c) 1.6×10^{13} years
    (e) 1.28 light seconds; reply time: 2.56 seconds; light takes an average of 8 min 19 s to reach Earth from the Sun.
    (f) 1 light year $= 9.46 \times 10^{12}$ km; Proxima Centauri is 4.05×10^{13} km distant;
    1 parsec $= 3.26$ light years; the melon would have to be placed 11 000 miles from the beach ball!
    (g) The supernova exploded around 5000 BC; we travel about 20 million km through the Galaxy each day; the 'cosmic year' lasts about 220 million years.

26. In any single sequence, two numbers are always missing. In all sequences, one number never appears.

28. (a) The lighthouse is 1.7 km from Jaws Tower; trawlers must not approach nearer than 300 m to Tiger Rock.
    (b) The trawler was fishing illegally. Small craft may approach to within 1.06 km of the lighthouse.

30. Using C for clockwise, A for anticlockwise and X for exchange, there are two solutions:

    X C X A X C X C X C X A X A
    C X A X C X C X C X A X A X

Index